THE MOLECULAR BASIS
OF OPTICAL ACTIVITY

The Molecular Basis of Optical Activity

OPTICAL ROTATORY DISPERSION AND CIRCULAR DICHROISM

ELLIOT CHARNEY

Laboratory of Chemical Physics
National Institutes of Health

A Wiley-Interscience Publication
JOHN WILEY & SONS

New York · Chichester · Brisbane · Toronto

Library of Congress Cataloging in Publication Data

Charney, Elliot.
    The molecular basis of optical activity.

    "A Wiley-Interscience publication."
    Includes index.
    1. Optical rotatory dispersion. 2. Circular
dichroism. I. Title.

QD473.C48   541'.22   79-9705
ISBN 0-471-14900-4

Printed in the United States of America

10 9 8 7 6 5 4 3 2 1

To my parents

PREFACE

There is something tangible and real about a familiar physical model, and when it fails we tend to blame nature first and accuse her of being unreal, complicated, paradoxical and lacking in common sense.—Walter Kauzmann.

This monograph attempts to develop the rationale for observations that are made on the degree and nature of the polarization of circularly polarized light transmitted or emitted by chiral molecules. This rationale, of necessity, derives from theories of the interaction of electromagnetic radiation with matter. Rather than develop these rigorously, for this has already been done, I have tried to demonstrate the most significant aspects and to argue frequently by physical analogy. Unfortunately, as Walter Kauzmann noted, analogies are never complete and sometimes appear quite perverse. Surely most of those who have worked on chiroptical problems must have felt that way on many occasions. I have tried to incorporate some feeling for why this is so as an aid or guide to understanding the application of the physical theories to the experimental observations.

The plan of the book appears in the table of contents and needs no elaboration. The approach, however, is subjective and does need emphasis. I have taken the point of view that the methods of quantum mechanics and the sophistication of computers is only now beginning to approach the point where successful calculations of the chiroptical properties of molecules is possible. Physically intuitive conclusions have been more easily discernable when optical activity is considered to be the result of an asymmetric perturbation by other atoms and bonds in the molecule chirally disposed about a symmetric chromophore. For these reasons, the basis of perturbation treatments is discussed first. An intrinsic part of the problem for both the dissymmetric and the asymmetrically perturbed symmetric chromophores is the description of the structural parameters of the nuclear and electronic configurations of the molecules. For this problem, concepts of symmetry as they have been developed in the twentieth century turn out to be very helpful; to emphasize this, two chapters are used to discuss the utility of these concepts.

Although those who bore with me while I wrote this book may think otherwise, it has been a pleasure, not least because of the fascinating insights gleamed from original papers, many of which I had read before but with far less attention than they deserved. A number of papers and books led me on to paths that had nothing to do with the subject but which were fascinating in their own right. I might mention in this respect, *Men of Mathematics*, the delightful book by E. T. Bell. There are several places where I have leaned very heavily on treatments in the literature. Hopefully, it will be sufficient recompense to the authors that I found it unnecessary to improve on what they have done and am grateful that their efforts made mine easier.

I want to thank my many colleagues who provided stimulation and help. I should like to acknowledge, particularly, my debt to Ulrich Weiss and Herman Ziffer who aroused my interest in optical activity and to Albert Moscowitz with whom the early work on diene optical activity was done. I am especially grateful to Thomas Bauman, Victor Bloomfield, Edward Bunnenberg, Albert Burgstahler, Cherie Fisk, Gary Gray, Robert Hexter, James Hofrichter, Joan Rosenfield, Ignacio Tinoco, Lin Tsai, and Herman Ziffer for their valuable comments on all or part of the book. To Theodore Hoffman of John Wiley my thanks for the patience of a saint and more; also to the National Institutes of Health and most particularly to E. D. Becker and J. E. Rall for support and encouragement. My thanks also to Walter Stockmayer and the other members of the Department of Chemistry who made my stay at Dartmouth so delightful for a part of the time while I was writing this book. I wish also to thank Sally Aigner and Agnes Brummett for the care and patience with which they typed a large part of the first draft and Helen Orem for help and special attention with the graphics. Finally, my gratitude to Mary Lou Miller who skillfully typed the remainder of the manuscript and the final draft, and whose sharp eye and intelligence saved the reader from many an awkward turn of phrase, and saved me from committing more than a few errors.

ELLIOT CHARNEY

March 1979
Bethesda, Maryland

CONTENTS

The application of perturbation models, quantum mechanics, and group theory to the optical rotatory dispersion and circular dichroism associated with some specific chromophores: ketone, hydroxyl, olefin, disulfide, diene, phenyl and amide chromophores.

With special emphasis on helical polymers, the theory of excitonic interactions between subunits and its applications as developed by Moffit and others.

The special nature of the optical activity of partially oriented systems and of crystals, and the role of symmetry in the latter.

Each of the three phenomena and their potential utility briefly described.

THE MOLECULAR BASIS
OF OPTICAL ACTIVITY

CHAPTER ONE

DISCOVERY, DEVELOPMENT, AND NOMENCLATURE

1.1   DISCOVERY OF OPTICAL ACTIVITY AND DEVELOPMENT OF THE THEORY

In September 1960, Ignacio Tinoco, Jr., was able to write, "I know of no successful published theory of Optical Activity." While it would be difficult to defend the same statement today, in view of the intensive research of the last decade and a half, it is still not possible to predict confidently the chiroptical activity of a molecule from a priori considerations and a knowledge of its structure. Nevertheless, enormous theoretical and experimental advances have been made in our understanding of the phenomenon in the last few years and in the century and a half since the first observation by the French astronomer Arago and its appreciation by Jean Baptiste Biot in 1811 and 1812. In this part of the nineteenth century optical observations could be counted on to arouse intense interest and excitement. The ink on Biot's memoir (1) could hardly have been dry before Augustine Fresnel (2), Herschel (3), Faraday (4), and Pasteur (5) started their theoretical and experimental explorations. T. M. Lowry, whose excellent monograph was for many years the only available treatise on optical activity in the English language, has pointed out that Biot's memoir was nearly 400 pages long. Biot's descriptions are not models of conciseness, but the length does speak for the difficulty of explaining a very complex phenomenon in classical, nonmathematical terms, as well as for the depth and extensiveness of his work. The history of the initial rapid developments, the subsequent slower but steady progress, and the explosion of interest in the mid-twentieth century make a fascinating story whose outline we will scarcely sketch. The facts appear distributed in a few monographs and symposia reports, but no coherent history has been written, Lowry's book being the best available compilation to 1935. In March 1966, in his closing remarks (6) to the Royal Society Discussion on

1

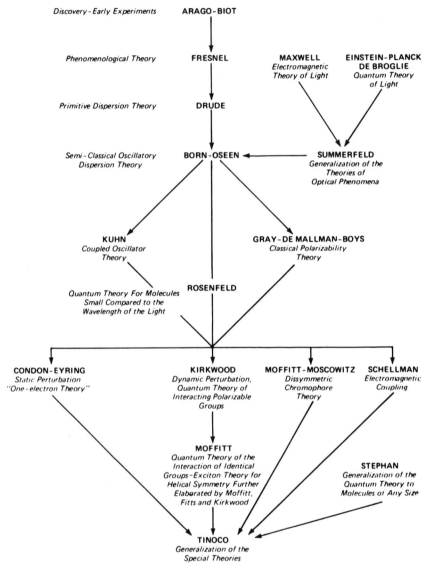

Fig. 1.00 The genealogy of the development of the theories of chiroptical activity. Some of the subsequent specialized treatments of the component theories are discussed later in the text, as are semi-empirical approaches based on these theories. Some recent parallel developments, such as scattering theory, should be especially valuable for the elucidation of second order chiral phenomena, but have not been extensively used for the chiral observations discussed in this book.

Circular Dichroism, Sir Christopher Ingold summarized the 150-year his-tory of chiral-optical phenomena by distinguishing six definitive periods approximately 30 years apart. The first was that of Arago and Biot, and also of Herschel, who identified the different but related crystals of quartz with their optical activity of opposite sign, and of Fresnel, who developed the first phenomenological theory. The second period, identified by Ingold with Pasteur, is also the time when James Clerk Maxwell formulated the fundamental equations of electrodynamics on which the entire classical structure of the theories of optical phenomena is based. Both Newton and Einstein have pointedly noted that all of scientific endeavor stands on the shoulder of its predecessors, but perhaps nowhere is this more visible, or at least clearly delineable, than in the field of optical chirality. In Fig. 1.00, the geneology of the theories of optical rotation and differential rotatory absorption, as we know it, is illustrated. Included are most of the major developments, but a few which do not fit precisely in the pattern of development have been omitted, and none at all of the experimental work of the last two decades is included. Despite this, it is clear that an almost continual thread has been woven into the fabric of our comprehension of these phenomena. The advances have been gradual and the interest con-tinuous. As a typical measure of this interest and the importance attributed to it, the initial paper in the first issue of the *Journal of the American Chemical Society* (1879) was entitled "A Method for the Detection of Artifical or Dextro-Glucose in Cane Sugar and the Exact Determination of Cane-Sugar by the Polariscope" (7). No single work is responsible for our present knowledge, but if we were to choose the most significant contribu-tion to our current understanding, it would have to be Rosenfeld's devel-opment of the quantum origin of optical activity (8). A large part of this book (almost all of it beyond the classical electromagnetic theory of Chapter 2) is devoted to relating the observations of chiral activity to Rosenfeld's identification of it with the electric and magnetic "transition moments" that are produced by the absorption and refraction of circularly polarized light.

If wave mechanics and the Schrödinger equation may be considered as exact formulations of physical phenomena, Rosenfeld's description of the molecular origin of the rotational strength is an exact theory of optical activity for chromophores which are small compared to the wavelength of the measuring radiation. To that extent Tinoco's statement is incorrect. Any modern student, however, will quickly recognize that there is a vast gulf between Rosenfeld's statement of the relation between quantum mechanics and rotational strength and the actual calculation of rotational strengths from theoretical quantum treatments of molecular eigenstates. It is this gulf that has not been completely bridged. Nevertheless, Tinoco's statement was perhaps too pessimistic in its implications, although quite accurate in its formulation.

The first significant postquantum advances came with the extension of Rosenfeld's quantum theory by Edward U. Condon and his colleagues (9) and by John Kirkwood (10) to particular models. In Condon's theory the effect of the asymmetric field of the remainder of the molecule on an electron in an otherwise symmetric chromophore was introduced. Condon described this as a one-electron theory. The choice of nomenclature is unfortunate, since it could equally well apply to subsequent developments, including, for example, the quantum description of optical activity in absorbing media by William Moffitt and Albert Moscowitz (11). However, as Condon et al. explained, the name was intended to distinguish their description from the earlier theories of Max Born, Hans Kuhn, and others, whose development was cast in terms of coupled oscillators. Kirkwood's theory dealt with an entirely different aspect—not with the effect of the *static* asymmetric field on the chromophoric electron, but rather with the asymmetric *dynamic coupling* of that electron with the electronic transitions of other electrons. Tinoco, starting in 1960 with the paper (12) from which the opening quotation of this chapter is taken and continuing over a period of several years, reformulated and extended the quantum theory of optical activity to include both static and dynamic coupling effects. In all of these theories, the radiation field with which the optically active molecules interacted was treated classically, so that even Rosenfeld's basic theory which gives the quantum origin of the molecular optical activity is semi-classical. Completion of the quantum treatment by quantizing the radiation field, as well as the molecular states, leads to exactly the same result (13, 14) but has the added feature that it does not, as did Kirkwood's theory and the exciton theory of Moffitt (15), require the approximations associated with the assumptions that the wavelength of the light is long compared to the length of the molecules, an approximation that could obviously lead to difficulties in the treatment of the optical activity of large polymers. In addition to classical radiation theory, all these developments have in common the use of quantum mechanical perturbation theory, which, as we shall see, permits important approximations to be made when the energy of the chromophoric electron is perturbed only slightly by the radiation field or by the field of the surrounding atoms. A nonperturbation approach has been taken in more recent work by Rhodes (16) and Loxsom (17), who have also shown that supposed difficulties in the Kirkwood-Moffitt approach are due to the strict application of the Rosenfeld theory with its restriction to chromophores small compared to the wavelength. The essential conclusions of the Kirkwood-Moffitt theories are not, however, in question. The origin of the optical activity has also been described by Ying-Nan Chu (18) in terms of two-photon processes in quantum electrodynamics. In this treatment the examination of higher order processes shows how optical activity of small magnitudes could arise in symmetric systems for which the first-order optical activity is forbidden. The convergence of the theoretical developments in the Tinoco generalizations illustrated in Fig. 1.00 is therefore not intended to imply that the theoretical developments were completed with this generalization. Developments continue as advances are made both in quantum mechanics and quantum electrodynamics or radiation theory. But with the Tinoco generalization, it became possible to bring together all the classical and quantum treatments and to make predictions of the effects of structural and electronic parameters on chiroptical properties for molecules of any size, with some confidence that at least the major interactions would be accounted for.

This brief historical introduction characterizes the approach of this book, which is directed toward providing the semiclassical and quantum description of the phenomena of optical rotation and circular dichroism, largely in the formalism of the original work. The extensive theoretical work of other investigators will be discussed in the context of the basic electromagnetic and quantum theories. So also, the experimental work of investigators too numerous to mention here has formed the inspiration and the testing ground for the theoretical developments. Particular samples of these will be chosen to demonstrate the relationship between the theories and the observations. One early investigation was, however, critically important for the theoretical developments. This was the work of André Cotton (19), who demonstrated that the special shape of the wavelength dependence in the optical rotation curves of optically active solutions at absorption bands was due to the differential absorption of left and right circularly polarized light, that is, to circular dichroism. This effect and the curves that measure it, to which Cotton's name is now attached, were fundamental clues for establishing the origin of chiroptical activity in the transition moments associated with optical transitions between molecular states of different energies.

The extension of quantal descriptions to include the details of vibronic interaction (20, 21) is another aspect to be dealt with as measurements become more sophisticated. The measurements themselves and the development of instruments for making them require a long story, better left for another discussion. The optical activity of optically anisotropic media is still another problem that has received only scant attention despite the fact that the first measurements—and many subsequent ones—were made on such media, crystals in particular. Max Born (22) was the first to deal with this phenomenon on a theoretical basis, and several authors, especially

Shubnikov (23), have treated the phenomenological optics as applied to crystals in terms of symmetry. More recent treatments will be discussed in the text.

A symmetry approach of great generality has been developed by Ernst Ruch and his associates (24). In this treatment chirality, a property that differentiates an object from its mirror image, is treated as an algebraic phenomenon in which the pseudoscalar properties appear through the use of *chirality functions*. This approach appears to have important direct application to the predictivity of the results of stereoselective reactions where analogies independent of quantum mechanical considerations can be established. It is unfortunate that limitations of time and space will not permit its exposition here. Finally, and not so briefly, the all-important aspect of symmetry, which forms the foundation for optical activity, must be extensively explored. For this we depend heavily on two sources, the basic application of symmetry considerations to determine selection rules for electric and magnetic transition moments and the very important contributions of symmetry considerations to optical activity by John A. Schellman (25, 26).

No introduction to the relationship between symmetry and optical activity would be complete without reference to the original statement of the subject by Pasteur (5). Here is the way in which he described his fascinating experiment:

In this first section of this report I will establish...the connection between crystalline form and the direction of rotatory polarization....When I discovered the hemihedrism of [the crystals of] all the tartrates, I hastened to study carefully double sodium and ammonium paratartrate; I saw that tetrahedral facets, for corresponding isomorphic tartrates, were placed, with respect to the main surfaces of the crystal, sometimes rightward, sometimes leftward, on the various crystals I obtained. With great care I separated the right-hemihedral from the left-hemihedral crystals; I observed their *solutions* separately in Mr. Biot's polarization apparatus, and saw, to my surprise and pleasure that the [solutions of] right-hemihedral crystals deviated the plane of polarization rightwards while the [solutions of] left-hemihedral crystals deviated the plane of polarization leftwards. Thus, starting from chemical paratratric acid, I obtained in the ordinary manner, double sodium ammonium paratartrate and the solution, after standing for a few days, redeposited crystals which all had exactly the same angles and the same appearance; there could be no doubt that the molecular arrangement was quite different from one to the other group. The rotatory power thus attests the mode of asymmetry of the crystals. The two kinds of crystals are isomorphic, and isomorphic with the corresponding tartrate; but the isomorphism appears there with a hitherto unexampled peculiarity; *the isomorphism of two dissymmetrical crystals looking at each other in a mirror.*

Thus as early as 1848 Pasteur recognized that the inherent repository of the optical activity of dissymmetric crystals were the separate molecules and that in these molecules lay the origin of the dissymmetry of the crystals themselves. He moreover recognized that a chiral crystal structure could arise from the inherent handedness of the molecules and that the resultant optical activity could have contributions both from the inherent molecular activity and the chirality of the crystals (see Section 9.3). Pasteur's statement can, of course, be related to modern theoretical discussion of the relationship between symmetry and optical activity. This will be discussed in considerable detail in Chapter 6. Here we want only to make some preliminary observations on the relationship between symmetry and optical activity and present some of the consequences.

We will see that optical activity arises from the vectorial properties of the interaction of polarized radiation with molecules along symmetry-fixed directions. Optical activity itself is not a vectorial property but instead a property that has both sign and magnitude but not direction, i.e., a pseudoscalar. Thus an optically active substance that has a positive sign for rotation at a particular frequency of the incident radiation will rotate plane-polarized light of that frequency in a clockwise direction regardless of the direction of illumination, with respect to axes fixed in the sample, provided only that the system is optically isotropic for unpolarized radiation—a liquid or gas, for example (27). (By convention, when an observer looking toward the oncoming radiation observes a clockwise rotation, the rotation is designed as positive.) It is apparent therefore that the optical rotation of the incident polarized beam would be exactly canceled if it were returned through the same path by reflection from a mirror after passing through an optically active material (28).

The symmetry considerations that fix the direction for polarized absorption provided the basis for "selection rules" familiar in spectroscopic nomenclature. These rules, in turn, arise from a property of the quantum mechanical Hamiltonian operator which requires that it be invariant to *symmetry operations*, that is, to operations that interchange identical particles. It may be intuitively apparent that for molecules with Pasteur's requirement of mirror image, that is, nonsuperimposability, only certain *symmetry operations* are possible. We will find that those symmetry operations that interchange like atoms or groups of a molecule by reflection through a plane or by inversion through a center of symmetry are incompatible with optical activity. Thus molecules that contain these particular *symmetry elements* do indeed have superimposable mirror images and are therefore optically inactive. Let us examine a few particular examples.

For optically inactive molecules, we can easily come up with the trivial cases of diatomics and triatomics, since these by definition must have planes and in some cases centers of symmetry. With four or more mass

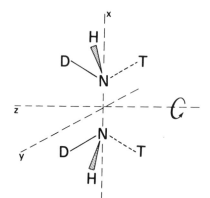

Fig. 1.01 The simplest real molecule that could exhibit optical activity if its optical isomers were resolvable—ammonia with two of its hydrogen atoms substituted by deuterium and tritium.

points, it is possible to construct geometric figures that do not have planes or centers of symmetry. Perhaps the simplest such example would be ammonia in which the nitrogen atom is bonded to one proton, one deuterium atom, and one tritium atom. Begging for a moment the question of the magnitude (the optical activity of this molecule would be exceedingly minute), we may use this molecule to illustrate two features of optically active molecules which are not under symmetry control, namely, their absolute configuration and the property of racemization. For example, in Fig. 1.01 one such molecule and its image reflected through the *yz* plane are illustrated. It is apparent that rotation of the molecule about the *z* axis, which is the only operation that results in the superposition of the nitrogen atoms, superimposes the tritium (T) atom and deuterium (D) atom as well, but not the hydrogen (H). There is no way of superimposing all the equivalent atoms in these two molecules without changing the absolute configuration of one.

The term *absolute configuration* refers to the spatial coordinates of the mass points in a defined set of three mutually orthogonal axes. Thus the two ammonia molecules illustrated in the figure have different absolute configurations, and their specification requires not only the coordinates of the mass points, but the definition of the coordinate system. In some cases it is possible for a molecule of a given absolute configuration to change into its mirror image by electron redistribution, which may or may not require the rupture of a chemical bond. This phenomenon is termed *racemization*. For the purpose of illustrating racemization, we can look again at the ammonia molecule which has a low-frequency umbrellalike inversion vibration in which the nitrogen atom moves through the plane formed by the other three atoms. It is thereby converted into its mirror image with the frequency of this inversion, which is 2×10^{10} cycles/sec.

Consequently, even if we were to prepare a quantity of mono-deutero-tritioammonia molecules, they would convert from one optical isomer to that of their mirror image; that is, the sample would *racemize* within 10^{-10} sec. The two optical isomers are known as *enantiomorphs*. Absolute configuration is, however, a factor in determining whether optical isomers are enantiomorphs, for not all the optical isomers of more complex molecules need be enantiomorphs; if there are no symmetry elements whatsoever (symmetry elements other than the *identity*, of course) associated with the coordinates of the atoms of the molecule, it may have no enantiomorphs among its optical isomers.

The ammonia molecule was chosen to illustrate the properties of absolute configuration and racemization, but historically and more practically, these properties become active problems with molecules containing more than four atoms. This will be discussed in Chapter 7, where the problem of *conformation* will also be examined.

The powerful tool for discussing, analyzing, and predicting the existence of optically active absorption bands is group theory which, as applied to spectroscopic and optical problems, provides the apparatus for dealing with symmetry. In Chapters 5 and 6 we will discuss some of the postulates, methods, and conclusions of group-theoretical treatments applied to problems of absorption and rotation of plane-polarized light. We will see then, among other things, that the requirements that a molecule not have a center or plane of symmetry are only the special cases of a more general symmetry restriction, albeit perhaps the most important in terms of the vast majority of molecules. In Chapter 5 we present only sufficient discussion of group theory to provide some apparatus with which to work. There are many excellent discussions (29) of group theory and its spectroscopic uses, which the unfamiliar reader is well advised to consult if a more comprehensive understanding is required.

In still another way symmetry has entered the field of optical activity through the characterization of optically active "chromophores" as *dissymmetric* or asymmetric. These terms were used as early as 1848 by Pasteur (5) to describe the difference between crystals that lack certain symmetry elements (asymmetric) and those that lack specifically centers or planes of symmetry and are thus not superimposable on their mirror images (dissymmetric). In modern terminology these terms have come to have different, although related meanings. We will see that in order for a molecule to be optically active, Rosenfeld's theory leads to the requirement that there exist parallel components of the electric and magnetic dipole transition moments associated with its absorption bands. If for any given optical transition, these components exist in zeroth order, without perturbation by the asymmetric field of the remainder of the molecule not involved in that particular transition, then the molecule or chromophore responsible for the transition is said to be *dissymmetric*. If, on the other hand, in zeroth order no such parallel components exist, but the optical transition nevertheless contributes to the optical activity, the molecule or chromophore is *asymmetric*. This distinction has been useful if not quite rigorous. It should be obvious then that all optically active transitions could be derived from dissymmetric chromophores if exact wave functions for the eigenstates involved in these optical transitions can be formed. Because we have not been able to do this but have had to depend instead on approximate wave functions, the distinction has been a useful one. It has given rise to some of the principal models and mechanisms of optical activity, since there are many more optically active transitions that do not have zeroth-order parallel electric and magnetic dipole transition moments than do. Schellman has distinctively characterized these models and examined (26) how symmetry determines those properties of the perturbing field of the asymmetric environment which produce the activity from an otherwise symmetric (in zeroth order) chromophore.

1.2   STEREOCHEMICAL CONCEPTS AND NOMENCLATURE

It is the asymmetric stereochemical architecture of molecules that is responsible for the phenomena of optical rotation and circular dichroism, as well as for a myriad of other physical and chemical properties. Molecular stereochemistry, which is dependent on this architecture, is a wide and well-documented subject (30). The basic rules by which most of the current methods for specifying the stereospacial arrangements of the atoms and bonds in a molecule were derived were laid down in a series of elegant papers by Cahn, Ingold, and Prelog in 1951 (31), 1956 (32), and 1966 (33) and have been formalized by the International Union of Pure and Applied Chemistry (34). We will only outline briefly some of the principal aspects of this architecture and its nomenclature, hoping in this way to pay tribute to J. H. van't Hoff and J. A. LeBel, whose insight and fortitude in the face of misguided criticism (35) made clear the existence of the dependence of optical rotation on the stereostructural arrangements of atoms in the molecule (36, 37).

Concepts of structural stereochemistry can be illustrated with examples using tetrahedral carbon atoms, which provided the basis of the van't Hoff-LeBel ideas. The basic notions are, of course, not that restricted; the configuration of any molecule is specified by its three-dimensional arrangement of atoms, and no carbon atoms at all need be involved. The necessary and sufficient condition that a molecule be chirally active is that it have an absolute configuration that is not superimposable on its mirror image. In Section 1.1 we pointed out that the specification of configuration requires a statement or reference to the sense of the coordinate system. Alternately some referral unit such as a right- or left-handed helix, which itself requires a statement of the sense of the coordinate system, can be used. Once that is done, it is trivial to understand that two molecules containing the same atoms and bonds in the same sequential arrangement, but differing with respect to the relation of the sequence to the axes of the coordinate system, are structurally related. The relationship between the sequential arrangement and the axes is designated *handedness*, and the relationship between the molecules is termed *structural isomerism*. If the isomers are exact mirror images of each other, then as we have seen earlier, they are *enantiomers* or antipodes. Enantiomers, or for that matter any structural isomers, maintain their separate integrity as long as an energy barrier exists to prevent them from turning into their antipodes or into another nonantipodal structural isomer if they are not enantiomers. Depending on the external conditions and the height of the barriers, this conversion of one structural isomer into another (racemization or epimerization) may occur over a period of time that can range from infinitesimally small, as in the case of the mono-deutero-tritioammonia molecule, to infinitely long at or below room temperature, as, for example, in a very sterically hindered biphenyl molecule (Fig. 1.02). The term racemization refers to the conversion of enantiomers to a nonoptically active equilibrium mixture, and the term epimerization is used to describe the similar conversion of one or more (but not all) optical centers of nonenantiomeric optical isomers.

Structural optically active isomers which are not enantiomers are designated diastereoisomers and must contain more than one molecular grouping responsible for the chiral activity—otherwise only two such isomers could exist, and they would be enantiomers. These structural isomers are nicely illustrated by the ephedrine molecule, which can exist in four

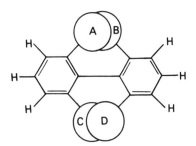

Fig. 1.02 A sterically hindered substituted biphenyl molecule which can exist in two different absolute configurations of opposite chiroptical properties. Racemization about the single bond connecting the aromatic rings is slow at room temperature or below.

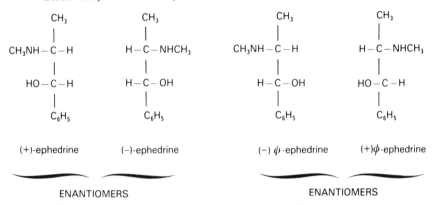

Fig. 1.03   The diastereoisomers of ephedrine (in the Fischer convention).

diastereoisomeric forms of which each of two pairs are in fact enantiomers (Fig. 1.03). A simple rule specifies the relation between the number of optical centers and the number of possible stereoisomers; there are 2^n stereoisomers in a molecule with n different optical centers. Van't Hoff and LeBel's concept of an optical center is easily generalized, requiring only that a fixed spacial nonlinear and nonplanar arrangement of four or more mass points (liganded atoms) can be specified. While our modern understanding of optical centers in terms of the electronic transitions attributable to their nuclear and electronic makeup is much more sophisticated, van't Hoff and LeBel's simple structural concepts alone are sufficient to have produced vast quantities of information about molecular structure and molecular conformations. The more theoretical underpinnings we will be discussing in this book are not designed to replace the structural ideas, but to supplement them.

Returning to the nomenclature of structural stereochemistry, the distinction to be made between conformation and configuration is important, although admittedly sometimes blurred. The *configurations* of ephedrine are shown in Fig. 1.03, but not the *conformation* of which there are many more possibilities. For example, in each of the stereoisomers the rotation of the methyl group about the carbon-nitrogen bond produces conformations in which a carbon-hydrogen bond is eclipsed either with the nitrogen-hydrogen bond or with the nitrogen bond with the carbon atom of the backbone (Fig. 1.04). The barrier between these conformations is low so that almost free rotation occurs, and as a consequence there is no observable effect of these different conformations on the optical activity. The effect of metastable conformations is much more pronounced in ring structures and in polymers. We will examine these effects, respectively, in Chapters 7 and 8.

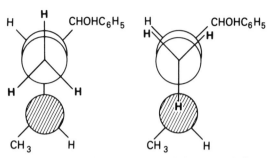

Fig. 1.04 The conformations (rotamers) of the methyl group of the (+)-ephedrine di-astereoisomer. In this figure the large forward empty circle represents the carbon atom of the methyl group and the crosshatched circle the nitrogen atom. Other rotamers, for example, of the methyl group directly bonded to the nitrogen atom are not shown in this representation.

Despite the IUPAC rules, neither the configurational notation nor the pictorial representation of optical isomers is completely uniform. However, leaving aside molecules with pseudoasymmetric or prochiral centers (which we define below) and molecules whose dissymmetry derives from a sense of twist of the chromophoric bond or bonds, the generally acceptable notation is one in which each asymmetric atom or central group is designated by the letters R and S, depending on the sequence in which surrounding liganded atoms or groups of atoms are arranged. R *(L. rectus)* stands for a clockwise and S *(L. sinister)* for a counterclockwise sense of the sequence in which each liganded atom is assigned a priority according to an established system (30). It is important to note that these designations refer to the absolute geometric arrangements about the optical centers and not to the resultant optical activity. Thus the designations (−)-R-(compound) and (+)-R-(compound) refer respectively to levorotating and dextrorotating forms of compounds having R arrangements of an optical center. More specifically than is usually noted, the (−) and (+) refer to the sense of the optical rotation at the wavelength of the Na-D line, 589 nm. Truly enantiomeric compounds could not, of course, have the same sense of optical activity with opposite absolute configurations, but it is quite possible for diastereoisomers with the same absolute configuration at one of the optical centers to have rotations of opposite sign at the Na-D line, viz., (+)-ψ-ephedrine and (−)-ψ-ephedrine which in the R, S designations are, respectively, (+)-(1S : 2R)-ephedrine and (−)-(1R : 2R)-ephedrine. As in the case of rotational isomerism, various schemes for pictorial representation are utilized. One of the more common and quite useful, especially for visualizing dihedral angles, is the Newman projection (38), Fig. 1.05. Another very useful representation for acyclic stereoisomers is the Fischer convention (39) used in Fig. 1.03. In this convention the

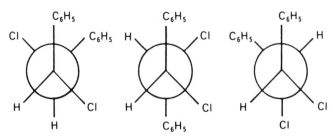

Fig. 1.05 The Newman projections of one of the conformations of 1,2 diphenyl-1,2 dichloro-ethane. Note that in this case since the optical centers of the two carbon atoms are identically substituted, there are only two configurationally different optical isomers (enantiomers) and not four ($=2^n$) as in ephedrine, which has two differently substituted optical centers.

main chain of bonds, positioned vertically, is imagined to be in the plane of the paper or projecting behind it, while the bonds to side atoms or groups project toward the observer.

In addition to the more general rules, some families of compounds have had special systems of characterization. In the α-amino acids, for example, the letters D and L are used for the absolute configurations illustrated in Fig. 1.06. Once again, these are not related to the (+) and (−) or lowercase *d* and *l* that designate dextro- and levorotatory. In the polypeptides formed from these amino acids, diastereoisomers become possible despite the fact that the monomers themselves are enantiomeric. Thus a D-amino acid may polymerize to form either or both right-handed and left-handed α helices depending on the free-energy difference between the resulting helices. Helices may or may not have an element of symmetry resulting from the nature of the monomer from which they were polymerized. If the monomer of a single-stranded helical polymer has a center of symmetry, or if there is formed a two-stranded helix in which the direction of the atom sequence in each strand is opposite, then the atomic or group sequences in both directions are identical and the helix is said to be palindromic (*Gr. palindromos*, meaning running back). Single-stranded and

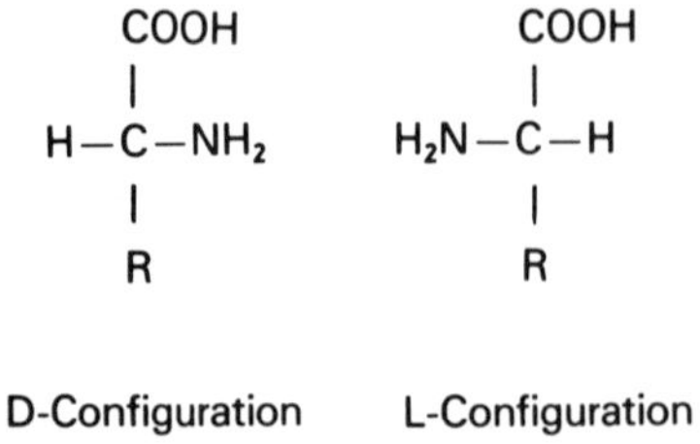

COOH
|
H—C—NH₂
|
R

D-Configuration

COOH
|
H₂N—C—H
|
R

L-Configuration

Fig. 1.06 The D and L conventions for α-amino acids.

in many cases double-stranded palindromic helices derive their chiral activity entirely from the helical structure, as we will see in Chapter 8.

Finally, we wish to touch very briefly on the definitions of pseudochirality and prochiral centers. The development of concepts and nomenclature in this field is still in flux (40, 41). A prochiral center is most generally perceived as a nonoptically active center with a plane of symmetry existing in a molecule with other optically active centers, so that stereochemical reactions involving the prochiral center are able to distinguish between otherwise equivalent atoms or groups on each side of the symmetry plane. A pseudoasymmetric center is one that has a nominal plane of symmetry by virtue of the fact that two (in the case of a tetrahedral atom) of its ligands have the same chemical structure but are enantiomorphic. The subtleties and systems of nomenclature in compounds of these types are discussed in references 40 and 41.

In the chapters that follow these stereochemical concepts will enter only after we have come to some understanding of the theory of the interactions between light and matter that are responsible for the fact that centers of molecular structural asymmetry or dissymmetry delineate the sign of the optical activity. This, then, is the central problem with which this book is concerned. We will begin with the classical description of light as electromagnetic radiation in preparation for a discussion of the phenomena of optical activity and circular dichroism, which are generically referred to as chiral and chiroptical phenomena. Optical activity is defined as the observed rotation of plane-polarized light when it is passed through an "optically active" substance. Circular dichroism is the observed elliptical polarization of right and left circularly polarized light produced by an optically active medium in the vicinity of its absorption bands. The magnitude of the optical rotation is measured in degrees (ϕ) or radians (δ) and is found to be proportional to the difference between the indices of refraction for left and right circularly polarized light:

$$\phi \sim (\eta_1 - \eta_r). \tag{1.200}$$

The magnitude of the circular dichroism, which is usually measured as the ellipticity θ, in degrees, is proportional to the difference in extinction of the left and right circularly polarized beams:

$$\theta \sim (\varepsilon_1 - \varepsilon_r). \tag{1.201}$$

These chiral parameters are properties of matter in bulk. Our problem is to examine how these properties are derived from the structure and properties of the individual molecules. We will find that it is the molecular reaction to the electric *and* magnetic fields of the radiation which gives rise to the differential response of the optically active and circularly dichroic media. Using Maxwell's description of these fields, the magnitude of the chiral activity turns out to be proportional to a parameter g, the factor of proportionality for optical rotation (in degrees) being $4\pi^2c/\lambda^2$:

$$\phi = \frac{4\pi^2c}{\lambda^2}g.$$ (1.202)

In this expression c is the velocity of light and λ is the wavelength of the measuring radiation. Then we will see that Rosenfeld (8) found from quantum mechanical perturbation theory that g is related to the electronic structure of the individual molecules through

$$g = \frac{c}{3\pi h}\sum_b \mathrm{Im}\,\boldsymbol{\mu}_{ab}\cdot\mathbf{m}_{ba}\frac{\lambda^2-\lambda_{ab}^2}{\lambda^2\lambda_{ab}^2}.$$ (1.203)

In this equation h is Planck's constant, Im stands for "Imaginary part of" $\boldsymbol{\mu}_{ab}$ and $\mathbf{m}_{ba}$ are, respectively, the electric and magnetic dipole transition moments associated with an optical transition between electronic states a and b of the molecule, and λ_{ab} is, of course, the wavelength associated with the energy difference between those states. $\sum_b \mathrm{Im}\,\boldsymbol{\mu}_{ab}\cdot\mathbf{m}_{ba}$ is called the *rotational strength*, and the task we set for ourselves is to examine the way in which the structural and electronic properties of molecules in their ground and excited states determine the *sign* and the *magnitude* of the rotational strength. We should find this a hearty problem.

REFERENCES AND NOTES

1. J. B. Biot, *Mem. Inst. (de Fr.)*, L, 1–372 (1812). Since Biot's memoirs are not everywhere obtainable, a two page description with extracts is to be found in Lowry's *Optical Rotatory Power*, published by Longmans, Green and Co., London, 1935, now available as a Dover Publishing Co. reprint.
2. A. Fresnel, *Ann. Chim.*, **28**, 147 (1825).
3. J. Herschel, *Trans. Cambridge Phil. Soc.*, **1**, 43 (1820).
4. M. Faraday, *Phil. Mag.* [iii], **28**, 294 (1846).
5. L. Pasteur, *Ann. Chim. Phys.*, *III*, **24**, 442 (1848).
6. C. K. Ingold, *Proc. Roy. Soc. (London)*, A **1448**, 171 (1967).
7. P. De P. Ricketts, *J. Am. Chem. Soc.*, **1**, 1 (1879).
8. L. Rosenfeld, *Z. Phys.*, **52**, 161 (1928).
9. E. U. Condon, W. Alter, and H. Eyring, *J. Chem. Phys.*, **5**, 753 (1937).
10. J. G. Kirkwood, *J. Chem. Phys.*, **5**, 479 (1937).
11. W. Moffitt and A. Moscowitz, *J. Chem. Phys.*, **30**, 648 (1959).

12. I. Tinoco, Jr., *J. Chem. Phys.*, **33**, 1332 (1960).

13. M. J. Stephen, *Proc. Cambridge Phil. Soc.*, **54**, 81 (1958).

14. H. F. Hameka, *J. Chem. Phys.*, **36**, 2540 (1962); *Ann. Phys.*, **26**, 122 (1964).

15. W. Moffitt, *J. Chem. Phys.*, **25**, 467 (1956).

16. W. Rhodes, *J. Chem. Phys.*, **37**, 2433 (1962); **53**, 3650 (1970).

17. F. M. Loxsom, *J. Chem. Phys.*, **51**, 4899 (1969); *Int. J. Quantum Chem.*, III S, **147** (1969).

18. Y. N. Chu, *J. Chem. Phys.*, **50**, 5336 (1968).

19. A. Cotton, *Ann. Chim. Phys.*, *VII*, **8**, 347 (1896).

20. O. E. Weigang, Jr., *J. Chem. Phys.*, **43**, 71 (1965).

21. O. E. Weigang, Jr., and E. C. Ong, *Tetrahedron*, **30**, 1783 (1974).

22. M. Born, *Dynamik der Krystalgitter*, B. G. Teubner, Leipzig, Germany, 1915; *Z. Phys.*, **8**, 390 (1922).

23. A. V. Shubnikov, *Principles of Optical Crystallography*, English language ed., Consultants Bureau, New York, 1960.

24. E. Ruch and A. Schonofer, *Theor. Chim. Acta (Berlin)*, **19**, 225 (1970); E. Ruch, *Acc. Chem. Res.*, **5**, 49 (1972).

25. J. A. Schellman, *Acc. Chem. Res.*, **1**, 144 (1968).

26. J. A. Schellman, *J. Chem. Phys.*, **44**, 55 (1966).

27. In the case of crystalline solids, the same principles hold, but the situation is much more complex. This will be discussed in Chapter 9 and even more extensively in the references given there.

28. E. U. Condon, *Rev. Mod. Phys.*, **9**, 432 (1937).

29. Treatises on group theory and its applications are subjectively more useful with various preconditions of understanding and depending on the particular purpose for which they are being consulted. Without prejudice as to their utility or quality, listed below are some of those which the author has found useful: L. H. Hall, *Group Theory and Symmetry in Chemistry*, McGraw-Hill, New York, 1969; M. Tinkham, *Group Theory and Quantum Mechanics*, McGraw-Hill, New York, 1964; E. P. Wigner, *Group Theory*, Academic, New York, 1959; the latter is a revised and translated edition, the original having been published in 1931 by Fried, Vieweg and Sohn, Brunswick, Germany.

30. There are a number of excellent texts and articles, many of which are referenced in a particularly concise and convenient short monograph by K. Mislow entitled *Introduction to Stereochemistry*, W. A. Benjamin, 1965.

31. R. S. Cahn and C. K. Ingold, *J. Chem. Soc. (London)*, **612**, (1951).

32. R. S. Cahn, C. K. Ingold, and V. Prelog, *Experientia*, **12**, 81 (1956).

33. R. S. Cahn, C. K. Ingold, and V. Prelog, *Angew. Chem. (int. ed.)*, **5**, 385 (1966).

34. IUPAC 1968 Tentative Rules, Section E. Fundamental Stereochemistry, *J. Org. Chem.*, **35**, 2849 (1970).

35. F. G. Riddell and M. J. F. Robinson, *Tetrahedron*, **30**, 2001 (1974).

36. J. H. van't Hoff, *Arch. Neer.*, **9**, 445 (1874); *Bull. Soc. Chim. Fr.*, **23**, 295 (1875).

37. J. A. LeBel, *Bull. Soc. Chim. Fr.*, **22**, 337 (1874).

38. M. S. Newman, *J. Chem. Educ.*, **32**, 344 (1955).

39. E. Fischer, *Berichte*, **24**, 2683 (1891).

40. V. Prelog and G. Helmchen, *Helv. Chim. Acta*, **55**, 2581 (1972).

41. H. Hirschmann and K. R. Hanson, *Tetrahedron*, **30**, 3649 (1974).

CHAPTER TWO

THE MOLECULAR RESPONSE TO LIGHT

2.1 BEING AN INTRODUCTION TO THE INTERACTION OF POLARIZED RADIATION WITH MATTER

Unlike other chapters in this book, this chapter starts with the index 2.0 rather than 2.1 to emphasize the significance of the introductory comments. In these comments we will try to establish intuitively some of what is required to understand the way in which light interacts with matter so that it becomes possible to permit the development that follows to fit comfortably in that part of our brain reserved for abstract notions. Lest the reader think that this is an impudent or insulting exercise, consider the following quotation which appears just after one of the most beautiful expositions of the classical formulation of electric and magnetic fields in Feynman's *Lectures on Physics*:

> I have asked you to imagine these electric and magnetic fields. What do you do? Do you know how? How do I imagine the electric and magnetic field? What do I actually see? ...you say, "Professor please give me an approximate description of the electromagnetic waves, even though it might be slightly inaccurate. ..."
>
> I'm sorry I can't do that for you. I don't know how. I have no picture of this electromagnetic field that is in any sense accurate. I have known about the electromagnetic field a long time—I was in the same position 25 years ago that you are now, and I had 25 years more of experience thinking about the wiggling waves. When I start describing the magnetic field moving through space, I speak of the **E** and **B** fields and wave my arms and you may imagine that I can see them. I'll tell you what I see. I see some kind of vague shadow, wiggling lines—here and there is an **E** and **B** written on them somehow, and perhaps some of the lines have arrows on them—an arrow here or there which disappears when I look too closely at it. When I talk about the fields swishing through space, I have a terrible confusion between the symbols I use to describe the objects and the objects themselves. I cannot really make a picture that is even nearly like the true waves. So if you have some difficulty in making such a picture, you should not be worried that your difficulty is unusual. (1)

Feynman was, of course, both accurate and modest. The picture of electromagnetic waves which derives from Maxwell's equations and from subsequent modifications to accommodate new phenomena and new observables is extremely accurate and has been tested time and time again. The principal difficulties to which he was alluding involve the problem of relating the mathematics to the visual picture our nature demands when we deal with physical phenomena. Even this is not completely hopeless because, as Feynman observes, we can indeed draw little lines and associate electric and magnetic fields **E** and **B** with them. And we can treat various parameters of these fields as vectors having magnitude and direction and describe both visually and mathematically what happens to these vectors as they travel through space or through matter which reduces their size by absorbing the energy associated with them, or changes their direction by reflecting or diffracting them, or rotates them about the line along which they are moving because of the asymmetry of the matter through which they are moving—or indeed all of these simultaneously.

We have the notion of a light wave. All of us are familiar with that. We know that if it is monochromatic, it has a definite wavelength, which means that the mathematical description of its propagation in space is a trigonometric function, say the $\sin\theta$ or the $\cos\theta$ where θ is a function of that wavelength. The drawing of that function is the familiar curved line of Fig. 2.00. Is that all there is to the light wave? A little more sophistication tells us that the line represents the time and/or space behavior of the tip of an arrow drawn from the horizontal coordinate; that the height of the arrow, that is, the value of the line projected on the vertical x coordinate, represents the magnitude of some property of the wave. It was in the mid-nineteenth century that James Clark Maxwell described the electric field of the light wave; we label it with the letter **E** whose numerical value is the magnitude of **E** and whose boldface type indicates that it has a direction in space relative to a set of coordinates fixed in space—in this case, parallel to the x axis and in the xz plane. The wavelength of the radiation represented by this line is λ, which is just the distance along z

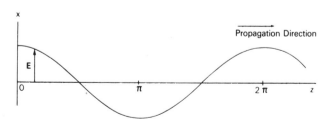

Fig. 2.00   Magnitude of the electric field component **E** of an electromagnetic wave.

between a full cycle, that is, between values θ and $\theta + 2\pi$. The wavelength is related to the frequency of the light by $\omega = c/\lambda$ if λ is measured in vacuum and c is the velocity of light in vacuum (or free space as it is more frequently referred to). This brings up an important point. In dense material (even in air for that matter, although the effect is naturally smaller), the wavelength of light changes from its value in a vacuum, and it changes because of a property of the material which we know as its index of refraction. In terms of the macroscopic properties of matter, the index of refraction is a well-defined quantity. For example, for a transparent medium with thickness longer than the wavelength of light (and with light of say 5000 Å, 5×10^{-5} cm, that does not have to be very thick), it is exactly given by Snell's law which specifies how the sines of the angles of incidence and refraction of the light are related to the refractive indices on both sides of the surfaces between the media.

Up to now most of us are on familiar ground. Perhaps not so familiar is the fact that the magnitude of the index of refraction is strongly responsive to the extent of the absorption of the radiation by the material through which it is traveling. If the index in a transparent region is designated by the symbol η, this response is measured by the absorption index k and the value of the refractive index by $\hat{\eta} = \eta + ik$. If the material is transparent at the wavelength of the radiation, $k = 0$ and the refractive index is just η. The index of refraction is therefore a response to the light wave, for there is no property of the material related to the index of refraction that can be measured without the light, that is, without using electromagnetic radiation of some wavelength.

We will have other occasions to use and discuss a complex quantity that consists of a real part and an "imaginary" part, designated here, respectively, by η and ik. In order that unfamiliarity with the use of complex quantities in this context should present no barriers, we interpose here that the connotation becomes obvious from a comparison between the refractive index and absorption index in a region of optical absorption and sine and cosine functions (Fig. 2.01). The values of the sine and cosine are seen to be 90 degrees out of phase with each other; that is, at 0, 90, 180, 270, and 360 degrees, or indeed at any angle, the sine has the values that the cosine had 90 degrees earlier. In a very similar, *but not identical way*, the absorption index k lags behind the real part of the refractive index η. The quantity i ensures that the mathematical description of the two observables will fit this description. Additional insight into the meaning of i will be found in the discussion of the polarization of a light beam and of the magnetic moment associated with an optical transition between two molecular states.

Although it does not manifest itself except in the presence of the radiation field, the refractive index is therefore a property of the medium. We will see that it is intimately related to microscopic properties, that is, to

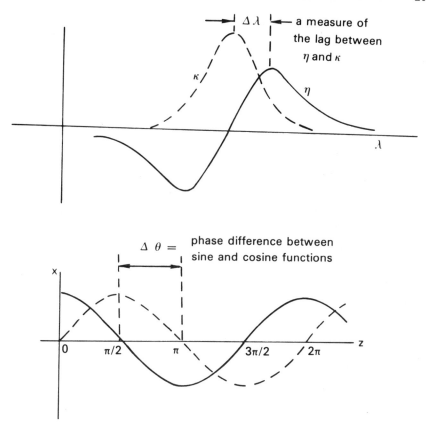

Fig. 2.01 The relation between the refractive index η and the extinction coefficient κ in the vicinity of a resonance absorption compared with the relation between sine and cosine functions.

the electronic configuration of the molecules in their ground and excited states. But let us go back to our light waves about which we know yet so little. Just a little more intuition and perhaps we can justify going to Maxwell's equations. Maxwell's equations can be used to rigorously derive the classical equation for optical activity. We will not reproduce the derivation but will examine Maxwell's equations for the physical insight they give into the interaction of light with matter.

The wave in Fig. 2.00 is a plane wave; the vector $\mathbf{E}$ points always in the x direction (alternating $+$ and $-$ as it goes through space in intervals of πz or πt when z and t are whole numbers) and is in the xz plane. If it is disturbing that the field of the light wave goes alternately positive and negative, while light as we know it visually is either present or not present but does not seem to have a sign, wait until Chapter 4 where the not unsurprising result will be reached that light intensity is proportional not to E but to $|E|^2$. The propagation of the wave is along the line z. But since there is no component of the vector E along the z axis, $(E_z = 0)$, neither can there be a variation of E_z with respect to the distance along this axis. Symbolically,

$$\frac{\partial E_z}{\partial z} = 0. \tag{2.000}$$

Similarly, in the curve we have drawn there is no component of the vector E along the y axis, so that

$$\frac{\partial E_y}{\partial y} = 0. \tag{2.001}$$

If the wave emanated from a source at the origin, propagating in the z direction with the velocity c, then each location on the z axis represents a point in time, and at any given time t, its amplitude, $|E| = E_x$, is a constant. Since the derivative of a constant is zero, it follows also that

$$\frac{\partial E_x}{\partial x} = 0. \tag{2.002}$$

This is an interesting result, because if the vector E is defined in terms of Cartesian components,

$$E = iE_x + jE_y + kE_z, \tag{2.003}$$

where i, j, and k are unit vectors, respectively, along the x, y, z coordinates, and a vector ∇, "del" in terms of its components,

$$\nabla = i\frac{\partial}{\partial x} + j\frac{\partial}{\partial y} + k\frac{\partial}{\partial z}, \tag{2.004}$$

then since $i \cdot i = j \cdot j = k \cdot k = 1$ and $i \cdot j = j \cdot k = k \cdot i = 0$,

$$\nabla \cdot E = \frac{\partial E_x}{\partial x} + \frac{\partial E_y}{\partial y} + \frac{\partial E_z}{\partial z}. \tag{2.005}$$

It follows from Eqs. 2.000–2.002, that for this representation of a plane wave in terms of a vector E parallel to the x axis (for which we have just seen each of the partial derivatives is zero),

$$\nabla \cdot \mathbf{E} = 0. \tag{2.006}$$

This is the first of Maxwell's equations for an electromagnetic field in the absence of free charges or currents. It is the mathematical symbolism for the statement that the magnitude and direction of the electric field of a plane light wave does not depend on its position in space. The only assumption we have made, based on a rather sketchy understanding of a light wave, is that the wave can be represented by a vector which specifies the amplitude and direction of the electric field of the wave at points in time or space. So we are no closer to answering Feynman's question, but we should now have some intuitive feeling for the basis of describing, by Maxwell's first equation, the electric field of that wave at a fixed point, and for the fact that matter interacts with that field in a manner regulated by the specific properties that give rise to its refractive index.

We should, and will, consider in more detail the response of the medium that gives rise to the refractive index. In fact, one might almost say that this is what this book is all about—if we restrict our thinking to the response when the light is circularly polarized. However, first let us continue with our preliminary look at the properties of light because even in our state of relative ignorance, we know enough to realize that we have hardly touched the surface. When we think of a wave, we rarely think of it as static. Even if we think of a standing wave produced by a "continuous" coherent source, that is, one in which the magnitude of the **E** vector remains constant with time at a fixed point in space, we still think of the wave as propagating in space (the "continuous" source emits the radiation as discrete quanta). So there must be a time variation of the electric field. The derivative of the field with respect to time is not equal to zero:

$$\frac{\partial \mathbf{E}}{\partial t} \neq 0. \tag{2.007}$$

Before invoking Maxwell's equation which relates this time variation of the electric field to a magnetic field, recall what we know from everyday experience, namely, that passing a current through a coil surrounding an iron core produces a torque on the coil, which, if it is mounted so that it is free to rotate, will do so—a simple electric motor. What has happened? The current was turned on to produce a flow of electrons. The electric field created by the current changes from zero to some finite value creating a magnetic field in the iron core which produces a torque on the coil. The motion of the coil, in turn, varies the electric field on the iron core leading to the continuous production of a magnetic field in the core. The motor runs until the current is shut off, whereupon the electric field drops to zero, and the motor stops as soon as friction of the bearings overcomes the momentum imparted by the torque produced during the time the field was reduced to zero. A child's example? Perhaps, but it was Faraday (2) in 1840 who discovered the essential ingredient—a *changing* electric field.

Well, our light wave has such a field, $\partial E/\partial t \neq 0$, and it was Maxwell (3) who described how this time-dependent electric field produced a magnetic field. The equation that describes it (in the absence of currents or charges),

$$c^2 \nabla \times \mathbf{B} = \frac{\partial \mathbf{E}}{\partial t}, \tag{2.008}$$

tells something very important if we only remember that the cross product $\mathbf{L} \times \mathbf{M}$ of any two vectors is a vector at right angles to both. So Maxwell's second equation says that the electric field $\mathbf{E}$ and the magnetic field $\mathbf{B}$ of a light wave are at right angles to each other.

What about the direction of the magnetic field $\mathbf{B}$? We already know that the electric field $\mathbf{E}$ is perpendicular to the direction of propagation of the light, and now we know that $\mathbf{B}$ is perpendicular to $\mathbf{E}$. Is $\mathbf{B}$ in the direction of propagation z, or is it perpendicular to both $\mathbf{E}$ and the direction of propagation, that is, in the y direction? (Fig. 2.02).

There are several ways of answering this question, but at this point we are prepared to invoke our belief in electromagnetic theory and call upon a third of Maxwell's equations, which like the second, tells us that the magnetic field is perpendicular to the electric field:

$$\nabla \times \mathbf{E} = -\frac{\partial \mathbf{B}}{\partial t}. \tag{2.009}$$

Thus $\nabla \times \mathbf{E}$ is a vector and it could have components along the $x, y,$ and z

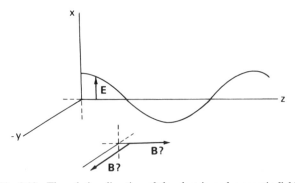

Fig. 2.02   The relative direction of the electric and magnetic fields.

axes. However, from the definition of the cross product of two vectors, $\mathbf{L} \times \mathbf{M} = (L_z \cdot M_y - L_y \cdot M_z)\mathbf{x} + (L_x \cdot M_z - L_z \cdot M_x)\mathbf{y} + (L_y \cdot M_x - L_x \cdot M_y)\mathbf{z}$ and since we have defined the light wave as plane polarized, having a component of $\mathbf{E}$ along x only,

$$(\nabla \times \mathbf{E})_x = \frac{\partial E_z}{\partial y} - \frac{\partial E_y}{\partial z} = 0 = -\frac{\partial B_x}{\partial t},$$

$$(\nabla \times \mathbf{E})_y = \frac{\partial E_x}{\partial z} - \frac{\partial E_z}{\partial x} = \frac{\partial E_x}{\partial z} = -\frac{\partial B_y}{\partial t},$$

$$(\nabla \times \mathbf{E})_z = \frac{\partial E_y}{\partial x} - \frac{\partial E_x}{\partial y} = 0 = -\frac{\partial B_z}{\partial t}. \tag{2.010}$$

Therefore only the B_y component of $\mathbf{B}$ exists, and *the magnetic, as well as the electric, field of the light wave is transverse to the direction of propagation.*

We have come pretty far in a few short pages. But we are still a long way from understanding chiral phenomena. We still have to tackle the molecular response to the radiation both from a macroscopic and molecular point of view, the subject that will occupy us for the remainder of the book. We will deal with the macroscopic response rather rapidly in this chapter. But before that, let us consider just a few more aspects of electromagnetic fields. For one, for completeness, we set down another equation usually included in the set of Maxwell's equations, one that is fairly obvious from what we already know. From Eqs. 2.009 and 2.010 we know that B_z and B_x are zero and the E_x and B_y must be similar functions (except perhaps for a constant). So we are led by analogy to Eq. 2.006 to Maxwell's fourth equation, which is just that

$$\nabla \cdot \mathbf{B} = 0, \tag{2.011}$$

which, like the statement $\nabla \cdot \mathbf{E} = 0$, just says that the magnitude and direction of the magnetic field of the light is also independent of its position in space.

You will have noticed that except for the discussion of the refractive index, all of our comments on the electric field have been with respect to free space; especially, nothing has been included in Maxwell's equations which would account for any effect of matter on the electric field of the light waves. This is not quite true for the corresponding magnetic field $\mathbf{B}$, which is induced by the electric field and for which we have begged the question of the response to a material medium by casting the equation in the form of this field. $\mathbf{B}$ may or may not be the same as the free-space field $\mathbf{H}$. We will comment on this again.

The most exquisite way of looking at the interaction of light with matter is to leave classical theory completely and go to quantum electrodynamics where both the light and the material are treated as quantum phenomena. This is also the approach least amenable to intuition and visualization. Classically, we know that forces are exerted by these fields, and we have the example of the electric motor. We are also aware from dielectric theory that if a dielectric material has incident on it a constant electric field $\mathbf{E}$, then the field $\mathbf{D}$ inside the medium is related to $\mathbf{E}$ by a quantity known as the dielectric constant ϵ, which is a property of the dielectric material:

$$\mathbf{D} = \epsilon \mathbf{E}. \tag{2.012}$$

That is a pretty simple relation. We will find that the dielectric constant is just the square of the index of refraction. In the absence of free charges, ϵ is always greater than unity, so that the field $\mathbf{D}$ in a nonconducting medium is always greater than the applied field $\mathbf{E}$. Therefore ϵ is another way of expressing the response of the medium to the field, to a static rather than a time-dependent field. What *is* the nature of this response and where does it arise? Let us stay with the static field for a moment and consider a very simple system of a neutral molecule held fixed in space. Regardless of its shape or size, if it is a neutral molecule, the center of positive (nuclei) and negative (electrons) charges are fixed with respect to each other. If the centers do not coincide, we call the molecule polar. If they do, it is nonpolar. We all know this, of course, but the purpose of repeating it here is to emphasize that in the presence of a field, these centers are displaced, and the amount by which they are displaced depends linearly on the electric field, the factor of proportionality being termed the polarizability. Thus a new polarity or dipole moment, μ_{ind}, is induced. The charge displacement occurs, of course, because the field exerts a force, moving the center of positive charge in the direction of the negative direction of the field and the center of negative charge in the positive direction. The induced moments are additive for a system of many molecules:

$$\mathbf{P} = \sum^{N} \mu_{\text{ind}} = N\mu_{\text{ind}}, \tag{2.013}$$

N being the number of molecules in unit volume. The induced moments depend on the charge separation induced by the field and, in fact, are proportional to the field strength, $\mu_{\text{ind}} = \alpha\mathbf{E}$, where the constant of proportionality α is termed the molecular polarizability. Thus the total moment is

$$\mathbf{P} = N\alpha\mathbf{E}. \tag{2.014}$$

In order to see how the light affects the polarization $\mathbf{P}$ of the medium through which it travels, we need only consider the nature of the electric field, which we have associated with the light wave. But first let us stay with the static field in order to establish how the field **D** in the medium depends on the incident field, that is, what the response of the medium as expressed in the dielectric constant ϵ has to do with the intrinsic electronic and nuclear structure of the molecules that make up the medium on which the field **E** is acting. Without going through an extensive derivation, it is sufficient to note that the induced polarization's being linearly proportional to **E** may be considered to be an additional field. When we account for the solid angle over which the field is working by multiplying by 4π, the field inside the medium is the sum of the applied and induced fields:

$$\mathbf{D} = \mathbf{E} + 4\pi\mathbf{P}. \tag{2.015}$$

Substituting 2.014 in 2.015 and rearranging:

$$\mathbf{D} = \mathbf{E} + 4\pi N\alpha\mathbf{E}$$
$$= (1 + 4\pi N\alpha)\mathbf{E}. \tag{2.016}$$

Compare Eq. 2.016 with 2.012; the dielectric constant turns out to be related to the proportionality between the induced moment per molecule and the field $\mathbf{E}: \epsilon = 1 + 4\pi N\alpha$. In free space, therefore, $\epsilon = 1$ and $\mathbf{D} = \mathbf{E}$. The denser medium, however, contributes $4\pi N\alpha\mathbf{E}$ to the field, and the polarizability α represents the molecular response of the medium. In a static field that is all there is to say except to define how α depends on the molecular electronic structure. In a light wave, however, there are two other considerations: both α and **E** are frequency dependent. We will have more to say about this later in this chapter, as well as in subsequent ones. For the moment we are content to note it and call attention to the fact that Eq. 2.015 is a corollary of Maxwell's equations.

All of the same arguments can be used to obtain the relation between the magnetic field in the medium, the applied magnetic field **H**, and the induced magnetic moment per unit volume **I**:

$$\mathbf{B} = \mathbf{H} + 4\pi\mathbf{I}, \tag{2.017}$$

where $\mathbf{I} = N\mathbf{m}$, and **m** is the induced magnetic moment per molecule.

It will not have escaped notice that Maxwell's equations have been written in a rather unsymmetrical way in that we used the free-space field **E** and the material magnetic field **B** to specify the transverse fields of the light wave in Eqs. 2.008 and 2.009, respectively. If we had written Maxwell's equation that describes how a changing electric field gives rise to the magnetic field (Eq. 2.008) in terms of the free-space magnetic field intensity **H**, we should have had to replace $c^2\nabla\times\mathbf{B} = \partial\mathbf{E}/\partial t$ by $c^2\nabla\times\mathbf{H} = \partial\mathbf{D}/\partial t$ and introduced the concept of the material response immediately in order to define **D**. By writing it in terms of $\partial \mathbf{E}/\partial t$, we were able to beg the question until a few paragraphs later where the response of matter comes up more naturally. This is a classical pattern, but it conforms to an intuitive picture in which a moving charge induces a magnetic field. No confusion should arise since the relation 2.017 sets straight the analogy between **B** and the material electric field **D**, usually termed the "magnetic induction" and "electric displacement," respectively. The relation between **B** and **H** which characterizes the overall response of matter to the magnetic field is $\mathbf{B} = \zeta \mathbf{H}$, where the dimensionless ζ is called the magnetic permeability of the medium, a parameter analogous to the dielectric constant.

We have come a long way from considering a light wave to be a line representing a trigonometric function on a piece of paper. With just a little more exploration of the relation between the response parameters and the structure of matter, we should be prepared to raise the question of how this response manifests itself in chiral activity.

The question which we have delayed coming to grips with is how the polarization **P**, which is related to the response parameter ϵ and therefore to the refractive index through Eqs. 2.015 and 2.016, is, in turn, related to the microscopic properties of the molecule. We are not going to give a mathematical derivation, either classical or quantum mechanical. It is beautifully derived in many places, the Feynman *Lectures* for one. Rather, let us describe it and then write down some resulting expressions that will help us understand it. We know that an electric field exerts a force on a charge. Coulomb's law describes the force between two charges—obviously due to the action of the field of each on the other. We also used the concept in discussing the polarization of the charges in a molecule which give rise to the field component $4\pi \mathbf{P}$. We shall suppose now that the charge that our field is going to act on is an electron bound in some equilibrium relationship with the nucleus of an atom or atoms. The field may either be a static field or the dynamic field of a light wave. If the field **E** is static, then the displacement of the bound electron $\mathbf{x} - \mathbf{x}_0$, remains constant until the field is turned off and the electron relaxes to its original equilibrium position $\mathbf{x}_0$. However, if the field is a time-varying field, say a sinusoidal field, the displacement of the electron is a function of the frequency of the field and of the retarding force that keeps the electron from responding with the frequency of the applied field. In that case, an equation of motion for the electron which gives the displacement can be written (4):

$$(\mathbf{x} - \mathbf{x}_0) = \frac{q^2/M}{\omega_i^2 - \omega^2 + i\gamma\omega} \cdot \mathbf{E}. \qquad (2.018)$$

The charge on the bound electron is q; M is its mass; ω is the frequency of the field (light wave); γ is a measure of the retarding force that keeps the electron bound and oscillating about x_0 with the natural oscillating frequency ω_i. The latter is a concept relatively easy to understand by just thinking of a plucked string which oscillates with a frequency depending on its length (ergo, a harp). Also just as a plucked string slows down as a result of dissipating its energy by "friction" of the molecules in the string and against the molecules in the air in which it is vibrating, so too does the bound electron dissipate some of its energy; the dissipation is represented by the term $i\gamma\omega$, γ being the so-called dissipative factor. We will see later how this dissipative factor manifests itself in chiral phenomena by shaping the dispersive curves of the well-known Cotton effects in optical rotation and circular dichroism. The displacement of the electron also determines the magnitude of the dipole moment induced by the field, $\mu_{ind} = q(x - x_0)$. Combining these equations for the displacement, we find that the induced electric dipole moment per molecule is

$$\mu_{ind} = \frac{q^2/M}{\omega_i^2 - \omega^2 + i\gamma\omega} \cdot \mathbf{E}. \qquad (2.019)$$

But from Eqs. 2.013 and 2.014, we know that the moment induced by field $\mathbf{E}$ is just $\alpha\mathbf{E}$. Therefore the polarizability α is given by

$$\alpha = \frac{q^2/M}{\omega_i^2 - \omega^2 + i\gamma\omega}. \qquad (2.020)$$

We have not written an equation yet for the electromagnetic wave; that is reserved for the next section. But if we had, we would have seen that the velocity of the wave is represented by the velocity of light divided by the refractive index. This is just another way of saying that the wavelength or frequency changes as the light goes from one medium to another. The response of the electron is related to the frequency of the applied field; Eq. 2.020 is an expression of this response in terms of the molecular polarizability. So there must be a relation between the polarizability α, which is the molecular response, and the index of refraction η, which characterizes the bulk response of the electric field of a light wave. We have to sum the polarizability per molecule over all the molecules in unit volume by multiplying by the number of such molecules, N. When this is done, it can be shown that

$$\eta = 1 + \frac{N\alpha}{2} \qquad (2.021)$$

More accurately $\eta^2 = 1 + N\alpha$, but for small $N\alpha$, this is equivalent to $\eta = 1 + N\alpha/2$. N, the number of molecules per cubic centimeter, is of the order of 10^{23}, while α (in cubic centimeters) is between 10^{-25} and 10^{-24}, so that $N\alpha$ is of the order 0.01 to 0.1. For such values, η determined from $\eta^2 = 1 + N\alpha$ and $\eta = 1 + N\alpha/2$ are almost equal.

Substituting Eq. 2.020 in Eq. 2.021 and accounting for the fact that each of the charges q_i of the molecule contributes to the polarizability, we find that the refractive index is

$$\hat{\eta} = 1 + \sum_i \frac{Nq_i^2}{2M} \frac{1}{\omega_i^2 - \omega^2 + i\gamma_i\omega}. \tag{2.022}$$

We will examine equations related to or derived from Eq. 2.022 in more detail later, but there are already obvious implications from this expression. Notice, for example, that the mass enters in the denominator of the second term. We have been talking about electrons, but nothing in the "derivation" is really limited to electrons; so M really refers to any particle of charge q. Since the nuclear charge is just the sum of the protonic charges, which for many atoms of interest (carbon, oxygen, and nitrogen, for example) is less than ten times the charge of the electron, while the mass of the nucleus is at least a thousand times that of the electron, contributions to the refractive index and, as we shall see in Chapter 10, to chiral phenomena from nuclear motions are likely to be two to three orders of magnitude smaller than those from electronic motion.

Notice also the frequency dependence. In regions where the medium is transparent, the dissipative factor—the real part of $i\gamma\omega$—is very small compared to $\omega_i^2 - \omega^2$. The frequency dependence is therefore $(\omega_i^2 - \omega^2)^{-1}$. In fact, we need only recall that we have previously defined the refractive index phenomenologically as $\hat{\eta} = \eta + ik$, where k is the ordinary absorption index. Here we see that $k = 0$ corresponds to the situation where the measuring frequency ω is very far from the resonance frequency ω_i. The resonance frequency therefore corresponds to absorption; that is, with every absorption band we associate a resonance frequency ω_i. As ω approaches ω_i, the second term in Eq. 2.022 tends to get very large, that is, to approach positive or negative infinity (see Fig. 2.03). Of course the closer ω gets to ω_i, the less can we ignore $i\gamma_i\omega$; nevertheless, just from the frequency dependence alone, the second term can contribute either positively or negatively to the refractive index. We shall see later in a similar expression for optical activity that besides the frequency dependence there are other parameters that control the sign. In this classical expression for the isotropic refractive index, the sign of the second term is controlled only

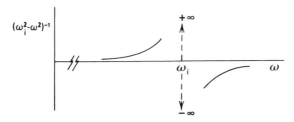

Fig. 2.03 The frequency dependence of the refractive index in transparent regions of the spectrum. ω_i is the frequency at a resonant absorption.

by a frequency dependence and is always positive for frequencies lower than a resonance frequency and negative at higher frequencies ($\omega_i > \omega$).

We have defined the response of the medium to an electric field in classical terms by describing the motion of bound electrons or nuclei as if they were balls on springs having resonance frequencies, ω_i acted on by a mechanical force. Although this kind of classical description can still give useful results, we have become accustomed to treat the molecule, if not the radiation field, quantum mechanically, using semiempirical methods to treat complex molecules. It is not really a bound electron we are discussing but quasi-elastically bound electrons and/or nuclei in discrete energy states. Associated with each state is an energy $h\omega$ ($= hc\nu$), where h is Planck's constant. Between two such states transitions may take place, provided symmetry considerations or changes in electron spin do not prevent it. Symmetry will be discussed in detail in Chapters 4 and 5, as well as in other places, and we shall have a little to say about spin considerations. If the energy difference is very small, then we are talking about nuclear motions, vibrations, or rotations having frequencies as small as $\nu = 1$ cm^{-1} or as large as about 4000 cm^{-1}, corresponding, respectively, to energy differences of 1.24×10^{-4} electron volts (eV) and 0.5 eV. These are energies associated with absorption of electromagnetic radiation in the infrared region of the spectrum. If the energy differences are much larger and range from 4000 cm^{-1} up to perhaps 80,000 cm^{-1} (10 eV), then we are talking about transitions between electronic states, and the energies are those that are associated with absorption in the visible, ultraviolet, and vacuum ultraviolet regions. Our discussion up to now has barely hinted at the fact that the fields of the light wave would induce such transitions, but we know that they do. Furthermore, we know that quantum considerations impose restrictions on the natural resonance frequencies associated with these transitions. Also their strengths vary enormously. If we apply quantum theory to derive an equation of motion of the particles excited by a light wave, then the resulting expression for the refractive index differs from the classical description (Eq. 2.022) in only two respects. The polarizability is associated with each transition between discrete states $(a \rightarrow b)$, so that contributions to the polarizability must be summed over transitions to all states. Also, since each transition has a distinct strength, a factor called the oscillator strength, f_{ab}, specifies the magnitude of the contribution of each transition to the polarizability:

$$\alpha(\omega) = \frac{q^2}{M} \sum_b \frac{f_{ab}}{\omega_{ab}^2 - \omega^2 + i\gamma_{ab}\omega}, \qquad (2.023)$$

where classically $q_i = q f_{ab}$ and q is the electronic charge. The refractive index therefore is given by

$$\hat{\eta}(\omega) = 1 + \frac{Nq^2}{2M} \sum_b \frac{f_{ab}}{\omega_{ab}^2 - \omega^2 + i\gamma_{ab}\omega}. \qquad (2.024)$$

We have finally, albeit by assuming or ignoring quite a bit of physics and its mathematical description, reached some understanding of the response of the system to the electric field of the light wave. We find that the response has its origin in the fact that the radiation induces transitions between the states. According to the classical picture, these transitions change the nuclear and electronic configurations of the molecule so as to produce an induced dipole moment. As a consequence, they are called *electric dipole transitions*. In exactly the same way, the magnetic field may produce *magnetic dipole transitions*. The latter are very important in chiral phenomena but produce only negligible effects in ordinary absorption. If, in fact, we went into detail about the interaction of the field and the electrons, we would find that quadrupole, octopole, etc., transitions may be induced. Equations 2.021–2.024, for the refractive index and polarizability, would contain additional terms. In general, these are negligible in ordinary absorption processes, but may enter into the chiral properties.

Equation 2.024 does tell us one thing more, namely, that *each* transition contributes to the refractive index, which, in turn, means that the *specific* response of the medium arises from the nature of the excited states of the molecules that comprise it. Since chromophoric groupings of atoms in many molecules are very similar, there are of course generalities that can be made, and some of these generalities will form the basis for later chapters.

Our objective now is to apply these notions to a phenomenological description of chiral activity and to describe how Rosenfeld's supplement to Maxwell's equations produces a description of the response parameters that give rise to optical activity and circular dichroism. We will then return in Chapter 3 to a somewhat more rigorous description of how the states of a system are modified by the electric field of the light wave and how this understanding is used in the quantum theory of optical activity.

2.2 THE INTERACTION OF POLARIZED RADIATION WITH MATTER WITH SPECIAL EMPHASIS ON CHIRAL PHENOMENA

Optically active media exhibit all the optical properties of inactive media. They refract, absorb, and scatter light, but in addition, they respond differently to radiation depending on whether it is left or right circularly polarized. The differential absorption of light of these two polarizations is known as circular dichroism, and the differential refraction as circular birefringence or, more commonly, as optical rotation. The term optical rotation arises from the fact that the measurement is frequently made by observing the angle of rotation of one of a pair of initially crossed linear polarizers required to return the transmission of the radiation to zero after the optically active medium has been placed between them. In the remainder of this chapter, we will outline the theoretical basis for these phenomena in terms of the interaction of the radiation with the molecules of the media. Before doing so, however, it will be helpful to examine the optical phenomenon itself as a macroscopic process. We will treat the radiation field classically and will not, in fact, develop the theory in the detail of Rosenfeld's 1928 paper (5). Instead, we will limit the discussion to the way in which Maxwell's concepts are utilized to arrive at the result that optical activity arises from the interaction of electric and magnetic moments associated with optical transitions between molecular states of different energy. For practical reasons, in the exposition of particular models as distinct from the general theory, we will be examining only the contributions from transitions between the ground state and those of higher energy.

2.3 THE OPTICAL PHENOMENON AS AN EXPRESSION OF A BULK PROPERTY OF THE MEDIUM

Unpolarized light can be described as an array of plane waves that are randomly oriented with respect to a plane perpendicular to the directions of propagation. The electric field of each wave in this array is represented by a vector whose amplitude is given by

$$\mathbf{E} = \mathbf{E}_0 \cos 2\pi\omega\left(t - \frac{\eta z}{c}\right) \tag{2.201}$$

This mathematical description is that of a wave with periodicity of 2π, exactly like that illustrated in Fig. 2.00 except that the effect of the refractive index of the material through which the wave is propagating has now been included. If we chose to place the traveling wave in a right-handed Cartesian axis system such that z is the axis of propagation, then the plane of the wave may be arbitrarily chosen parallel to one of the two other perpendicular axes, say the x axis. In this case, if $\mathbf{i}$ is a unit vector in the direction of the x axis, the amplitude of the wave is given by

$$\mathbf{E} = E_0 \mathbf{i} \cos\theta \qquad\qquad (2.202)$$

where $\theta = 2\pi\omega[t - (\eta z/c)]$, and E_0 is the maximum amplitude of the wave, that is, $\mathbf{E} = E_0\mathbf{i}$ when $\theta = 0°$. The plane wave may be decomposed into two circularly polarized waves propagating in the z direction. A simple description of these two waves is given by

$$\mathbf{E_r} = E_0(\mathbf{i}\cos\theta_r - \mathbf{j}\sin\theta_r), \qquad\qquad (2.203a)$$

$$\mathbf{E_l} = E_0(\mathbf{i}\cos\theta_l + \mathbf{j}\sin\theta_l), \qquad\qquad (2.203b)$$

and their sum can easily be shown to be $\mathbf{E} = E_0\mathbf{i}\cos\theta$ when the phase angles $\theta_r = \theta_l = \theta$. That these equations do represent circularly polarized waves may be seen from the following considerations. Since the sine and cosine functions are complementary, the amplitudes $\mathbf{E_r}$ and $\mathbf{E_l}$ remain constant for all values of θ_r and θ_l. However, the directions that the two electric vectors trace about the z axis, that is, the xy plane, differ: to the right for the negative combination and to the left for the positive combination. Left and right are defined here as they have been historically; that is, from the point of view of the observer looking at the oncoming ray. This means that in a right-handed coordinate system (see Fig. 2.04), a left circularly polarized ray is traced out by a transverse vector going from positive x to positive y, while a right circularly polarized ray goes from positive x to negative y. This tends to be out of phase with our intuition because we are accustomed to looking along the direction of propagation rather than against it. Therefore, careful attention must be paid to coordinates when determining the sign of optical activity. The fact that the sense of direction requires a definition provides interesting and sometimes amusing parables about communication. For example, one may ask how to communicate in words alone the definition of direction to an experimenter who has the same scientific and technological capabilities, but who either does not know the words left and right or who has no calibration of parity or directional sense with our own. One of many methods that require a familiarity with physical phenomena and only one assumption, namely,

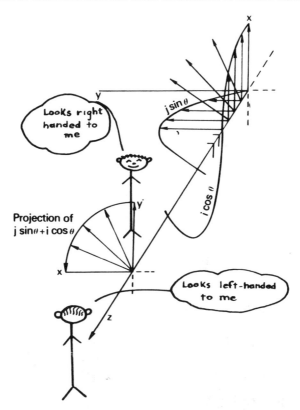

Fig. 2.04 Circularly polarized light in an optically active medium: the definition of right and left chirality with respect to the direction of propagation of the radiation.

that both the experimenters have powers of observations that allow them a sense of direction relative to themselves, is the following: Build a loop of wire such that the plane of the loop is perpendicular to a line going from the observer to the loop. Now cause a current of electrons to flow around the loop in either direction. Measure the direction of the resulting magnetic field in the loop. If the field points away from the observer, the current is flowing in a right-handed or clockwise sense in the loop; if toward the observer, the current is flowing in a left-handed or counterclockwise direction. To demonstrate how precarious is our ability to communicate precisely, however, consider if one of the experimenters is using one of the steeple clocks and the other is using the other clock (Fig. 2.05) on the City Hall of the old Jewish ghetto in Prague, Czechoslovakia!

Consider a set of simple experiments that may be performed with either the plane-polarized beam or the two corresponding circular polarized

Fig. 2.05 The clocks on the City Hall of the old Jewish Ghetto in Prague, Czechoslovakia, read the same time. Which one goes *clockwise?*

beams of equal amplitude and phase. The polarized beam is produced by interposing a perfect linear polarizer between an unpolarized source of radiation and the remaining elements. Between this polarizer and a detector capable of measuring the radiation intensity, we place an analyzer (actually another linear polarizer identical with the first). If we ignore reflection losses, the analyzer can be rotated so that 50 percent of the radiation that would be measured by the detector in the absence of both the polarizer and analyzer is transmitted by the two elements. In this case the analyzer and polarizer are said to be parallel. Rotating the analyzer by precisely 90 degrees will then reduce the transmission to zero. The polarizer and analyzer are now said to be crossed or perpendicular.

We continue with an experiment designed to examine optical rotation. An optically active transparent material is placed between the crossed polarizer and analyzer. Immediately the detector records the presence of some light. Rotating the analyzer by some number of degrees $\pm\phi$, the light at the detector is again reduced to zero. Furthermore, an additional rotation of ± 90 degrees, permits 50 percent of the total incident power to be transmitted again. So we conclude that the optically active material has rotated the plane of the polarized light by ϕ degrees in a positive or negative (right- or left-handed) sense depending on the nature of the optically active medium. Furthermore, the sense of the rotation (i.e., clockwise or counterclockwise) bears a fixed relation to the direction of propagation of the light. If the light traverses the same medium in each of two opposite directions—as when it is returned through the medium by reflection at a mirror, the net rotation vanishes.

If we inquire what has happened to rotate the plane of polarization of the light in the optically active material, we can look for an answer at two levels. The first level is macroscopic and this is relatively easy. The representation of the linear vector that traces out the plane wave is replaced by the representation for two circularly polarized beams traveling, respectively, clockwise and counterclockwise. Then if the material in the beam exhibits different properties with respect to the two beams, say, for example, that its indices of refraction are different for right circularly polarized light than for left circularly polarized light, one of these components will lag behind the other—the lag being proportional to the difference between the two indices of refraction and the length of the path through the material and inversely proportional to the wavelength of the light. To examine the meaning of the statement that one component will lag behind the other, we will determine the values of $\mathbf{E}_r$ and $\mathbf{E}_l$ at a time t after the wave has entered the active medium, at which time the values of θ_r and θ_l (initially identical) will differ because the refractive indices η_r and η_l are unequal (Fig. 2.06). From the definition of θ (see Eq. 2.202):

$$\theta_r = 2\pi\omega\left(t - \frac{\eta_r z}{c}\right), \tag{2.204a}$$

$$\theta_l = 2\pi\omega\left(t - \frac{\eta_l z}{c}\right), \tag{2.204b}$$

$$\theta_r - \theta_l = \frac{2\pi\omega z}{c}(\eta_l - \eta_r). \tag{2.205}$$

The difference, which is known as the phase lag, is just twice the angle by

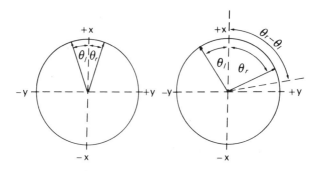

Entering the Active Medium   Leaving the Active Medium

$$\eta_l - \eta_r = 0 \qquad\qquad\qquad \eta_l \neq \eta_r$$

$$\theta_r = \theta_l \qquad\qquad\qquad \theta_r \neq \theta_l$$

$$\delta = \frac{(\theta_r - \theta_l)}{2} = 0 \qquad\qquad \delta = \frac{\theta_r - \theta_l}{2} > 0$$

Fig. 2.06   The phase difference between left and right circularly polarized components as the radiation leaves an optically active medium is twice the magnitude of the observed angle of rotation of the linearly polarized radiation.

which the incident plane-polarized wave has been rotated by the optically active medium. This can be seen when we represent the two circularly polarized waves specified by Eqs. 2.203a and 2.203b (which are now out of phase because $\theta_r \neq \theta_l$) by a plane-polarized wave:

$$\mathbf{E} = \mathbf{E}_r + \mathbf{E}_1 = E_0\mathbf{i}(\cos\theta_r + \cos\theta_1) - E_0\mathbf{j}(\sin\theta_r - \sin\theta_1)$$

$$= 2E_0 \cos\frac{(\theta_r + \theta_1)}{2}\left[\mathbf{i}\cos\frac{(\theta_r - \theta_1)}{2} - \mathbf{j}\sin\frac{(\theta_r - \theta_1)}{2}\right]. \qquad (2.206)$$

We now see that because $\theta_r \neq \theta_1$, $\mathbf{E}$ must now have an x *and* y component ($\mathbf{i}$ and $\mathbf{j}$). $\mathbf{E}$ is therefore no longer parallel to the x axis. When we refer to Fig. 2.06, E is now a plane-polarized wave rotated by $(\theta_r - \theta_1)/2$ from its original orientation in the xz plane. From Eq. 2.205 we see that this rotation results from the passage of the incident radiation through the optically active medium whose index of refraction for left and right circularly polarized radiation differs by $\eta_l - \eta_r$. The relation between the rotation and the difference in the indices of refraction is

$$\delta = \frac{(\theta_r - \theta_1)}{2} = \frac{\pi\omega z}{c}(\eta_1 - \eta_r), \qquad (2.207)$$

or since the wavelength $\lambda = c/\omega$,

$$\delta = \frac{\pi z}{\lambda}(\eta_1 - \eta_r);\tag{2.208}$$

δ is thus the angle (in radians) by which the plane-polarized wave has been rotated in traversing the optically active medium of thickness z. More commonly, the rotation is specified in degrees per decimeter:

$$\phi = \frac{1800}{\lambda}(\eta_1 - \eta_r),\tag{2.209}$$

obtained by replacing π rad by 180 degrees and z cm by 1 dm. Some indication of the sensitivity of optical rotation measurements in terms of molecular parameters can be obtained by substituting a reasonable value of ϕ in Eq. 2.209. A small but easily measurable value of ϕ is 10^{-3} degrees at a wavelength of 600 nm (measurements of optical activity are frequently made at the wavelength of the sodium D line at 589 nm), in which case the refractive index difference $(\eta_1 - \eta_r)$ is $(10^{-3}\ \text{deg/dm})(6 \times 10^{-6}\ \text{dm})/1800\ \text{deg} = 0.3 \times 10^{-11}$ or less than one-thousand millionth of the mean refractive index of most materials.

It is important to remember that we have been discussing transparent optically active material for which k_1 and $k_r \approx 0$. Otherwise the amplitudes of the components $E_0\mathbf{i}$ and $E_0\mathbf{j}$ of the circularly polarized radiation would have to be modified not only by the phase angle θ which describes the velocity difference with which they travel through the medium, but also by an exponential factor $\exp(-4\pi k_1 z/\lambda)$ and $\exp(-4\pi k_r z/\lambda)$ which describes how each of the circularly polarized beams is differentially absorbed as it travels through z cm of the active material (circular dichroism).

Before proceeding to examine the phenomenon at the molecular level, we continue with the experiment to examine circular dichroism. In doing so, we will ignore the fact that the actual experimental method must be modified so as not to measure the simultaneous effects of rotation and dichroism. If the optically active medium is not transparent at the wavelength of the incident radiation, then the transmitted intensity is substantially reduced, but not by precisely the same amount as the isotropic absorption measured under the same conditions using unpolarized light. Now, not only do the two circularly polarized beams leave the material out of phase, but also one of the components is absorbed more strongly than the other. This situation is illustrated in Fig. 2.07 where it is observed that the resulting components trace an ellipse with major axis at the angle of rotation. The *ellipticity* ψ, which is a measure of the eccentricity of the ellipse results, of course, from differential absorption of the circularly

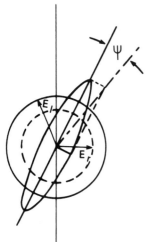

Fig. 2.07   Ellipticity, a measure of circular dichroism.

polarized component beams as the complex part ik of the refractive index contributes to the optical properties of the medium. For the circularly polarized light

$$\hat{\eta}_l = \eta_l + ik_l, \tag{2.210}$$

$$\hat{\eta}_r = \eta_r + ik_r. \tag{2.211}$$

Referring to the figure, we see that the ellipse may be characterized by the angle ψ whose tangent is equal to the ratio of the major to the minor amplitudes of the ellipse. The minor amplitude is just $|\mathbf{E}_l - \mathbf{E}_r|$ and the major $|\mathbf{E}_l + \mathbf{E}_r|$. Therefore,

$$\tan\psi = \frac{|\mathbf{E}_r| - |\mathbf{E}_l|}{|\mathbf{E}_r| + |\mathbf{E}_l|}. \tag{2.212}$$

$\mathbf{E}_l$ and $\mathbf{E}_r$ are of course no longer equal when the extinction for left and right circularly polarized light are different. When we express them in terms of the extinction difference, the function on the right-hand side of Eq. 2.212 can be shown (6) to be equal to the hyperbolic tangent, $\tanh \pi(k_l - k_r)z/\lambda$. The hyperbolic tangent of ψ is very nearly equal to ψ for values of ψ less than 1. The measured ellipticity is rarely greater than 1 even for very short wavelengths because the extinction difference $k_l - k_r$ is rarely greater than 10^{-6} or 10^{-7}. Consequently,

$$\psi = \frac{\pi z}{\lambda}(k_l - k_r). \tag{2.213}$$

In the more commonly used units for absorption, the extinction coefficient ε is related to the absorption index by

$$\varepsilon = \frac{4\pi k}{2.3\lambda C} \qquad (2.214)$$

where C is the concentration in moles per liter. Substituting in equation 2.213, the ellipticity in radians is

$$\psi = \frac{2.3\,Cz}{4}(\varepsilon_l - \varepsilon_r) \qquad (2.215)$$

or in degrees = square centimeters per mole:

$$[\theta] = \frac{18{,}000 \times 2.3}{4}(\varepsilon_l - \varepsilon_r) \approx 3300(\varepsilon_l - \varepsilon_r) \qquad (2.216)$$

In the next section we turn to an examination of optical rotation and circular dichroism at the molecular level and therefore to an investigation into the molecular parameters that give rise to the inequalities between $\hat{\eta}_l$ and $\hat{\eta}_r$ and between ε_l and ε_r.

Sometimes one gets the impression from descriptions of optical activity or circular dichroism that the observed activity results from a statistical difference between the number of molecules that rotate or absorb the circularly polarized radiation in one direction and those that exhibit the opposite behavior. This is nonsense. At a given frequency the optically active transition of a pure optical isomer is identical for each molecule; it does differ from the achiral absorption. The ambiguity results from the *description* of the phenomena in terms of these differences, not from statistically different populations of the molecules or transitions. Perhaps the most intuitive way of seeing this is in terms of scattering theory. When optical activity is approached from this point of view, the description is in terms of the differential rate at which forward scattered photons with angular momenta positive and negative with respect to the direction of propagation are produced. If the momenta of the incoming photons are equally distributed between the two phases, the differential scattering reflects the fact that those photons that did interact with an optically active molecule were either changed in sign (phase) according to the chirality of the molecule or given additional momentum in the same phase. It does not reflect a differential concentration of molecules with different chirality. Of course, if the material is not optically pure, that is, if it consists of a mixture of two or more optical isomers, then the observed chiral activity is indeed a statistical average resulting from a predominance of chiral absorption or refraction in one direction.

2.4 ELECTROMAGNETIC THEORY RELATING THE BULK REFRACTIVE INDICES TO MOLECULAR PARAMETERS

In this section we examine the response of optically active media to circularly polarized electromagnetic radiation using Maxwell's description of light waves as time-varying electric and magnetic fields. We will find that the response is governed by a molecular parameter β, which is proportional to the difference in the refractive indices for left and right circularly polarized light. We will be making two important assumptions. First, we will assume that associated with the absorption of radiation by a system of bound electrons (i.e., the electrons in stationary states of molecules), there are changes in the charge distribution that give rise to electric and magnetic "transition moments." In Chapter 3 we will see that this is not an assumption but rather a consequence of quantum theory.

The other assumption is the classical assumption that electromagnetic radiation has associated with it an electric field that is capable of inducing a moment proportional to it in a medium with which it interacts, just as does the static field that was described in Eqs. 2.013 and 2.014:

$$\mu_{ind} = \alpha E \tag{2.301}$$

This electric moment and the moment induced by the magnetic field of the radiation

$$m_{ind} = \kappa H \tag{2.302}$$

are part of the requirements of Maxwell's description of electromagnetic radiation; κ is the magnetic susceptibility analogous to α.

We start by accepting Maxwell's description of the interaction of the electric and magnetic fields of a light wave with matter. From the relationships between the applied fields, E and H, and the internal fields, D and B, we have arrived at a series of equations that express the dependence of the total induced electric moment P and magnetic moment I on the respective fields:

$$D = E + 4\pi P \tag{2.303}$$

$$D = \epsilon E \tag{2.304}$$

$$B = H + 4\pi I \tag{2.305}$$

$$B = \zeta H. \tag{2.306}$$

In these equations the field D originates only from the applied electric field

E directly, and less directly through the induced electric polarization $P = N\alpha E$; the field B originates only from the magnetic field H directly, and through the induced magnetic polarization $I = N\kappa H$. But Maxwell's equations, Eqs. 2.008 and 2.009, had already shown that a time-varying magnetic field produces an electric field, and vice versa; yet no corresponding terms appear in these equations. Thus there must be missing parts of P and I. Rosenfeld recognized that these electromagnetic field equations are therefore inadequate to describe the effect of radiation on optically active media. In order to obtain suitable expressions, the contributions of a moving charge to the magnetic field in the form of the time variation of the electric field and the time variation of the magnetic field must also be included. Methods of doing this are discussed in Section 3.4 and of course in Rosenfeld's original paper (5). The final result for optically active media is that the equivalents of Maxwell's equations in inactive media are

$$D = \epsilon E - g \frac{\partial H}{\partial t}, \tag{2.307}$$

$$B = \mu H + g \frac{\partial E}{\partial t}, \tag{2.308}$$

where g is a constant, and $\partial H / \partial t$ and $\partial E / \partial t$ are the time derivatives of the applied magnetic and electric fields. These fields are those that act on the electrons in the medium; D represents the effective *electric* field which consists of (1) the applied field E, enhanced by the medium in proportion to its isotropic dielectric constant, and (2) a field arising from the variation of the magnetic field of the light wave with time; similarly for B, the effective magnetic field. Correspondingly, the classical moments induced by the light wave microscopically, that is, in each molecule, are no longer given by Eqs. 2.301 and 2.302 but instead by

$$\mu_{ind} = \alpha E - \frac{\beta}{c} \frac{\partial H}{\partial t} + \cdots + \text{smaller terms}, \tag{2.309}$$

$$m_{ind} = \kappa H + \frac{\beta}{c} \frac{\partial E}{\partial t} + \cdots + \text{smaller terms}. \tag{2.310}$$

The total electric moment or polarization in a system containing N molecules per unit volume is just N times the moment induced in each molecule (Eq. 2.013), so that in a chiral substance

$$P = N\mu_{ind} = N\left(\alpha E - \frac{\beta}{c} \frac{\partial H}{\partial t}\right). \tag{2.311}$$

Let us examine this expression. We see that the total polarization, that is, the total response of the medium to the electromagnetic wave, consists of one part proportional to the electric field and another part proportional to the rate at which the magnetic field varies, the factors of proportionality, aside from the constants N and c, being, respectively, the polarizability α and the as yet unspecified parameter β. We have seen that the polarizability is related to a microscopic property of the molecules that make up the medium, namely, the change in their static charge distribution which results from the application of a static field and more important from our point of view, from the changes in the charge distribution under the influence of nonstatic radiation fields. Similarly, β, which will turn out to be the rotatory parameter that specifies the magnitude of the chiral activity, is related to microscopic molecular properties; since it arises in response to the variation of $\mathbf{H}$ with t, we will be concerned only with its time-dependent properties. The response parameters α and β can now be related to the macroscopic properties of the medium through Maxwell's relations; substituting Eq. 2.311 in Eqs. 2.303 and 2.304, we get

$$\mathbf{D} = \mathbf{E} + 4\pi N\alpha\mathbf{E} - 4\pi N \frac{\beta}{c} \frac{\partial \mathbf{H}}{\partial t}, \tag{2.312}$$

or

$$\mathbf{D} = (1 + 4\pi N\alpha)\mathbf{E} - 4\pi N \frac{\beta}{c} \frac{\partial \mathbf{H}}{\partial t}. \tag{2.313}$$

If we compare Eq. 2.307 with Eq. 2.313, it is apparent that

$$\epsilon = 1 + 4\pi N\alpha, \tag{2.314}$$

$$g = 4\pi N \frac{\beta}{c}. \tag{2.315}$$

Equation 2.314 is the already familiar expression (see Section 2.0) for the dielectric constant in terms of the polarizability of the molecule and is related to the ordinary refractive index through $\eta^2 = \epsilon$ (7). In order to determine the nature of g, it is necessary to obtain the relationship between the field inside the medium $\mathbf{D}$, and the applied fields $\mathbf{E}$ and $\mathbf{H}$. The relations between g and the indices of refraction for circularly polarized light are found to be (6)

$$\eta_r = \sqrt{\mathbf{E}} - 2\pi\omega g, \tag{2.316a}$$

$$\eta_l = \sqrt{\mathbf{E}} + 2\pi\omega g. \tag{2.316b}$$

The rotation, which has been defined phenomenologically (Eq. 2.209) in terms of the difference between η_r and η_1, is then proportional to g:

$$\delta = \frac{\pi z}{\lambda}(\eta_1 - \eta_r) = \frac{4\pi^2 z}{\lambda^2}g. \tag{2.317}$$

Substituting for g the molecular parameter β from Eq. 2.315,

$$\delta = \frac{16\pi^3 Nz}{\lambda^2 c}\beta. \tag{2.318}$$

Equation 2.318 relates the rotation to β. Both α and β thus describe the response of the molecule to the radiation field, the former that of an optically inactive medium to unpolarized radiation, the latter that of an optically active medium to circularly polarized light. Both are related through the bulk refractive index to measurable optical phenomena, the former to ordinary isotropic absorption and therefore a simple scalar, the latter to optical rotation which may be positive or negative and therefore a pseudoscalar. Rosenfeld's theory of the quantum origin of β shows that the rotation arises from simultaneous electric and magnetic dipole moments connecting the ground and excited states. We shall examine this result in detail. The quantum expression for β is obtained by identifying the coefficient of $\partial \mathbf{H}/\partial t$ in the expanded form of Maxwell's equation for circularly polarized light in an optically active medium, with the coefficient of $\partial \mathbf{H}/\partial t$, which appears in the appropriate quantum perturbation theory treatment of the interaction of radiation with matter. For this treatment the perturbation is expanded to include the operators for both the electric and magnetic dipole moments. In Chapter 3 we will examine how this is done, but in the next section we will take a preliminary look at the end result. Our aim is to raise the questions and then to describe (1) the nature of the optical transitions that give rise to circular dichroism, (2) how these transitions give rise to chiral activity both at the absorption frequencies and at frequencies far removed from measurable absorption, and (3) generalizations that can be made about the types of chromophores or transitions that are optically active.

2.5 CHARACTERISTICS OF THE OPTICAL ROTATORY PARAMETER β

The title of this section might more appropriately be "How a Very Small Equation Controls an Optical Phenomenon" or, alternately, "All About $R_{ab} = \mu_{ab} \cdot \mathbf{m}_{ba}$." The object of this discussion is to establish the relationship between the "microscopic" or molecular properties that determine the rotational strength R_{ab} of an optically active material in terms of the electric μ_{ab} and the magnetic $\mathbf{m}_{ba}$ dipole "transition moments" between electronic states a and b, and the observed chiral activity. In the last section we saw how the classical refractive index difference $(\eta_l - \eta_r)$ for optically active media is related to a molecular parameter β. In this section we will look at Rosenfeld's formula (5) for β in terms of the quantum mechanically derived induced electric and magnetic dipole transition moments and relate these latter quantities to the observed optical rotation. In principle, this is all that is required in order to calculate the optical activity from theoretically derived quantities. We will see, however, in Chapters 3 and 4, that while a great deal of insight into the origin of optical activity may be gained in this way, as a practical matter, more frequently it is necessary to pursue circuitous routes in attempting to predict or calculate the sign and magnitude of the optical activity or circular dichroism of real molecules.

The concept of transition moments arises out of quantum mechanical theory for the "expectation values" of physical quantities. We cannot develop that theory here any more than we have "developed" electromagnetic theory. It is, in fact, frequently given as a postulate with the understanding that it is self-consistent with the fundamental expressions of quantum mechanics, the time-dependent Schrödinger equation, and its corollaries. For example, Kauzmann gives it as the third "law of quantum mechanics":

When a great many measurements of any dynamical variable are made on an assembly of systems whose state function is ψ, the average result obtained will be $\bar{a} = \int \psi^* a \psi \, d\tau / \int \psi^* \psi \, d\tau$ where a is the operator corresponding to the dynamical variable and the integration is over all configurations accessible to the system. (8)

In this book, we will be interested in the average value (expectation value) of the electric or magnetic moment associated with the changes between two states ψ_n and ψ_m. The operator in the case of the electric dipole moment will be the dynamical variable $\mathbf{r}$, the distance of the electron from some chosen center of coordinates in a molecule. Alternately, the dynamical variable, by a suitable change of units can be the operator ∇, which is defined as the rate of change with respect to the coordinate. We are in the same dilemma here as we are in the case of the classical description of electromagnetic fields. The expectation value, in this case of the transition moments, is a precise measurable quantity derivable, in fact, from classical Hamiltonian and Newtonian mechanics, with the addition of the DeBroglie-Planck quantum conditions and the Heisenburg uncertainty relations; but our physical concepts are not quite adequate to the task of visualizing it. Let us try, however, to get a slightly better picture of it with the help of Werner Heisenberg's original (1930) formulation (9). Heisenberg pointed out that the classical dipole moment of a charge q, located at the position r, distant from a center of coordinates, is just qr. In quantum mechanics, the probability that the electron will occupy the available space τ is $\int \psi^* \psi \, d\tau$ where ψ is the mathematical function describing the space available to the electron in a given state. The integral thus describes the charge distribution ($d\tau = dx \, dy \, dz$ in Cartesian coordinates). An electron with a negative charge $q = -e$, therefore, has a charge distribution $-e \int \psi^* \psi \, d\tau$, and the dipole moment μ associated with it, if the charge density is described with respect to the electron at a position r in the molecule, will be $\mu = -e \int \psi^* r \psi \, d\tau$. At this point, Heisenberg suggested that if the electron can occupy two states, ψ_n and ψ_m, it could be considered to be a "virtual oscillator" with a dipole moment $\mu_{nm} = -e \int \psi_n^* r \psi_m \, d\tau$. This virtual oscillator has nothing really to do with quantum mechanics except that the distribution of space permitted to the electron ψ is given by the solutions (eigenstates) of the Schrödinger equation. In principle, any description of the charge distribution would do, but the quantum distribution was Heisenberg's choice. This virtual oscillator therefore gives rise to the "transition dipole moment," in this case the electric dipole transition moment, which is a molecular property. A similar case can be made for the magnetic moment. In the discussions in this book we will replace the notation of the integral by a notation due to Dirac (10) which has some operational advantages but the identical meaning: $-e \int \psi_n^* r \psi_m \, d\tau = \langle n | er | m \rangle = \langle n | \mu | m \rangle$. In this chapter these transition moments will be introduced ad hoc as the product of Rosenfeld's result for the rotational strength parameter. In the next chapter we will see how quantum mechanical perturbation theory leads to the conclusion that the intensity of radiation absorption by an achiral molecule is proportional to the square of the electric dipole transition moment, and that the intensity difference for the absorption of right and left circularly polarized light by a chiral molecule is proportional to the product of the electric and magnetic dipole transition moments.

Rosenfeld's important conclusion is that the chiral response parameter is

$$\beta = \frac{c}{3h} \sum_b \frac{\text{Im}\{\langle a|\mu|b\rangle \cdot \langle b|m|a\rangle\}}{\nu_{ba}^2 - \nu^2}, \tag{2.401}$$

where Im stands for "imaginary part" of the terms between the curly brackets in terms of complex number notation. The two factors between the curly brackets are integrals connecting the two molecular states $\langle a|$ and $\langle b|$ by an "operator" in the usual quantum mechanical notation; in this case the operators are the electric (μ) and the magnetic (**m**) dipole moment operators; ν_{ba} is the frequency of the radiation associated with the transition between the states a and b, and h is Planck's constant. "Molecular states" will be discussed and more precisely defined in Section 3.2. At this point, however, we emphasize that the state of a molecule designated by $\langle a|$, $\langle b|$, etc., is the Dirac description of the charge distribution of electrons and nuclei having the fixed energies $\mathcal{E}_a$, $\mathcal{E}_b$, etc., appropriate to the specific orbital, angular momentum, and spin quantum numbers for those energies. For the purpose of investigating *chiral* transitions between states, that is, events that carry a molecule from state $\langle a|$ to state $\langle b|$ via the absorption or emission of circularly polarized radiation, it suffices to *approximate* the state; to characterize $\langle a|, \langle b|$, etc., by just the description of the distribution of charge associated with the particular electron or electrons whose energy is changed by the transition. This description we know as a wavefunction or orbital (ψ_a, etc.) because of the wave properties of the electron. The transition moments $\langle a|\mu$ or $\mathbf{m}|b\rangle$ are integrals over all space but in practice only over all values of the coordinates x, y, and z, for which the amplitude $\psi_a \mu \psi_b$ is sufficiently large to contribute at least 1 or 2 percent of the overall value of the integral. For most practical purposes, using spherical (r, θ, ϕ) rather than Cartesian coordinates, the integration only has to be performed over the radial coordinate r (11). A typical wavefunction is the hydrogenlike function for a $1s$ electron, $\psi_{1s} = (1/\sqrt{\pi})(z/a_0)^{3/2} \exp -[(z/a_0)r]$ where z is nuclear charge ($z = 1$ for hydrogen), a_0 is a constant (the radius of the first Bohr orbit), and r is the distance of the electron from the center of the coordinates (12). The function is not mysterious; its shape is shown in Fig. 2.08, and a naive orbital diagram of the hydrogen $1s$ electron would consist of a sphere (or a circle cut along a diameter) drawn about the nucleus at the radius corresponding to the maximum in the curve of the probability distribution obtained from this function. Similarly, the integral has the usual mathematical significance, namely, the area under the curve produced by integrating the function $\psi_a \mu \psi_b$. But, of course, it has much more profound significance, as we see from Eq. 2.401 and the equations that follow. In Chapter 3 we will see how it is that this integral (and its magnetic counterpart) come to be the controlling factors in chiral phenomena. There is another important property of the Dirac designation of states by the symbol $\langle a|$, etc., and of the transition moments by $\langle a|\mu|b\rangle$; when used in matrix algebra formalism to carry out symbolic operations, not only μ, but also the states $\langle a|$, etc., have vector properties. Thus, if $\langle a|$ and $\langle b|$ are orthogonal, then $\langle a|b\rangle$ is zero. This will become important when vibronic contributions to optical activity are considered. Since the integral $\langle a|\mu|b\rangle$ is real when $\langle a|$ and $\langle b|$ are real wavefunctions, and the integral $\langle b|\mathbf{m}|a\rangle$ is

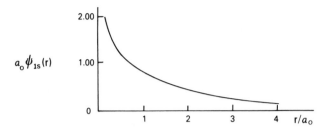

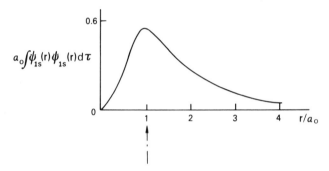

Fig. 2.08 The dependence of the ψ_{1s} wavefunction of the hydrogen atom on the distance r from the nucleus. Spherical coordinates are used for hydrogenlike wavefunctions. The lower curve is a plot of the charge density of the $1s$ electron radially from the nucleus.

imaginary because the magnetic moment operator **m** contains the constant $ih/\pi c$, where i is the imaginary number $\sqrt{-1}$, the entire term in the curly brackets (Eq. 2.401) is imaginary. If we replace **m** by **m'** where $\mathbf{m'} = -\mathbf{m}\sqrt{-1}$, then we can rewrite β, which can be thought of as the chiral polarizability, in terms of real quantities only:

$$\beta = -\frac{c}{3\pi h} \sum_b \frac{\langle a|\boldsymbol{\mu}|b\rangle \cdot \langle b|\mathbf{m'}|a\rangle}{\nu_{ba}^2 - \nu^2}. \qquad (2.402)$$

For comparison the expression for the isotropic or achiral polarizability α is

$$\alpha = \frac{2}{3h} \sum_b \frac{\nu_{ba}|\langle a|\boldsymbol{\mu}|b\rangle|^2}{\nu_{ba}^2 - \nu^2}. \qquad (2.403)$$

We recall that we are limiting the analysis to optical phenomena associated with transitions from the ground state only, so that in Eq. 2.401 and 2.402,

$\langle a|$ is the wavefunction for the ground electronic state and the sum is over all $\langle b|$ excited states. The isotropic absorption intensity is proportional to $|\langle a|\boldsymbol{\mu}|b\rangle|^2$ and consequently is not signed, although α is signed because the denominator may be positive or negative depending on whether the measuring frequency ν is lesser or greater than the frequency ν_{ba}. This same denominator appears in β. The rotation (in units of radians) at any given frequency is related to β and is obtained by substituting Eq. 2.402 in 2.318:

$$\delta = -\frac{16\pi^2 N\nu^2 z}{3hc} \sum_b \frac{\langle a|\boldsymbol{\mu}|b\rangle \cdot \langle b|\mathbf{m}'|a\rangle}{\nu_{ba}^2 - \nu^2}. \qquad (2.404)$$

Since the integrals $\langle a|\boldsymbol{\mu}|b\rangle$ and $\langle b|\mathbf{m}'|a\rangle$ may be of the same or opposite sign, the sign of δ is controlled by the signs of these transition moments. The dot product of the two transition moments is called the rotational strength, R_{ab}, so that Eq. 2.404 may be abbreviated as

$$\delta = \frac{16\pi^2 Nz}{3hc} \sum_b \frac{\nu^2 R_{ab}}{\nu_{ba}^2 - \nu^2}. \qquad (2.405)$$

Equations 2.404 and 2.405 predict a "resonance catastrophe" at the central frequency of absorption bands where ν_{ba}^2 equals ν because as ν approaches ν_{ba}^2, the difference approaches zero and the rotation approaches plus or minus infinity; the predicted shape of the dispersion of the rotation is very similar to that of the refractive index illustrated in Fig. 2.03. The experimental observation, however, is that the rotation of isolated optically active bands goes smoothly through zero near or at the resonant frequency ν_{ba}^2. In order to accord with experiment, we again invoke the classical theory of bound electrons; the electron is assumed to obey a Hooke's law damped motion when it is excited by the light wave. The motion and its resultant dispersion can be simulated by inserting a parameter $i\nu\gamma_{ba}$ in the denominator of Eq. 2.405 to account for the damped motion:

$$\hat{\delta} = \delta - i\delta' = \frac{16\pi^2 Nz}{3hc} \sum_b \frac{\nu^2 R_{ab}}{\left(\nu_{ba}^2 - \nu^2\right) + i\nu\gamma_{ba}}. \qquad (2.406)$$

The damping factor γ_{ba} to a reasonable degree of accuracy, is approximated by the width of the optically active band at half its maximum height. In this equation, the rotation $\hat{\delta}$ is written in complex form because of the arbitrary introduction of the complex number i in the damping factor $i\nu\gamma_{ba}$. While introducing this factor is ad hoc, it is not without either experimental or theoretical justification. From the experimental point of

at first it was easier to measure the monotonic curves in trans-
[...] nearly transparent regions of the spectrum. Since the correspond-
[...] curve in absorbing regions is generally named after its discoverer,
[...] Cotton (16), we prefer to apply the same name to the "anomalous"
[...] curve—as has sometimes been done. The *Cotton* effect is illustrated
[...] ig. 2.09. In this book, therefore, a Cotton effect is the chiroptical
[...] ctrum in an absorption band regardless of whether it is measured as
[...] ical rotation or circular dichroism.

[...] In most cases for measurements made in the visible or ultraviolet
[...] ortion of the spectrum, the term $\gamma_{ca}^2 \nu^2$ in the denominator of the summa-
[...] ion term of Eq. 2.502 can be discarded. For example, if the rotation is
[...] measured at 500 nm ($\nu = 20,000$ cm^{-1}) and we are concerned about the
contribution of an absorption band at 250 nm ($\nu_{ca} = 40,000$ cm^{-1}) having a
usual half-bandwidth of about 10 nm ($\gamma_{ca} = 1600$ cm^{-1}), then the ratio of
[...] ν^2 to $(\nu_{ca}^2 - \nu^2)^2$ is less than 0.001. So as a good approximation over most
[...] he [...] quency range, we can rewrite Eq. 2.502 as

$$\delta = \frac{16\pi^2 Nz}{3hc}\left[\sum_{c \neq b} \frac{R_{ac}\nu^2}{(\nu_{ca}^2 - \nu^2)} + \frac{R_{ab}(\nu_{ba}^2 - \nu^2)\nu^2}{(\nu_{ba}^2 - \nu^2) + \gamma^2 \nu^2}\right]. \qquad (2.503)$$

[...] m is no longer a resonance term; that is, the frequency-
[...] art does not contain a factor in the numerator that goes to
[...] measuring frequency approaches the transition frequencies. It
[...] infinity as ν approaches any other resonant frequency, but then,
[...] it should not be part of this term but should instead contribute a
[...] the second term. Its behavior is that of a monotonically increas-
[...] creasing function of generally gentle curvature and with a sign
[...] be positive or negative, depending on the magnitudes and signs
[...] contributing rotational strengths and, for each contribution, on
[...] ther the measurement is made at a higher or lower frequency than the
[...] responding transition frequency. The dispersive behavior for five rather
[...] cal situations is plotted in Fig. 2.10. Here we see that in the vicinity of
[...] optically active electromagnetic transition between molecular states, the
[...] optical rotation fluctuates through a mean value that is dependent on the
contribution of other transitions and may be positive (if the lobe at longer
[...] avelengths is positive relative to the mean value at the transition
[...] equency ν_{ab}) or negative. In discussing dimeric and polymeric systems
[...] er, we will see that still other types of dispersive behavior are possible.
[...] so, accidental near degeneracies, for example, chirally active optical
[...] sitions with similar energies occurring from a single "chromophore" or

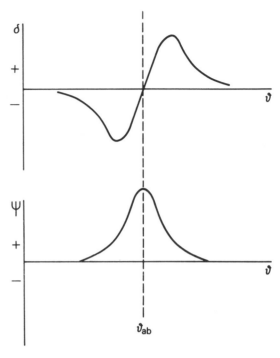

Fig. 2.09 The Cotton effect arising from the absorption of electromagnetic radiation by a chiral molecule. The frequency dependence of the optical rotation is shown above and of the ellipticity or circular dichroism, below.

view, it is justified by the fact that the resulting Eq. 2.407 removes the "resonance catastrophe" and reproduces qualitatively and often quite accurately the magnitude of the experimental optical rotatory dispersion (Fig. 2.09). Theoretically it arises (Section 2.0) from the classical equation of motion of an electromagnetically perturbed electron. While this classical picture may be far from reality, many of its consequences, including this one, correspond closely to the somewhat more realistic quantum picture.

When we separate the real and imaginary parts of Eq. 2.406, the measurable rotation is given by

$$\delta = \frac{16\pi^2 Nz}{3hc}\sum_b \frac{\nu^2(\nu_{ba}^2 - \nu^2)R_{ab}}{(\nu_{ba}^2 - \nu^2)^2 + \nu^2\gamma_{ba}^2}. \qquad (2.407)$$

In the course of reaching this expression for the optical rotation, a number of questions have been begged. Two of these are of principal importance and will be dealt with more completely in Chapter 3. One is the reasoning behind the identification of the parameter β, which appears in the macroscopic field equations derived from classical electromagnetic theory with the scalar product of the quantum mechanical electric and magnetic dipole transition moments—that is, how Rosenfeld arrived at Eq. 2.401, which permits us to reach the conclusion that the rotation is given by Eq. 2.404 or 2.407. The second question of importance that we have ignored relates to the origin of the frequency dependence, except that we have shown that electromagnetic theory requires that there be a time variation of the electromagnetic field of the light in order to produce a chiral response. Let us examine this first, at the phenomenological level.

2.6 THE DISPERSIVE BEHAVIOR OF THE OPTICAL ACTIVITY

The frequency dependence or *dispersion* of the optical activity comes from two sources: one is the definition of the rotation in terms of the wavelength or frequency of the light used to observe it (Eq. 2.209) and the second is the frequency dependence of the index of refraction (Eq. 2.022). The latter arises classically from Maxwell's equations or quantum mechanically from perturbation theory (Chapter 3). The dispersion is not independent of the nature of the interaction between neighboring groups, of atoms or molecules as we shall see in Chapters 4 and 8. If for the moment, however, we restrict our attention to isolated systems (by which we mean chromophores or molecules containing chromophores that are either or both spatially and energetically separated from each other), then the frequency dependence of the optical activity is given by Eq. 2.407. Several properties of this expression can be seen at a glance. It is a sum of terms, each of which arises from an optical transition so that at any frequency the optical activity will be a complex sum of contributions from all the electromagnetically induced changes in state from the ground state. Fortunately the contributions from transitions whose natural "Bohr" or resonance frequency is near the measuring frequency will usually be dominant; at most, finite contributions from only a very few transitions need be taken into account. This is a result of two factors: (1) The energy separation between optical transitions in the same molecule is often large. As we will soon see, the frequency dependence of the magnitude of the optical activity is such as to make each contribution decrease rather slowly with the difference between the frequency of the measuring radiation and that of the optical transition. Nevertheless, the large separation between many optical transitions tends to make contributions from each of the many possible higher energy transitions rather small at any given frequency. (2) There exists sum rules for optical rotation (6, 13–15), the most basic of

which specifies that the rotational streng... a given molecule must sum to zero:

$$\sum_{b \neq a} R_{ab} = 0.$$

In order that this be so, many of the terms in the ... must cancel each other, so that at any particular obs... only a very few nearby transitions will contribute s... overall value of the rotation. The right-hand side of ... separated into two terms, one for the contribution of a ... $a \rightarrow b$, and the other for the contributions of all the other tra...

$$\delta = \frac{16\pi^2 Nz}{3hc} \left[\sum_{c \neq b} \frac{R_{ac}(\nu_{ca}^2 - \nu^2)\nu^2}{(\nu_{ca}^2 - \nu^2)^2 + \gamma_{ca}^2 \nu^2} + \frac{R_{ab}(\nu_{ba}^2 - \nu^2)\nu^2}{(\nu_{ba}^2 - \nu^2)^2 + \gamma_{ba}^2 \nu^2} \right]$$

For the moment it will suffice to look only at the term for... transition. Since the $(\nu_{ba}^2 - \nu^2)$ appears in the numerator as a fa... obvious that at this level of approximation this transition will... nothing when the measuring frequency ν is precisely the same... the optical transition ν_{ab}. More important, even when the... frequency is far from the transition frequency where the terms (... the denominator get very large thereby decreasing the magnitud... rotation, the latter decreases rather slowly because of the ν^2 fact... numerator. The general shape of this contribution is given in Fig. ... is, of course, the familiar form of an isolated rotatory dispersion c... so-called *anomalous* rotatory dispersion near an optical transi... distinguish it from the monotonically decreasing tail, which is obs... the measurements are made in a spectral region far from an abso... band).

There is some variety in the literature with respect to the nom... for optical rotation and circular dichroism spectra. The monoto... rotatory dispersion curves observed on the low-frequency sid... tion bands are more or less uniformly referred to as "plain"... curves and extend well beyond the wavelengths at whic... dichroism (CD) is observed with ordinary concentrations... of optically active materials. In the vicinity of an optic... tion band, the optical rotatory dispersion (ORD) ex... inflection points and is most frequently but not uni... "anomalous." This nomenclature is, however, ar...

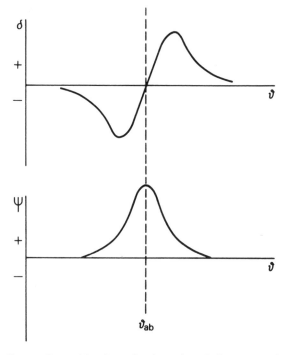

Fig. 2.09 The Cotton effect arising from the absorption of electromagnetic radiation by a chiral molecule. The frequency dependence of the optical rotation is shown above and of the ellipticity or circular dichroism, below.

view, it is justified by the fact that the resulting Eq. 2.407 removes the "resonance catastrophe" and reproduces qualitatively and often quite accurately the magnitude of the experimental optical rotatory dispersion (Fig. 2.09). Theoretically it arises (Section 2.0) from the classical equation of motion of an electromagnetically perturbed electron. While this classical picture may be far from reality, many of its consequences, including this one, correspond closely to the somewhat more realistic quantum picture.

When we separate the real and imaginary parts of Eq. 2.406, the measurable rotation is given by

$$\delta = \frac{16\pi^2 Nz}{3hc} \sum_b \frac{\nu^2 (\nu_{ba}^2 - \nu^2) R_{ab}}{(\nu_{ba}^2 - \nu^2)^2 + \nu^2 \gamma_{ba}^2} . \tag{2.407}$$

In the course of reaching this expression for the optical rotation, a number of questions have been begged. Two of these are of principal importance and will be dealt with more completely in Chapter 3. One is the reasoning behind the identification of the parameter β, which appears in the macroscopic field equations derived from classical electromagnetic theory with the scalar product of the quantum mechanical electric and magnetic dipole transition moments—that is, how Rosenfeld arrived at Eq. 2.401, which permits us to reach the conclusion that the rotation is given by Eq. 2.404 or 2.407. The second question of importance that we have ignored relates to the origin of the frequency dependence, except that we have shown that electromagnetic theory requires that there be a time variation of the electromagnetic field of the light in order to produce a chiral response. Let us examine this first, at the phenomenological level.

2.6   THE DISPERSIVE BEHAVIOR OF THE OPTICAL ACTIVITY

The frequency dependence or *dispersion* of the optical activity comes from two sources: one is the definition of the rotation in terms of the wavelength or frequency of the light used to observe it (Eq. 2.209) and the second is the frequency dependence of the index of refraction (Eq. 2.022). The latter arises classically from Maxwell's equations or quantum mechanically from perturbation theory (Chapter 3). The dispersion is not independent of the nature of the interaction between neighboring groups, of atoms or molecules as we shall see in Chapters 4 and 8. If for the moment, however, we restrict our attention to isolated systems (by which we mean chromophores or molecules containing chromophores that are either or both spatially and energetically separated from each other), then the frequency dependence of the optical activity is given by Eq. 2.407. Several properties of this expression can be seen at a glance. It is a sum of terms, each of which arises from an optical transition so that at any frequency the optical activity will be a complex sum of contributions from all the electromagnetically induced changes in state from the ground state. Fortunately the contributions from transitions whose natural "Bohr" or resonance frequency is near the measuring frequency will usually be dominant; at most, finite contributions from only a very few transitions need be taken into account. This is a result of two factors: (1) The energy separation between optical transitions in the same molecule is often large. As we will soon see, the frequency dependence of the magnitude of the optical activity is such as to make each contribution decrease rather slowly with the difference between the frequency of the measuring radiation and that of the optical transition. Nevertheless, the large separation between many optical transitions tends to make contributions from each of the many possible higher energy transitions rather small at any given frequency. (2) There exists sum rules for optical rotation (6, 13–15), the most basic of

which specifies that the rotational strengths of all the optical transitions of a given molecule must sum to zero:

$$\sum_{b \neq a} R_{ab} = 0. \qquad (2.501)$$

In order that this be so, many of the terms in the sum term of Eq. 2.407 must cancel each other, so that at any particular observational frequency only a very few nearby transitions will contribute significantly to the overall value of the rotation. The right-hand side of Eq. 2.407 can be separated into two terms, one for the contribution of a given transition $a \rightarrow b$, and the other for the contributions of all the other transitions:

$$\delta = \frac{16\pi^2 Nz}{3hc} \left[\sum_{c \neq b} \frac{R_{ac}(\nu_{ca}^2 - \nu^2)\nu^2}{(\nu_{ca}^2 - \nu^2)^2 + \gamma_{ca}^2 \nu^2} + \frac{R_{ab}(\nu_{ba}^2 - \nu^2)\nu^2}{(\nu_{ba}^2 - \nu^2)^2 + \gamma_{ba}^2 \nu^2} \right]. \qquad (2.502)$$

For the moment it will suffice to look only at the term for the $a \rightarrow b$ transition. Since the $(\nu_{ba}^2 - \nu^2)$ appears in the numerator as a factor, it is obvious that at this level of approximation this transition will contribute nothing when the measuring frequency ν is precisely the same as that of the optical transition ν_{ab}. More important, even when the measuring frequency is far from the transition frequency where the terms $(\nu_{ba}^2 - \nu^2)^2$ in the denominator get very large thereby decreasing the magnitudes of the rotation, the latter decreases rather slowly because of the ν^2 factor in the numerator. The general shape of this contribution is given in Fig. 2.09 and is, of course, the familiar form of an isolated rotatory dispersion curve, the so-called *anomalous* rotatory dispersion near an optical transition (to distinguish it from the monotonically decreasing tail, which is observed if the measurements are made in a spectral region far from an absorption band).

There is some variety in the literature with respect to the nomenclature for optical rotation and circular dichroism spectra. The monotonic optical rotatory dispersion curves observed on the low-frequency side of absorption bands are more or less uniformly referred to as "plain" or "normal" curves and extend well beyond the wavelengths at which any circular dichroism (CD) is observed with ordinary concentrations and path lengths of optically active materials. In the vicinity of an optically active absorption band, the optical rotatory dispersion (ORD) exhibits two or more inflection points and is most frequently but not uniformly referred to as "anomalous." This nomenclature is, however, archaic and arose simply

because at first it was easier to measure the monotonic curves in transparent or nearly transparent regions of the spectrum. Since the corresponding CD curve in absorbing regions is generally named after its discoverer, André Cotton (16), we prefer to apply the same name to the "anomalous" ORD curve—as has sometimes been done. The *Cotton* effect is illustrated in Fig. 2.09. In this book, therefore, a Cotton effect is the chiroptical spectrum in an absorption band regardless of whether it is measured as optical rotation or circular dichroism.

In most cases for measurements made in the visible or ultraviolet portion of the spectrum, the term $\gamma_{ca}^2 \nu^2$ in the denominator of the summation term of Eq. 2.502 can be discarded. For example, if the rotation is measured at 500 nm ($\nu = 20{,}000$ cm^{-1}) and we are concerned about the contribution of an absorption band at 250 nm ($\nu_{ca} = 40{,}000$ cm^{-1}) having a usual half-bandwidth of about 10 nm ($\gamma_{ca} = 1600$ cm^{-1}), then the ratio of $\gamma_{ca}^2 \nu^2$ to $(\nu_{ca}^2 - \nu^2)^2$ is less than 0.001. So as a good approximation over most of the frequency range, we can rewrite Eq. 2.502 as

$$\delta = \frac{16\pi^2 Nz}{3hc} \left[\sum_{c \neq b} \frac{R_{ac}\nu^2}{(\nu_{ca}^2 - \nu^2)} + \frac{R_{ab}(\nu_{ba}^2 - \nu^2)\nu^2}{(\nu_{ba}^2 - \nu^2) + \gamma^2\nu^2} \right]. \tag{2.503}$$

The first term is no longer a resonance term; that is, the frequency-dependent part does not contain a factor in the numerator that goes to zero as the measuring frequency approaches the transition frequencies. It does go to infinity as ν approaches any other resonant frequency, but then, of course, it should not be part of this term but should instead contribute a term like the second term. Its behavior is that of a monotonically increasing or decreasing function of generally gentle curvature and with a sign that may be positive or negative, depending on the magnitudes and signs of the contributing rotational strengths and, for each contribution, on whether the measurement is made at a higher or lower frequency than the corresponding transition frequency. The dispersive behavior for five rather typical situations is plotted in Fig. 2.10. Here we see that in the vicinity of an optically active electromagnetic transition between molecular states, the optical rotation fluctuates through a mean value that is dependent on the contribution of other transitions and may be positive (if the lobe at longer wavelengths is positive relative to the mean value at the transition frequency ν_{ab}) or negative. In discussing dimeric and polymeric systems later, we will see that still other types of dispersive behavior are possible. Also, accidental near degeneracies, for example, chirally active optical transitions with similar energies occurring from a single "chromophore" or

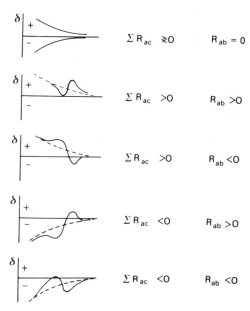

Fig. 2.10 Five typical optical rotation curves showing the effect of the appearance of a Cotton effect from an optical transition from the state a to the state b. In each of the curves, the transitions from the state a to the states c are at higher energies than plotted here (to the left of the vertical axis). The dashed lines are cumulative contributions to the rotation from the transitions $(a \rightarrow c)$ at higher energies.

from two or more different chromophores in a single molecule, will obviously overlap strongly and distort the dispersive behavior from the forms illustrated in Fig. 2.10.

2.7 EPILOGUE TO THE INTERACTION OF POLARIZED RADIATION WITH MATTER

Because almost the whole of the remainder of this monograph and especially Chapters 3, 4, and 6 are concerned with this subject, this section is written as an epilogue to, rather than a summary of, the interaction that produces the chiral optical activity.

Electromagnetic radiation, light, consists of electric and magnetic fields traveling in space. When these fields go from free space to a material medium, they produce a response in the medium that shows up in optical phenomena which can be measured because of a change in a property of the light, its intensity or its polarization, or both. The response manifests itself through the macroscopic or bulk property known as the refractive index, which in complex form includes the absorption index so that it specifies the behavior of the medium in spectral regions where the latter is transparent or absorbing. This is the classical way of looking at it.

Quantum mechanically it makes more sense to look at this response in terms of electromagnetically induced changes in stationary energy states, which are manifest primarily as an absorptive rather than a refractive process (in semiclassical quantum theory, but *not* in quantum scattering theory). However, these changes or transitions have a frequency and magnitude dependence that may be correlated with the optical polarizability and thus also with the refractive index. For achiral substances, these will produce ordinary absorptions and refraction (which can have polarization properties only if the medium is oriented). For chiral substances, using circularly polarized radiation as the inducing field (17), optical rotation and circular dichroism are the observable phenomena.

Let us go on to examine the modifications of the states of systems that produce these phenomena.

REFERENCES AND NOTES

1. R. P. Feynman, R. B. Leighton, and M. Sands, *The Feynman Lectures on Physics*, Vol. 2, Addison-Wesley, 1965.
2. M. Faraday, *Philos. Mag.*, **17**, 546 (1840).
3. J. C. Maxwell, *Philos. Trans.*, **155**, 459 (1865); *ibid.*, **158**, 643 (1865).
4. J. C. Slater and N. H. Frank, *Electromagnetism*, McGraw-Hill Book Co., New York, 1947, Chapter 9.
5. L. Rosenfeld, *Z. Phys.*, **52**, 161 (1928).
6. E. U. Condon, *Rev. Mod. Phys.*, **9**, 432 (1937).
7. P. J. Debye, *Polar Molecules*, Dover Publications, New York, 1945.
8. W. Kauzmann, *Quantum Chemistry*, Academic Press, New York, 1957, p. 159.
9. W. Heisenberg, *The Physical Principles of the Quantum Theory*, C. Eckert and F. C. Hoyt (Transls.), Dover Publications, New York, 1930, pp. 80–90.
10. P. A. M. Dirac, *Proc. R. Soc. (London)* **A123**, 714 (1929).
11. See the discussion of the hydrogen atom wavefunctions (pp. 210 ff.) in W. A. Kauzmann's *Quantum Chemistry*, Academic Press, New York, 1957. With cylindrical symmetry with respect to the angular coordinates, the integration with respect to these coordinates contributes values of the electric and magnetic transition moments that are independent of nuclear charge or differ only by a constant for different electronic states. For magnetically degenerate states, the situation is somewhat different, but then only if the transitions between the magnetic degeneracies are to be calculated.
12. H. Eyring, J. Walter, and G. E. Kimball, *Quantum Chemistry*, John Wiley & Sons, New York, 1944, Chapter 6; or any similar text on quantum mechanics.

13. W. Kuhn, *Z. Phys. Chem.*, **B4**, 14 (1929).

14. A. E. Hansen, *Mol. Phys.*, **33**, 483 (1977).

15. R. A. Harris, *Chem. Phys. Lett.*, **45**, 477 (1977).

16. A. Cotton, *C. R. Hebd. Seance Acad. Sci.* (*Paris*), **120**, 989, 1044 (1895).

17. Note that circularly polarized radiation of single handedness does not measure optical activity. While the radiation would suffer a phase lag in an optically active medium, no measurement would detect this except in relation to some other radiation traveling through the medium. The incident radiation must therefore be plane polarized; by a mathematical construction, we consider the plane-polarized radiation to decompose into two circularly polarized beams of opposite handedness and equal amplitude which suffer a phase lag relative to each other as they traverse the medium, and a phase lag and differential extinction if the medium is absorbing at the wavelength of the radiation.

CHAPTER THREE

THE ELECTRONIC AND VIBRONIC STATES OF MATTER AND THEIR CONTRIBUTIONS TO OPTICAL ACTIVITY

3.1 PERTURBATION OF STATIONARY STATES BY ELECTROMAGNETIC RADIATION: SEMICLASSICAL QUANTUM THEORY

In Chapter 2, using classical theory, we tried to understand how changes in molecular charge distributions are produced by the electric and magnetic fields of electromagnetic radiation. We are well aware that quantum theory or wave mechanics associates this same phenomenon with discrete optical transitions between stationary molecular states. The association is reasonably obvious, but a clear picture of it is essential to an understanding of the nature of absorption and refraction in both achiral and chiral molecular systems. It is formalized in the relation of the classically derived optical polarizability and achiral absorption intensity to the quantum mechanically derived *oscillator strength* f_{ab}, and of the classical optical activity to the *rotational strength* R_{ab}; the former is proportional to the square of the electric dipole transition moment, $|\langle a|\boldsymbol{\mu}|b\rangle|^2$, connecting the stationary states $\langle a|$ and $\langle b|$; the latter is proportional to the scalar or dot product of the electric and magnetic dipole transition moments, $\langle a|\boldsymbol{\mu}|b\rangle \cdot \langle b|\mathbf{m}|a\rangle$ (Eqs. 2.403 and 2.404). (Despite their juxtaposition here; f_{ab} and R_{ab} are not precisely comparable quantities because the rotational strength as defined by Eq. 2.405 is not dimensionless, while the oscillator strength is. Schellman has suggested that the truly comparable quantity be given the name chiral strength.) In this chapter we wish to develop some of the rationale for these statements by treating the interaction of radiation with matter as a perturbation of stationary energy states by the radiation. The result of the perturbation will be interpreted in terms of the fraction of molecules that reach a new stationary state when the incident electromagnetic radiation corresponds to the difference between the energy of the two states. If the radiation is unpolarized or if the molecules are optically inactive, or both, the fraction of molecules in the new state is found to be proportional to the Einstein coefficient of probability that such a transition will take place. Therefore, if the incident radiation flux and the number (or concentration) of the appropriate molecules in the path of the radiation are known, then the intensity of transmitted radiation or the absorption may be calculated. If the light is plane- or left and right circularly polarized *and* the molecules are optically active, the difference between the corresponding coefficients for left and right circularly polarized light are related to the difference in absorption for the two components of the radiation and thus to the circular dichroism. To do all of this, we will take a deep breath and assume all the foundations of quantum theory, so that we can write a small number of justifying statements without proof. We will write down just enough in the text to develop the ideas, relegating to original sources pertinent derivations and proofs. This semiclassical quantum theory will be discussed in Section 3.3, after a brief description of what we mean by the *states* of a molecule.

3.2 ORBITALS, CONFIGURATIONS, AND STATES

In the discussion so far, the interaction of radiation with matter has been said to induce or promote a transition from one energy state of the molecule to another. It will help to define somewhat more precisely what we mean by a state; our definition is a spectroscopic one, since the description of optical absorption and chiral activity will all be in terms of spectroscopic phenomena. What do we mean then by a spectroscopic state?

The arrangement of atoms into the structural units of molecules is accompanied by a rearrangement of the electron configuration, which we know as hybridization. The hybridized electrons (orbitals) take up the familiar patterns to which are given labels derived from angular momentum quantum numbering, such as the π orbitals formed from the slightly hybridized p electrons of two carbon atoms, the n orbitals of the nonbonding electrons on an oxygen atom, or the σ orbitals of the sp^3 hybrids of two carbon atoms or of the sp^3 hybrid orbital of a carbon atom and the hybridized d orbitals of metals. The complete nomenclature for a molecular orbital (excluding spin state designation) in any given state or "configuration" includes the number of electrons as a superscript. Thus, one of the nonbonding oxygen orbitals in the ground state of formaldehyde is $(n_1)^2$ where the subscript 1 designates a particular one of the nonbonding

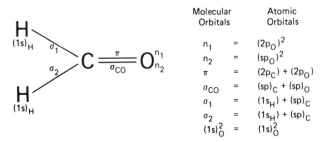

	Molecular Orbitals		Atomic Orbitals
	n_1	$=$	$(2p_O)^2$
	n_2	$=$	$(sp_O)^2$
	π	$=$	$(2p_C) + (2p_O)$
	σ_{CO}	$=$	$(sp)_C + (sp)_O$
	σ_1	$=$	$(1s_H) + (sp)_C$
	σ_2	$=$	$(1s_H) + (sp)_C$
	$(1s)_O^2$	$=$	$(1s)_O^2$

Fig. 3.00 Zeroth-order ground state molecular orbitals of formaldehyde; the atomic orbitals used in bonding are in parentheses. The ground state configuration (see text) contains the doubly filled bonding orbitals π, σ_{CO}, σ_1, and σ_2; the doubly filled nonbonding orbitals n_1 and n_2; and the doubly filled inner shell orbitals $(1s_C)$ and $(1s_O)$.

orbitals and the superscript 2 designates the number of electrons occupying the orbital. The formation of the n_1, n_2, σ_1, σ_2, π, and σ_{CO} molecular orbitals in formaldehyde is shown in Fig. 3.00.

Note that the highly stable inner shell electrons are usually indicated by their atomic designation $1s$, $2s$, etc., while the high-energy nonbonding electrons, which are easily promoted to still higher energies by radiation in the visible and ultraviolet range are designated by the letter n. Also note that the molecular orbitals of Fig. 3.00 are zeroth-order approximation but are sufficient to designate the symmetry.

Collecting all the electrons in the molecule, we write for the ground state *configuration*

$$(1s_C)^2(1s_O)^2(\sigma_1)^2(\sigma_2)^2(n_1)^2(\sigma_{CO})^2(\pi)^2(n_2)^2$$

in which the upper valence orbitals are written to the right and the lower energy, more stable orbitals to the left. If we ignore two- and three-electron promotions of low probability, then the lowest excited spectroscopic states are produced by one-electron promotions of an electron from the two or three highest energy-filled valence orbitals to the lowest unfilled or excited state orbital of the molecule—in this case to a π^* orbital. Thus we may write for configurations of the first excited states:

1. (core) $(\sigma_{CO})^2(\pi)^2(n_2)(\pi^*)$,
2. (core) $(\sigma_{CO})^2(\pi)(n_2)^2(\pi^*)$,
3. (core) $(\sigma_{CO})^2(\pi)^2(n_2)^2(\pi^*)$,

where (core) $= (1s_C)^2(1s_O)^2(n_1)^2(\sigma_1)^2(\sigma_2)^2$, and the promotions to the three excited configurations or valence states are designated, respectively, as

$n \rightarrow \pi^*$, $\pi \rightarrow \pi^*$, and $\sigma_{CO} \rightarrow \pi^*$. Each of these configurations may be given a state designation, commonly;

$\psi_0 =$ ground state,

$\psi_1 =$ first excited state (configuration 1),

$\psi_2 =$ second excited state (configuration 2), etc.

Alternatively the states may be designated by their symmetry such as ^1A_g for the ground state, as we will see in Chapter 6, or by the Dirac bracket notation $\langle 0|$, $\langle 1|$, $\langle 2|$, $\langle 3|$, etc. In this volume we use the alphanumeric and the italic letter designation $\langle a|$, $\langle b|$, etc., the latter of which is intended to avoid designating, except as may be indicated, relative state or orbital energies with respect to other states. Note that in this usage, "configuration" conveys no information about the three-dimensional structure of the atoms in the molecule except insofar as our chemical intuition specifies it from what we know about the orbitals, and in any event, it does not specify the absolute stereochemical configuration. Also, in this approximation, these configurations do not account for complete orbital rehybridization which may occur on excitation of an electron or for excitation to Rydberg orbitals. In some cases, as we shall see later, there are profound changes in the equilibrium geometry in excited states. These are therefore zeroth-order configurations. In modern quantum chemical calculations, it has become customary to attempt to approach a better description of the configuration in a number of ways, one of which consists of interacting the zeroth-order configurations. The simplest but not necessarily the most accurate way of doing this is to use as first-order configurations linear combinations of the zeroth-order configurations of other excited states of the same symmetry. In any event, when we speak of an optical transition between states, we are describing the promotion of an electron from an orbital in one configuration (usually the ground state) to an orbital in another configuration.

Configuration interaction (CI) methods are described in many places. For example, see references 1-4. One example of the marked changes in the orbital description of a chromophore from its zeroth-order description produced by configuration interaction methods will be found in Chapter 7.

The states as described here are pure electronic. Nothing has been said of the other quantized substates, vibrational, rotational, magnetic, etc. Later in this chapter we will discuss the vibrational states, the Born-Oppenheimer (5) approximation, the Franck-Condon (6) considerations, which permit us to separate the electronic part of the wavefunctions from the vibrational parts with considerable accuracy, and finally, the Herzberg-Teller (7) considerations, which demonstrate how vibrational interactions lead to optical transitions that would otherwise be forbidden by symmetry considerations. We are now prepared, however, to consider in more intimate detail the quantum mechanical theory of the perturbation of molecular states by electromagnetic radiation.

3.3   PERTURBATION OF STATIONARY STATES BY ELECTROMAGNETIC RADIATION: SEMICLASSICAL QUANTUM THEORY CONTINUED

All of the achiral absorptive and refractive phenomena, as well as the chiral phenomena, optical rotation, and circular dichroism, owe their origin to radiative transitions between stationary states. So the classical response (Maxwell's electromagnetic field laws) and the quantum response (Rosenfeld's transition moments) are corresponding descriptions. We will examine the ordinary isotropic or achiral absorption first (our object is to show that the absorption is proportional to the *square* of the electric dipole transition moment: $|\mu_{ab}|^2$).

The absorption of unpolarized light is usually measured in terms of the transmission (I/I_0), or more frequently, as the optical density (O.D.)= $\ln(I_0/I)$. If the latter is converted to the frequency-dependent molar extinction coefficient $\varepsilon(\nu)$ and the function $\varepsilon(\nu)/\nu$, plotted as a function of ν, then the total area under the absorption band between the limits where the absorption is measurably greater than zero is the integral $\int \varepsilon(\nu) d\nu/\nu$. To justify the statement that absorption is proportional to $|\mu_{ab}|^2$ and Eqs. 2.023, 2.024, and 2.403, we should obtain the result from quantum theory that the absorption intensity as measured by this area is proportional to the square of the electric dipole transition moment integral

$$\int \varepsilon(\nu) \frac{d\nu}{\nu} \sim |\langle a|\mu|b\rangle|^2. \tag{3.301}$$

Correspondingly, the circular dichroism as given by the area under the circular dichroism band is proportional to the rotational strength which we have already defined (Eq. 2.401) as the imaginary part of the scalar product of the electric and magnetic transition dipole integrals

$$\int [\varepsilon_1(\nu) - \varepsilon_r(\nu)] \frac{d\nu}{\nu} \sim \text{Im}\langle a|\mu|b\rangle \cdot \langle b|\mathbf{m}|a\rangle. \tag{3.302}$$

To start, we take for granted the quantum mechanical formulation of the time dependence of the unperturbed energy of one of the states, the time-dependent Schrödinger equation (8)

$$\mathcal{H}\psi = -\frac{h}{2\pi i}\frac{\partial \psi}{\partial t}, \tag{3.303}$$

and remind ourselves that this is a mathematical statement of the fact that there exist functions (ψ) that accurately describe (to within the Heisenberg uncertainty) the charge distribution of the nuclei and electrons at any time t, such that their variation with time ($\partial \psi/\partial t$) is proportional to the action on ψ of an "operator" $\mathcal{H}$, which is a function of the kinetic and potential energy of the nuclei and electrons. The values of ψ corresponding to *allowed* values of the energy found by solving the Schrödinger equation are called *eigenfunctions*. We need not be concerned here with the problem of solving the Schrödinger equation, and, in fact, investigations into chiral phenomena, as into other spectroscopic phenomena, frequently serve the useful purpose of helping to determine to what extent such solutions *are* reasonable representations of the states of molecules or of their chromophores. We need only accept the fact that in principle the equation is exactly soluble if the potential and kinetic energy of the system can be exactly specified. It follows that if the action of the electric and magnetic fields of the light wave can be correctly incorporated into the energy of interaction between the particles, we should be able to solve the equation for the way in which the state of the molecule changes under the influence of the light wave. Unfortunately, the complexities of the situation are such that a straightforward approach of this type is almost impossible using semiclassical theory, that is, without using quantum electrodynamics to describe the radiation field. Human ingenuity, however, turns out to be sufficient to transcend such difficulties. If the change in the potential energy introduced by these fields is small compared to the total energy of the particles, as indeed it almost always is, then perturbation theory (9) tells us we can write the equation as

$$(\mathcal{H}_0 + \mathcal{H}')\psi = -\frac{h}{2\pi i}\frac{\partial \psi}{\partial t}, \tag{3.304}$$

where $\mathcal{H}_0$ is the unperturbed energy operator and $\mathcal{H}'$ is a small perturbation introduced by the fields.

Not a requirement, but as an aid in thinking about the radiation-particle interaction, it will be well to recall that the wavefunctions ψ, which may be combined or separate functions of space and/or time, when they refer, for example, to the state of a particular electron, can be used to define those lines we draw (and call orbitals) to illustrate the area of confinement of 95 percent of the electronic charge about the nucleus or nuclei. They thus define for a given state or energy, the center of gravity of the charges, and for two different states, the changes in the center of gravity, if any, will define a change in charge distribution and so a change in the electric (and possibly magnetic) dipole moment. If the change in state is produced by the electric field of the radiation, the transition dipole moment $\langle \psi_a^* | \mu | \psi_b \rangle$ is not, in general, equal to the difference in the equilibrium molecular dipole moment in the two states, $\langle \psi_b^* | \mu | \psi_b \rangle - \langle \psi_a^* | \mu | \psi_a \rangle$, but is very closely related to it. In any event, the wavefunctions are not very mysterious; they are the mathematical expressions of charge distributions discussed in Chapter 2, and from them we draw the simple orbital diagrams with which we are all so familiar. Taking into account their vector properties, they may be multiplied and divided like any algebraic expression.

The solution of the differential equation 3.304 then becomes a straightforward matter whose results we can examine to help understand optical absorption. Let us look only at the essentials. The complete derivation of the perturbation theory treatment is found in many places (see for example, reference 8). If a system, an assembly of molecules let us say, starts out in a state designated by ψ_ℓ, having an energy $\mathcal{E}_\ell$, we want to know with what probability the radiation will induce a transition to a new state ψ_m, having an energy $\mathcal{E}_m$. We will be interested only in absorption, so that we are concerned with the case for which the energy of the perturbed state $\mathcal{E}_m$ is greater than $\mathcal{E}_\ell$. One way of thinking about this is to define an overall state of the system ψ, such that at any time after the system is placed in the field of the radiation,

$$\psi = a_\ell \psi_\ell + a_m \psi_m, \tag{3.305}$$

where the coefficients a_ℓ and a_m are functions of time, and since the system can have no other states, $a_\ell = 1$ when $a_m = 0$, and vice versa. If the state of the molecular system is defined in this way, then it can be shown (8) that a perturbation induced by the radiation field, which starts to produce the state ψ_m, changes the value of a_m from zero to some finite value at the rate of

$$\frac{da_m}{dt} = -\frac{2\pi i}{h} \exp\left[-2\pi i / h(\mathcal{E}_\ell - \mathcal{E}_m) t \right] \int \psi_m^* \mathcal{K}' \psi_\ell d\tau. \tag{3.306}$$

The exponential factor $[-2\pi i / h(\mathcal{E}_\ell - \mathcal{E}_m) t]$ represents the way in which the rate of decay of the ψ_ℓ state depends on the energy difference between the states. Equation 3.306 is the statement that no change will take place (da_m / dt will equal zero) unless the integral $\int \psi_m^* \mathcal{K}' \psi_\ell d\tau$ is greater than zero. Well, what is the integral? It is the function that defines the average or expectation value (9) of the energy of the system produced by the perturbation energy $\mathcal{K}'$.

What in turn is the perturbation energy? It is of course, the added energy or interaction energy of the system with the radiation fields. A complete treatment to account for both the circular dichroism and achiral absorption requires that the energy of interaction with *both* the magnetic and electric field of the radiation be accounted for. Since it is instructive to consider isotropic achiral absorption first, and since achiral absorption due to magnetically induced dipoles is extremely small (10), we can safely begin by neglecting the magnetic field of the light wave. Recall then that classical electrostatic theory defines the energy of a dipole in an *electric* field as $\mathbf{E \cdot \mu}$, where μ is the electric dipole moment which may consist of just the moment induced by the field (Eq. 2.019) or the algebraic sum of the induced moment and a preexisting permanent moment. Regardless of how the electric dipole moment arises in the field, once produced it is equal to the sum of the product of the charges and their distances from a chosen spatial origin: $\mu = \Sigma_i e_i \mathbf{r}_i$. This defines the dipole moment operator μ. Consequently, the interaction or perturbation energy of the molecule or, more particularly, of its chromophoric electrons is

$$\mathcal{H}' = \mu \cdot \mathbf{E} = \sum_i e_i \mathbf{r}_i \cdot \mathbf{E}, \qquad (3.307)$$

where $\mathbf{r}_i$ is the vector distance of the ith charge from the origin. $\mathbf{E}$ is just the electric field of the light wave defined in Section 2.2 as $\mathbf{E}_0 \cos\theta$ or its mathematical equivalent:

$$\mathbf{E} = \mathbf{E}_0 \frac{e^{2\pi i \nu t} + e^{-2\pi i \nu t}}{2}. \qquad (3.308)$$

with these definitions,

$$\mathcal{H}' = \mathbf{E}_0 \cdot \sum_i e_i \mathbf{r}_i \frac{e^{2\pi i \nu t} + e^{-2\pi i \nu t}}{2}. \qquad (3.309)$$

When this value of $\mathcal{H}'$ is substituted in Eq. 3.306, it becomes at once apparent that the increase in the contribution of the state ψ_m to the overall state of the system under the influence of the field is proportional to the magnitude of the electric field of the light wave (the intensity of the radiation) and to $\int \psi_m \mu \psi_\ell d\tau$:

$$\frac{da_m}{dt} \sim \mathbf{E}_0 \int \psi_m \mu \psi_\ell d\tau. \qquad (3.310)$$

The value of the *electric dipole transition moment integral* is $\int \psi_m \mu \psi_\ell d\tau \equiv \mu_{m\ell}$.

From this we draw the important conclusion that this integral must be finite, or the radiation will not induce a change in the state of the system. We will see later in this chapter, and at length in Chapter 6, the conditions, that is, the "selection rules," under which the integral is finite.

Early in the development of quantum theory, Einstein (11) related the intensity transmitted by an absorbing system to the incident radiation density $\rho(\nu)$, and to the probability $B_{m\ell}$, that the transition between the states m and ℓ would occur:

$$I = N_m B_{m\ell} \rho(\nu) h\nu, \qquad (3.311)$$

where N_m is the number of molecules in the path of the incident radiation. It was a straightforward matter for Slater (12) to identify the self-product of a_m with the Einstein coefficient $B_{m\ell}$.

The self-product is normally rigorously taken as the product of a_m with its complex conjugate because of the usual generalization of wavefunctions in complex notation, but for our purpose it is clear that we are dealing with real quantities only and we can ignore the generalization.

To obtain a_m and its self-product, Eq. 3.306 is integrated after substituting the value of $\mathcal{H}'$ from Eq. 3.309; for absorption, only terms for which $\mathcal{E}_\ell < \mathcal{E}_m$ are retained; for emission, only terms for which $\mathcal{E}_\ell > \mathcal{E}_m$. The resulting expression for the absorption intensity per unit volume is

$$I = \frac{8\pi^3 N_m \nu}{3hc} |\mu_{m\ell}|^2 \rho(\nu). \qquad (3.312)$$

Normalization of the intensity for the concentration of absorbing molecules and the path length of the light through the system yields an expression for the molar extinction coefficient $\varepsilon(\nu)$:

$$\varepsilon(\nu) = \frac{8\pi^3 \nu N_0}{6909ch} |\mu_{m\ell}|^2 \rho(\nu), \qquad (3.313)$$

where the incident radiation density $\rho(\nu)$ is related to the magnitude of the electric field of the light wave by $\rho(\nu) = [E_0(\nu)]^2/4$ and N_0 is Avogadro's number. Dividing both sides of Eq. 3.313 by the frequency ν and integrating over the entire range of the absorption band:

$$\int \frac{\varepsilon(\nu)}{\nu} d\nu = \frac{8\pi^3 N_0}{6909ch} |\mu_{m\ell}|^2 \int \rho(\nu) d\nu, \qquad (3.314)$$

where the integral $\int \rho(\nu) d\nu$ is proportional to the incident intensity or radiation flux in the same frequency range as the absorption band. Since the measurements are made relative to this intensity, it is reasonable to define it as conveniently as possible:

$$\int \rho(\nu)\, d\nu \equiv 1. \tag{3.315}$$

With this we arrive at the result we were looking for, the proportionality between the absorption of radiation and the electric dipole transition moments induced by the interaction between the radiation and the molecular system:

$$\int \frac{\varepsilon(\nu)}{\nu}\, d\nu = \frac{8\pi^3 N_0}{6909 ch} |\mu_{m\ell}|^2. \tag{3.316}$$

The numerical value of $8\pi^3 N_0/6909 ch$ in cgs units is 0.109×10^{39} with $\mu_{m\ell}$ in cgs $-$ esu units. If $\mu_{m\ell}$ is given in debyes, then since $1\, D = 10^{-18}$ cgs $-$ esu, $8\pi^3 N_0/6909 ch = 1.09 \times 10^2$.

The development of Eq. 3.316 required several simplifying assumptions. One of these is that the electric field does not vary over the molecule. This is equivalent to saying that the wavelength of the radiation is large compared to the molecular dimensions, a reasonably good assumption for ordinary molecules and visible or ultraviolet light, but it is not obvious that it will be the same for large polymers. With this assumption the field of the light wave need not be expanded to induce higher order moments (quadrupole and octopole) for ordinary isotropic absorption. However, we have seen in Chapter 2 that in order to distinguish chiral from achiral phenomena, it is necessary to include the time derivatives of the electric and magnetic fields in the relationships between **D** and **E** and between **B** and **H**. Thus, by Rosenfeld's criteria, Maxwell's macroscopic description of optically active media is associated with $\partial \mathbf{B}/\partial t$ and $\partial \mathbf{E}/\partial t$, the time variation of the fields passing through the media. Oosterhoff (13) has pointed out that this extension of Maxwell's equation has to do with the fact that "natural optical rotatory power implies that molecular dimensions cannot be neglected with respect to the wavelength of the light. In order to distinguish between left and right circularly polarized light, a molecule should not only feel the average field but also variations of the field inside the molecular volume." We will have more to say about this in the chapter on polymers.

Another assumption we have made is that the transition dipole $\mu_{m\ell}$ is invariant over the frequency range of the absorption band. But we have already called attention to the fact that an electronic absorption band consists of many subbands of vibrational, rotational, and, in solution, librational origin, many or all of which may in addition be in thermally excited states. Each contributes with varying degrees of probability to the absorption intensity. To treat each of these separately would not only require a degree of complexity far beyond our objective here, but would, in fact, obscure the interpretation of many important experimental results.

For this reason, the transition probability is divided into two factors, the invariant transition moment $\mu_{m\ell}$, theoretically calculable for the electronic state of the isolated molecule in its lowest vibrational-rotational state, and a shape factor for the absorption band. We have already seen the result of this in the classical expression for the polarizability (Eq. 2.023) and in the expression for the optical rotation (Eq. 2.407). We will see a similar separation for the circular dichroism. In discussing the dipole strength, however, it is apparent that we must be clear whether or not it includes all the factors—solvent interaction, vibronic factors, thermal excitation, etc.—that go into making up the shape factor. Thus, the dipole strength calculated from the integrated intensity under an experimental absorption band will include all these factors, while the dipole strength calculated from theoretical quantum mechanical considerations may or may not, depending on the manner of calculation.

3.4   CHIRAL PHENOMENA

Now we are prepared to look for a description of chiral phenomena in terms of the spectroscopic states of molecules. What has been accomplished so far can be grouped into two classifications, the classical physics of the interaction of light with matter and the quantum physics of the interaction of the electric field of the light with matter. In the former we ignored the Bohr condition that requires that energy be absorbed only in discrete quantities corresponding to the energy difference between allowed states, and in the latter we concentrated only on the electric field of the light; in the quantum theory we thereby avoided any interaction that would lead to optical activity or circular dichroism. We have, however, already anticipated the result of this interaction (Eqs. 2.404–2.407). Optical rotation is a dispersion phenomenon like the refractive index, and historically it was the first to be observed and discussed. If we keep in mind that the phenomenon originates in optical transitions, there is really nothing wrong with dealing with chiral phenomena as dispersion phenomena. This gives us an opportunity to discuss optical activity at frequencies far from any spectroscopic band where absorption is measured under normal conditions of concentration and sample thickness. However, conceptually it is easier to think of transitions occurring between states as absorption phenomena, and so we will cast our discussion in these terms. As a practical matter, it is also often advantageous to measure the circular dichroism rather than the optical activity. In Chapter 4 we will look at methods that have been developed for understanding chiral phenomena by relationships that either avoid direct connections with spectroscopic states or that veil the connection in terms of polarizabilities or other induced electromagnetic moments. These methods have been very powerful and remain a mainstay of the theoretical treatments of real systems which are not intrinsically dissymetric chromophores, but their utility stems from the problems we have in describing the spectroscopic states of matter accurately, problems resulting from the manipulative and computational difficulties in determining accurate wavefunctions and energy operators for molecules.

Consider some of the ideas that will be different in the treatment of circular dichroism from the treatment of achiral absorption. In the first place, we will have to specify the polarization of the light, that is, left and right circularly polarized. Indeed, it would appear that we should be able to obtain the magnitude of the circular dichroism simply by substituting expressions for the electric field of right and left circularly polarized light into the equation for the isotropic extinction, but, of course, this is not so because in the isotropic case we ignored the magnetic field of the radiation. So at the very least, the perturbation energy operator for circular dichroism will differ from that for isotropic absorption (Eqs. 2.303 and 2.309) by the inclusion of the energy of interaction with the magnetic field, as well as with the electric field:

$$\mathcal{H}' = -\boldsymbol{\mu}\cdot\mathbf{E} - \mathbf{m}\cdot\mathbf{H}. \qquad (3.401)$$

The relationship between the magnetic moment operator $\mathbf{m}$ and the magnetic transition moment $\mathbf{m}_{ab}$, or the classical induced magnetic moment per molecule $\mathbf{M}$, is the same as that between the electric dipole moment operator and the electric dipole transition moment:

$$\mathbf{M} = \sum_{b \neq a} \mathbf{m}_{ab} \equiv \sum_{b \neq a} \langle a|\mathbf{m}|b \rangle \qquad (3.402)$$

In principle, this definition and the expression of the interaction energy produced by the electromagnetic perturbation (Eq. 3.401) are all that is required. These expressions by themselves, however, provide little mathematical or physical intuition about the role of the magnetic moment in chiral activity, and it would be well to pause briefly to examine some aspects of the magnetic moment operator. Classically we know from the discussion in Chapter 2 that the magnetic moment arises from the time variation of the electric field or from the motion of a charge that produces that field. The quantum mechanical analog of that classical motion is the specification of the *orbital angular moment* $\mathbf{L}$ associated with the rapidly moving charge (the electron). The total magnetic moment is the sum of all sources of angular momentum and thus of the *spin moment* $\mathbf{S}$ as well:

$$\mathbf{m} = \frac{e}{mc}(\mathbf{L} + 2\mathbf{S}), \qquad (3.403)$$

but except insofar as excited state configurations can be considered to be combinations of even and odd electronic multiplets, spin angular momentum makes little or no contribution to chiral activity associated with ordinary optical absorption (14). Thus, no observations of chiral activity associated with transitions to the triplet states from ground singlet states of organic molecules have yet been made, although in principle small but finite activity should be associated with such transitions through the coupling of spin and orbital angular moment.

Focusing, then, on the orbital angular momentum only, the classical description of the momentum $\mathbf{L}$ of a charged particle moving in the curved path of an orbiting electron is that of a vector directed perpendicular to the motion of the particle. For a charged (charge $= e$) particle of mass $\mathbf{m}$, moving in a circular orbit, the angular momentum is equal to the vector cross product of the linear momentum $\mathbf{P}$, tangent to the orbit, and the radius vector $\mathbf{r}$, from the origin of the orbit:

$$\mathbf{m} = \frac{e}{2mc}(\mathbf{r} \times \mathbf{P}_r). \qquad (3.404)$$

Moving particles may have angular momentum components in three orthogonal space-fixed directions. Quantum mechanically, the linear momentum $\mathbf{P}_r$ is given by

$$\mathbf{P}_r = \frac{h}{2\pi i}\nabla, \qquad (3.405)$$

where ∇ is the operator defined in Eq. 2.004. Thus, by combining Eq. 3.403 and 3.405,

$$\mathbf{m} = \frac{eh}{4\pi imc}(\mathbf{r} \times \nabla). \qquad (3.406)$$

Inserting Eq. 3.406 into Eq. 3.402, we see that the magnetic dipole transition moment associated with the $a \rightarrow b$ transition is given by

$$\mathbf{m}_{ab} = \frac{eh}{4\pi imc}\langle a|\mathbf{r} \times \nabla|b\rangle \qquad (3.407)$$

or

$$\mathbf{m}_{ab} = -i\langle a|\mathbf{r} \times \nabla|b\rangle \beta_m, \qquad (3.408)$$

where $\beta_m = eh/4\pi mc$ is defined as the Bohr magneton and has the value 10^{-20} erg molecule^{-1} G^{-1} in cgs units.

Equations 3.407 and 3.408 already tell us quite a bit about the induced magnetic moment and its properties as they relate to chiral phenomena. Now we can see how the magnetic transition moment acquires its imaginary property (see Section 2.4), which, in turn, ensures that the rotational strength as specified by Rosenfeld's expression, Eq. 2.401, is a real quantity, as indeed it must be in order to be a physical observable (15).

Note that although we see "how" the magnetic transition moment is imaginary through the definition of momentum (Eq. 3.405), we do not see the "why." This would take us back to the fundamental laws of quantum mechanics.

We also saw earlier (Section 2.0 and especially Eq. 2.022) that in the classical view, the refractive response of a material medium to the electromagnetic field of the incident radiation is dependent on the mass of the particles. So too, we see now from Eq. 3.407, which is the quantum mechanical formulation of the induced magnetic response, that transitions with particles of large mass will make smaller contributions than those associated with smaller mass. Thus, infrared transitions that arise from nuclear vibration and rotations involving masses of about 10^3 times that of an electron will have optical activities of the order of 10^{-3} that of ultraviolet transitions associated with radiative promotions to electronically excited states.

In addition, by expanding $\mathbf{r} \times \nabla$ in Cartesian coordinates, we will see the source of rather severe symmetry restrictions that are imposed on chiral activity by the nature of the magnetic moment operator, just as such restrictions are imposed on absorption (as well as on chiral activity) by the electric dipole transition moment. If the radius vector $\mathbf{r}$ of the particle is expanded in terms of the unit vectors $\mathbf{i}, \mathbf{j}, \mathbf{k}$, parallel to the x, y, z coordinates:

$$\mathbf{r} = \mathbf{i}r_x + \mathbf{j}r_y + \mathbf{k}r_z; \tag{3.409}$$

then the vector cross product of $\mathbf{r}$ and ∇ (Eq. 3.408) is

$$(\mathbf{r} \times \nabla) = \mathbf{i}\left(r_y \frac{\partial}{\partial z} - r_z \frac{\partial}{\partial y}\right) + \mathbf{j}\left(r_z \frac{\partial}{\partial x} - r_x \frac{\partial}{\partial z}\right) + \mathbf{k}\left(r_x \frac{\partial}{\partial y} - r_y \frac{\partial}{\partial x}\right).$$
$$\tag{3.410}$$

Consider the hypothetical situation in which the electronic charge in both the ground and excited states is constrained to a plane—the xz plane, for example. Then r_y is null, and the partial derivative with respect to y, acting on the electronic wavefunction $|b\rangle$ in the transition moment integral

$\langle a|\mathbf{r} \times \nabla|b \rangle$, will similarly be null. When we refer then to Eq. 3.410, since r_y and ∂/∂_y are factors of $\mathbf{i}$ and $\mathbf{k}$, only the $\mathbf{j}$ component of the magnetic transition moment can have a finite value greater than zero. But recall that the Rosenfeld expression for the rotational strength (Eq. 2.402) involves the scalar or dot product of the electric and magnetic transition moments, so that in order for the rotational strength to be finite, there must also be a finite $\mathbf{j}$ component of the electric dipole transition moment ($\mathbf{j} \cdot \mathbf{j} = 1$ but $\mathbf{j} \cdot \mathbf{i} = \mathbf{j} \cdot \mathbf{k} = 0$). The electric dipole transition moment $\langle a|\boldsymbol{\mu}|b \rangle$ may also be written in terms of the ∇ operator as $(h/4\pi m \nu_{ba})\langle a|\nabla|b \rangle$ [exactly for exact wavefunctions and approximately for inexact wavefunctions (16)]. But the $\mathbf{j}$ component of ∇ (Eq. 2.004) is $\mathbf{j}\, \partial/\partial y$, which gives a zero component in operation on $|b \rangle$ under the restriction of a planar charge distribution in the xz plane. So the rotational strength in this case is zero. There is nothing startling about this, since we have known for a long time that molecules or chromophores with a plane of symmetry are optically inactive, but here, in a short mathematical statement, we see at least one reason why this is so, namely, that *if a planar charge distribution develops a magnetic transition moment in a given direction $\mathbf{j}$, this moment is of necessity perpendicular to the direction of any possible finite electric dipole transition moment*; $\mathbf{i}$ or $\mathbf{k}$. We will have more to say about this in Chapter 6 when we take up symmetry in detail and in subsequent descriptions of particular molecular systems. For the moment it is sufficient to note this result of the interaction of the magnetic field with matter and to return to our discussion of this interaction as a perturbation phenomenon.

Let us see what we have done. We started by examining how perturbation theory is used to demonstrate that achiral light absorption is proportional to the square of the electric dipole transition moment connecting two states $\langle a|$ and $\langle b|$ (Eq. 3.316). This was accomplished using the electric dipole interactions $\boldsymbol{\mu} \cdot \mathbf{E}$ as the sole perturbation. It was not necessary simultaneously to perturb the state wavefunctions, except in the sense of expanding the ground state wavefunctions to accommodate at least two possible states of the system under the influence of a given perturbation energy. We will see in Chapter 4 that the use of the perturbation theory to first order in the wavefunctions, as well as in the interaction energy, forms the basis of several methods of solving the principal problem in the quantum theory of chiral-optical phenomena, namely, the problem of explaining the rotational strength of transitions that occur in chromophores which have planes or centers of symmetry in the absence of asymmetric surroundings. This is not so strange since it is the equivalent of distorting the symmetry of the charge distribution. As originally formulated, Rosenfeld's treatment does not look for sources of asymmetric perturbation. It leads to equations for the contributions to the rotational strength from a single transition which has *intrinsically allowed* parallel electric and magnetic transition moment components and thus requires only the perturbation of the Hamiltonian energy $\mathcal{K}$, and not of the wavefunctions (except in the sense of the time dependence of the equilibrium concentration of two states). The basic quantal perturbation treatment of optical activity very much resembles that of the ordinary absorption. It differs, however, in one very fundamental way which we have recalled and discussed in the last several pages—in the necessity to include the magnetic perturbation energy $\mathbf{m} \cdot \mathbf{H}$ (Eq. 3.401), as well as the electric one. We examined some of the properties of the magnetic moment operator $\mathbf{m}$ related to chiral phenomena; now we wish to see how the inclusion of this parameter leads to the conclusion that the rotational strength is proportional to $\boldsymbol{\mu}_{ab} \cdot \mathbf{m}_{ab}$. This can be done several ways. One method is to expand the electric and magnetic fields in terms of potentials (vide infra) as Rosenfeld did in his original work (17). This treatment, repeated in a more modern fashion, is the basis of most discussions of optical activity [for example, see the monographs by J. P. Mathieu (18), D. W. Davies (19), and D. J. Caldwell and H. Eyring (20)].

Potentials have not been discussed earlier in this volume, and we will largely avoid doing so. However, it is worth taking the occasion of mentioning it to make some comments on concepts that enter into most descriptions of electromagnetic phenomena.

A field will exert a force, the magnitude of which is a function of the potential. Thus we are familiar with the concept of a voltage which is the potential difference between two points and with the fact that between two oppositely charged plates there exists an electric field, a potential or voltage difference, and a force that would pull them together in the absence of an equal and opposite restoring force. The greater the potential, the larger are the field and the force. The field and the potential have the same vectorial direction, and so in principle either may be used to describe the interaction between the field and some material medium. In the case of charged plates on which the charge is rapidly varied, there is both a vector and scalar potential, and the field is intimately related to the time rate of change of the vector potential as it is for an electromagnetic field. In some cases it is more convenient to use the potential; in descriptions of perturbation treatment of the origin of optical activity, the light wave is more frequently described in terms of the vector potential $\mathbf{A}$:

$$\mathbf{A} = \mathbf{A}_0 \cos^2 \pi \omega \left(t - \frac{\eta z}{c} \right), \tag{3.411}$$

rather than in terms of the fields as in Section 2.3. The vector potential has the advantage that a single interaction may be considered rather than separate interactions with the $\mathbf{E}$ and $\mathbf{H}$ fields. It is interesting to examine why this is so: The relation between the vector potential and the **E** and **H** fields are given (21) by

$$\mathbf{E} = -\frac{1}{c}\frac{\partial \mathbf{A}}{\partial t},\tag{3.412}$$

and

$$\mathbf{H} = \nabla \times \mathbf{A}.\tag{3.413}$$

Note from Eq. 3.412 that **A** is in the direction of **E**, but from 3.413 that **H**, being the *cross* product of the vectors ∇ and **A**, is perpendicular to **A**, a requirement since as we have seen, **H** and **E** are perpendicular to each other. Note also that **A** is a maximum when **H** is a maximum, and a minimum when **H** is a minimum. Similarly, when **E** is a maximum or a minimum, $\partial \mathbf{A}/\partial t$ is, respectively, maximum and minimum. If

$$\mathbf{E} = \mathbf{E}_0 \sin \omega t,\tag{3.414}$$

then

$$\frac{\partial \mathbf{A}}{\partial t} = -c\mathbf{E}_0 \sin \omega t.\tag{3.415}$$

Integrating Eq. 3.415, we find that

$$\mathbf{A} = -c\mathbf{E}_0 \cos \omega t + k,\tag{3.416}$$

or since it is immaterial where the wave starts, the constant k may be set equal to zero:

$$\mathbf{A} = -c\mathbf{E}_0 \cos \omega t.\tag{3.417}$$

Comparing Eqs. 3.414 and 3.417, we see that **A** follows a cosine curve while **E** follows a sine curve of the same argument (ωt). Thus by Eq. 3.412, **A** and **E** are in the same direction, and by Eqs. 3.414 and 3.417, they are 90 degrees out of phase with each other, while **A** and **H** are perpendicular, but in phase with each other. So **A** has the phase relationship of **H** and the directional property of **E**. We do not use the vector potential in the treatment that follows, but an understanding of the phase and directional relationship to **E** and **H** is a help in the analysis of optical phenomena in general and should certainly help to illuminate those discussions of chiral phenomena that do use it.

It is also appropriate at this point to remark on two aspects of an imaginary quantity. Perhaps physicists have made a mistake in borrowing the word "imaginary" from mathematics for the notation $i = \sqrt{-1}$. In any case, it is important to recognize that all measurable quantities are real, and that in working with fields or vector potentials, the separation of real from imaginary quantities must be done before squaring the complex sum of real and imaginary parts. With respect to the reality of measurable quantities, it is unfortunate from a conceptual viewpoint that the quantity i is woven into the fabric of quantum mechanics, but fortunately another property of quantum mechanical variables, the Hermetian property, ensures that whenever an experimental determination is made, it never contains i [see, e.g., the exposition of this by Kauzmann (10)]. The requirement for removing the imaginary part of a complex number to obtain a real observable is a fascinating one and may readily be seen by the following example. The intensity of a light beam is proportional to the square of the magnitude of the electric field, $\mathbf{E_0}$, associated with it. If we write the field in complex notation, and this is frequently a very useful or necessary thing to do,

$$\hat{\mathbf{E}} = \mathbf{E}_0 e^{i\theta}, \qquad \theta = 2\pi\omega\left(t - \frac{\eta z}{c}\right);$$ (3.418)

its square is just

$$|\hat{E}|^2 = |E_0|^2 e^{2i\theta}$$ (3.419)

which still has an imaginary part and so cannot be a real observable. On the other hand, we may achieve the same purpose of squaring the amplitude to obtain the intensity by multiplying the complex field $\hat{E}$ by its complex conjugate $\hat{E}^*$:

$$\hat{E}^* = E_0 e^{-i\theta},$$ (3.420)

since

$$|E|^2 = \hat{E} \cdot \hat{E}^* = E_0^2 (e^{i\theta} \cdot e^{-i\theta})$$ (3.421)

$$= E_0^2.$$ (3.422)

What has effectively been done by this procedure is to remove the complex description of the field in the expression for the intensity. Since the complex notation describes the state of linear or circular polarization, this is a simple demonstration of the fact that the intensity or flux of a radiation field is independent of its state of polarization.

Aside from the necessity for providing a real parameter to describe intensity by eliminating the imaginary part, we avoid describing the light wave as consisting of both electric and magnetic fields. Since the magnetic field is 90 degrees out of phase with the electric field, to describe both waves in a single equation, it is necessary to have trigonometric functions of the same parameter (θ) which are of necessity 90 degrees apart, that is, sine and cosine or secant and cosecant. In describing ordinary achiral absorption intensity, where the absorption is proportional to the square of electric dipole transition moment and the square of the electric field of the light wave, the magnetic field of the light wave may usually be ignored. When we come to optical activity, however, the problem is a little more complex. To specify the circular polarization of the light, we must write that the vector potentials of left and right circularly polarized light are

$$\hat{\mathbf{A}}_l = \mathbf{A}_0 e^{i\theta} = \mathbf{A}_0(\cos\theta + i\sin\theta), \tag{3.423}$$

$$\hat{\mathbf{A}}_r = \mathbf{A}_0 e^{-i} = \mathbf{A}_0(\cos\theta - i\sin\theta), \tag{3.424}$$

or the equivalent complex forms of the field $\hat{E}_l$, $\hat{E}_r$ and $\hat{H}_l$, $\hat{H}_r$. These quantities have to be carried through without squaring until the very end. This presents a problem because there are two other sources of i in Rosenfeld's equation, the imaginary part of the magnetic moment operator and the wavefunctions, which, for generality, are often treated as complex and therefore have imaginary parts. The problem is tractable, and a careful separation of the field-induced moments into real (Re) and imaginary (Im) parts does result in the expressions for the refractive and chiral parts of the moments, the isotropic polarizability α and the rotation parameter β.

Another approach which more closely resembles the perturbation treatment of achiral absorption is elegantly given by Kauzmann (10). Here the electric and magnetic fields are introduced separately. You will recall that Rosenfeld, using *classical electromagnetic theory*, derived the result that optical rotation will occur if electric and magnetic dipole moments are induced in the medium by changing electric and magnetic fields; the rotation parameter β appears as the coefficient of the derivatives of the magnetic and electric fields with respect to time. *Quantum mechanical* expressions for the induced moments have the same general form, and Rosenfeld identified the coefficients of $\partial H/\partial t$ and $\partial E/\partial t$ with those in the classical expressions β/c, and thus with the rotational strength. Kauzmann's approach to the quantum mechanical part of the derivation is slightly different from Rosenfeld's in that it does not use the vector potentials, but it is equivalent and perhaps somewhat easier to appraise in physically intuitive ways. No *new* physical interpretations are gained by going through the derivation in detail. We will examine it only closely enough to see the relationship between the classical and quantum descriptions.

Consider first the magnetic perturbation, the energy of which is

$$\mathcal{H}' = -\mathbf{m}\cdot\mathbf{H}, \tag{3.425}$$

where the magnetic field of the plane-polarized light is

$$\mathbf{H} = \mathbf{H}_0(e^{2\pi i\nu t}) = \mathbf{H}_0\frac{e^{2\pi i\nu t} + e^{-2\pi i\nu t}}{2}. \tag{3.426}$$

This results in an expression for the perturbation energy that has exactly the same form as that obtained for the electric field interaction with the electric dipole moment operator in the treatment of achiral absorption:

$$\mathcal{K}' = -\mathbf{H}_0 \cdot \mathbf{m} \frac{e^{2\pi i \nu t} + e^{-2\pi i \nu t}}{2}. \qquad (3.427)$$

Unsurprisingly, if the system is again described by an equation of the form of Eq. 3.305, the coefficient a_m, the square of which determines the probability that a magnetic dipole transition will take place, is found to be proportional to the magnetic transition moment integral $\langle \psi_m | \mathbf{m} | \psi_\ell \rangle$ and to the strength of the magnetic field $\mathbf{H}_0$. From this we draw the parallel conclusion to that for the electric field interaction, the system will not be perturbed—no magnetic dipole transition will be allowed unless the integral is finite. Even allowed magnetic dipole transitions by themselves are not readily measured by optical absorption, since their magnitude is extremely small, of the order of 10^{-5} of the intensity from an electric dipole transition. However, optical activity and circular dichroism are not proportional to the induced magnetic moment alone, but more importantly to the scalar product of the electric and magnetic moments. The induced electric dipole moment is determined using the value of a_m and, therefore, of the wavefunction obtained from the magnetic interaction. This moment, *averaged over all molecular orientations*, (we consider only liquids and gases), is

$$\mu = \frac{2}{3h} \sum_b \text{Re}[\mu_{ba} \cdot \mathbf{m}_{ab}] \frac{\nu_{ba}^2}{(\nu_{ba}^2 - \nu^2)} \cdot \mathbf{H}_0$$

$$- \frac{1}{3\pi h} \sum_b \text{Im}[\mu_{ab} \cdot \mathbf{m}_{ba}] \cdot \frac{1}{(\nu_{ba}^2 - \nu^2)} \frac{\partial \mathbf{H}_0}{\partial t} \qquad (3.428)$$

where Re and Im stand for real and imaginary parts, respectively.

By reversing the procedure and introducing the electric field interaction first to produce the electrically perturbed wavefunctions, a very similar expression results for the magnetic moment:

$$\mathbf{M} = \frac{2}{3h} \sum_b \text{Re}[\mu_{ba} \cdot \mathbf{m}_{ab}] \cdot \frac{\nu_{ba}^2}{(\nu_{ba}^2 - \nu^2)} \mathbf{E}$$

$$+ \frac{1}{3\pi h} \sum_b \text{Im}[\mu_{ba} \cdot \mathbf{m}_{ab}] \cdot \frac{1}{(\nu_{ba}^2 - \nu^2)} \cdot \frac{\partial \mathbf{E}}{\partial t}. \qquad (3.429)$$

Comparing these to Eqs. 2.310 and 2.311, we see that the coefficients of

$\partial\mathbf{H}/\partial t$ and $\partial\mathbf{E}/\partial t$ are identical (note that $\text{Im }\mu_{ba}\cdot\mathbf{m}_{ab} = -\text{Im }\mu_{ab}\cdot\mathbf{m}_{ba}$) to those of the classical theory; the value of the rotatory parameter β, therefore, is

$$\beta = \frac{c}{3\pi h}\sum_{b}\frac{\text{Im}\,\mu_{ab}\cdot\mathbf{m}_{ba}}{\nu_{ba}^2 - \nu^2}, \qquad (3.430)$$

and we have arrived at Rosenfeld's equation. The moments μ and $\mathbf{m}$ for the perturbed state are seen to each consist of two terms: one in the fields $\mathbf{E}$ and $\mathbf{H}$ and one in the time derivatives of the fields. From the point of view of deriving the rotatory parameter β, this is perfectly sufficient. Most other treatments that follow Rosenfeld's original procedure use the expanded form of the vector potentials of the fields rather than $\mathbf{E}$ and $\mathbf{H}$ and arrive, after dropping very small terms in the products of the coefficients a_m of the excited state or states, at three or four terms rather than two, of which one for the induced electric moment is proportional to $\mathbf{E}$ and identifiable with the polarizability α; another for the induced magnetic moment is proportional to $\mathbf{H}$ and identifiable as the magnetic susceptibility κ. These treatments are thus more general, but except in the case of measurements on oriented systems, for chiral phenomena we need only Eqs. 3.428 and 3.429. Indeed, even the first terms in these equations are superfluous, since with real wavefunctions $\langle a|$ and $\langle b|$, the magnetic transition moment, $\mathbf{m}_{ab} = eh/4\pi imc\cdot\langle a|\mathbf{m}|b\rangle$, is imaginary (see Eq. 3.407) and the electric transition moment, $\mu_{ab} = \langle a|\mu|b\rangle$, is real. The quantities $\mu_{ab}\cdot\mathbf{m}_{ba}$ and $\mu_{ba}\cdot\mathbf{m}_{ab}$ are thus pure imaginary and have no real part; the first terms in Eqs. 3.428 and 3.429 are therefore null from the point of view of a chiral observable. Only the second term in each expression has to do with optical activity.

We have reached a point at which we are very knowledgable. But it will take only a few more words to expose some still considerable gaps in that knowledge. The fact that there is still much to learn should not, however, prevent us from applying what we already know about the origin of optical activity and circular dichroism. In particular, we know that in examining ORD and CD curves we should look for the origin of anomalous dispersion or Cotton effects in some particular optical transition, and that this transition has to be at least partially electric and magnetic dipole allowed, that symmetry must not preclude the existence of parallel components of the moments, and that the underlying contributions to optical activity at any wavelength is the sum of contributions from a number of transitions; furthermore, only nearby transitions are likely to contribute significantly. This is patently useful information. We hope it will also be useful to have examined the physical origin of these statements.

There are some questions we have partly hidden under the rug, and there is the not-so-hidden question of how to perceive the optical activity of molecules that do not have planes (or centers) of symmetry, but in which the charge distribution in the chromophores that give rise to the chiral activity do appear to have such symmetry elements. In Section 3.6 we will discuss one mechanism for this, but in general, it forms the substance of the entire next chapter. Of the hidden problems, the related questions of the frequency dependence and vibrational contributions will form the substance of the next three sections. To focus on these problems, compare Eqs. 3.430 and 2.406. The frequency dependence of the rotatory parameter β, and therefore of the rotational strength and the optical activity, has the resonance catastrophe form $1/(\nu_{ba}^2 - \nu^2)$. In our earlier discussion of the dispersive behavior, the arbitrary introduction of a dissipative factor in the denominator of the expression for the optical activity (Eq. 2.406) was sufficient to remove this restriction and to describe the experimentally observed frequency dependence throughout the absorbing (resonance) region, as well as at frequencies far from the absorption.

3.5   FREQUENCY DEPENDENCE

In the discussion of the dispersive behavior of optical activity in Section 2.5, it became apparent that the frequency of the measuring radiation has a profound effect on the magnitude and sign of the measured rotation and, correspondingly, of the circular dichroism. Where do these effects arise? Is the behavior described in Section 2.5 the only behavior to be expected? Let us examine these questions separately, although, as we will see, their answers are related.

The first question may be approached with another—the answer to which is not restricted to chiral phenomena but applies to optical phenomena in general. Since according to quantum mechanics the difference in energy between states is discrete, why are ordinary absorption spectra, as well as the circular polarized spectra of chiral molecules, not infinitely sharp? There are two answers to this, and we are all familiar with them in a vague way. Let us try to be a little more precise. One is that under the appropriate conditions they can be almost infinitely sharp, at least no broader than the Heisenberg uncertainty limit or the Doppler width, the latter having to do with the relationship between the speed of light and the classical velocity of the "oscillating electron" with which the radiation is interacting.

More to the point, however, is that under normal conditions of measurement, for example, in solution at or near-ambient temperature, we are not dealing with a *single* discrete state, but with an assembly of molecules in a *multitude* of states whose energy is in the vicinity of, but not at the energy of, the unperturbed vibrationless pure electronic energy levels. Superimposed on this is the frequency distribution of the incident measuring radiation, which again has a distribution over energies in the vicinity of the nominal measuring frequency. The frequency of the radiation incident on the sample from an ordinary spectroscopic source, a monochromater illuminated by white light, for example, is distributed about the central frequency ν_0, with a width of $\pm \Delta \nu$ where $\Delta \nu$, for our purposes, can be considered to be one-half the frequency range over which the radiation is detectable ($2\Delta \nu$ is known as the spectral slit width). The multitude of states arise, both for chiral and achiral phenomena, from the vibrational, rotational, and environmental degrees of freedom available to the chromophore in question. The governance of which such states are available for interaction with the radiation produces the special band shape in each individual situation. In part, this governance is ruled by symmetry considerations. Since we do not take up symmetry control until Chapter 6, we will have to come back to this question once again. For now, suffice it to say that a molecule in a given electronic energy state may or may not undergo a radiation-induced transition to any particular vibrational and rotational state of another electronic state. Since there are many such possible substates, for example, $3N-5$ vibrational states where N is the number of atoms in the molecule, at least some of which are certain to be available (symmetry-wise), there will exist allowed electric and/or magnetic dipole transitions between a large number of different states, all in the region of the energy difference between the unperturbed pure electronic states. This is illustrated in Fig. 3.01. *Transitions between all of these states are not necessarily allowed*, but a sufficiently large number of them are; so that even for gaseous environmentally unperturbed polyatomic molecules, unless the incident radiation is very sharp, that is, $\Delta \nu$ is very small, transitions between the individual states are not necessarily resolved. This situation is illustrated in Fig. 3.02 where high resolution (small $\Delta \nu$) is observed to resolve at least some of the vibrational structure. This resolution can be obtained for molecules in solution only for particular solutes in noninteracting solvents (e.g., highly symmetric molecules in nonpolar solvents) or when a particular vibrational progression is much more strongly permitted than any of the others so that broadening resulting from rotational and environmental states is insufficient to completely smear the observed spectrum. We will see examples of this in the spectrum and circular dichroism of some ketones and olefins in Chapter 7. This situation can be strongly diagnostic for the nature of the electronic transition because Herzberg-Teller rules (vide infra) delineate those vibrations that

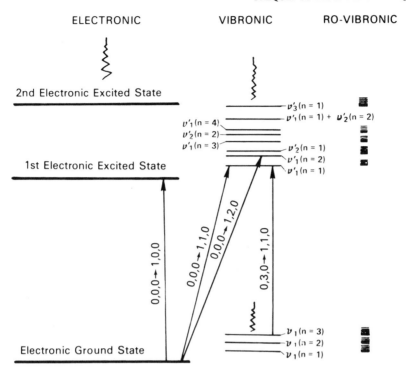

ELECTRONIC   VIBRONIC   RO-VIBRONIC

2nd Electronic Excited State

$\nu'_3(n = 1)$

$\nu'_1(n = 4)$ $\nu'_1(n = 1) + \nu'_2(n = 2)$
$\nu'_2(n = 2)$
$\nu'_1(n = 3)$ $\nu'_2(n = 1)$
                         $\nu'_1(n = 2)$
1st Electronic Excited State $\nu'_1(n = 1)$

$0,0,0 \rightarrow 1,0,0$

$0,0,0 \rightarrow 1,1,0$

$0,0,0 \rightarrow 1,2,0$

$0,3,0 \rightarrow 1,1,0$

$\nu_1(n = 3)$
$\nu_1(n = 2)$
Electronic Ground State $\nu_1(n = 1)$

Fig. 3.01 Molecular states and their optical transitions; the middle index of the quantum numbering on the transitions shown refer to the quantum number of the ν_1 vibration in the electronic state indicated by the first index. The last index, of course, refers to the rotational or ro-vibronic state and is rarely spectrally discernable except for very small molecules (chiefly di- and triatomics) in the gas or in low-temperature crystals.

can and cannot combine with the pure electronic states (7). Each of the quantized ro-vibronic (rotational, vibrational substate of an electronic state) states is further distributed in energy among molecules with almost a continuum of environmental substates in solutions or in crystalline solids, among a small set of environmental substates determined by the discrete asymmetry of the crystal. Depending on the strength of the coupling between the chromophoric groupings in the molecule and its environment, these may further broaden the energy distribution (22). Hydrogen bonding is perhaps the best known of the strong coupling situations, and its effect in broadening the transition arising from the OH stretching vibration in the infrared region of the spectrum is a well-known phenomenon. That same broadening, of course, appears in the electronic spectrum, and the diffuse nature of most absorption bands resulting from a carbonyl $n \rightarrow \pi^*$

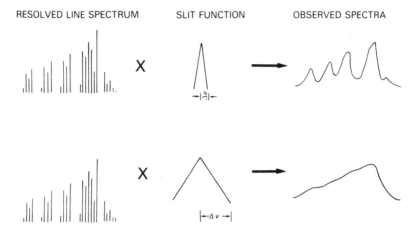

RESOLVED LINE SPECTRUM   SLIT FUNCTION   OBSERVED SPECTRA

Fig. 3.02 Resolution of a vibronic band; the effect of slit width (spectral resolution) on the observed spectrum of a band having very sharp transitions between different electronic vibrational or vibronic states. The slit function represents the intensity profile (frequency dependence) of the radiation from the spectrometer about the median frequency as the spectrum is scanned through the frequency range of the absorption band of the sample.

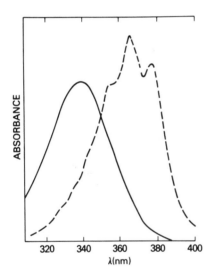

ABSORBANCE

320   340   360   380   400

λ(nm)

Fig. 3.03 The effect of hydrogen bonding on the spectral bandwidth of the vibrational components of an electronic transition. The many substates contributed by the varied conformations that simultaneously exist in a hydrogen-bonded complex of H_2O broaden and shift the spectrum of $(CH_3)_2NNO$ so that the spectra in H_2O (———) and in cyclohexane (----), taken at the same resolution, have completely different shapes. (The figure is adapted from Fig. 9.8 in H. H. Jaffe and M. Orchin, *Theory and Applications of Ultraviolet Spectroscopy*, Wiley, New York, 1962.)

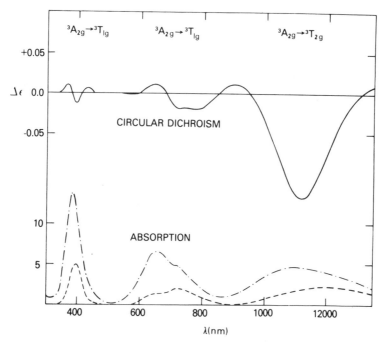

Fig. 3.04 Ligand field splitting of the degenerate states of Ni^{2+} ion. The very weak bands (note the value of the extinction coefficients) in the visible and near-infrared spectrum of the Ni^{2+} ion or its ammonia complex in aqueous solution (-----) is slightly enhanced when D-tartaric acid is added (-·-··-). In this case, the splitting of the degeneracy of the triple $d{\rightarrow}d$ transitions of the Ni^{2+} ion is, however, not observed in the isotropic absorption spectrum but stands out sharply in the circular dichroism spectrum (———); in particular, note the two positive and one negative CD bands corresponding to the isotropic absorption bands near 4000 Å and 11,000 Å. W. A. Eaton, unpublished spectrum.

transition in hydroxilic solvents is well known (Fig. 3.03). The splitting of the electronic sublevels of degenerate states of metal ions by ligands is another example (Fig. 3.04), this one usually characterized by a more discrete effect on the spectrum because of the magnitude of the splitting and the discrete geometrical arrangement of the ligands with respect to the metal ion.

There will be other occasions, in Chapter 6, where the symmetry of degenerate states is taken up, in Chapter 8, where we discuss repeating polymers, and in Chapter 10, in the discussion of magnetic circular dichroism, to bring up the effects of the degeneracy of molecular states. For the present, suffice it to say that any molecule or isolated chromophoric group in a molecule that has a threefold or greater axis of symmetry will have more than one state of the same energy in the absence of some perturbing field. Such states are known as degenerate states. The perturbing fields which break or "lift" the degeneracy may be external applied static electric or magnetic fields, or an asymmetric potential due to another part of the molecule—a ligand in a metal complex, for example, or indeed the interaction of two or more vibrations—rotations in the molecule itself. In addition it is possible for *transitions* to be degenerate even when there are no degenerate *states*. This can occur when transitions from occupied orbitals of different energy in the ground state of the molecule to different excited states happen to be of the same energy. These latter transitions are known as "accidentally degenerate" transitions, and different rules apply than those that apply to the "true" degeneracies.

Hydrogen bonding effects are quite different from those of ligand field splitting. Without going into a detailed description of the nature of the bonds, qualitatively it is sufficient to think of the continuum of possible geometric arrangements (bond angles and bond lengths) between the OH bond and the $C=O$ bond of a hydrogen-bonded carbonyl. Ligand splitting and hydrogen bonding are extreme effects of the environment; the entire range between is possible.

Transitions between each of these various states may contribute in varying degrees to optical phenomena. The extent to which they can contribute is dependent on a number of factors including the phenomenon itself. Since the dipole strength which controls the ordinary absorption and refraction of light is proportional to the square of the electric dipole transition moment, it is obvious that substates that are contributory to the magnetic transition dipole moment $\mathbf{m}_{ab}$ will contribute to the ordinary isotropic absorption only if they simultaneously contribute in the same way to $\boldsymbol{\mu}_{ab}$, and we will see that this is not a necessary condition (23). So the shape, or at least the frequency distribution of a circular dichroism band, can be different from that of an isotropic absorption band, and the optical rotation correspondingly different from the refraction. In addition to the frequency distribution, there is the question of sign, and here the effects become even more complex (23). We will return to examine these complexities after establishing the basis for being able to treat the question in terms of Herzberg-Teller-Weigang theories for intensity borrowing in chiral and achiral phenomena.

3.6 THE BORN-OPPENHEIMER APPROXIMATION AND VIBRONIC COUPLING IN CHIRAL PHENOMENA

In Section 3.3 where the quantum theory of optical activity is discussed, the Schrödinger equation (Eq. 3.303) for the time dependence of the wavefunction ψ was given in terms of the operator $\mathcal{H}$, called the Hamiltonian [after the mathematician and astronomer Sir William Rowan Hamilton (1805–1865), whose introduction of a "characteristic function" in the theories of mechanics and dynamics laid the foundation for the mathematical formulation of quantum theory in terms of the motions of particles]. $\mathcal{H}$ is a description of the kinetic and potential energy of the particles of the system whose distribution in space and/or time is described by ψ. We pointed out that perturbation theory permits the Hamiltonian operator to be considered as the sum of two parts, $\mathcal{H}_0$ and $\mathcal{H}'$, where $\mathcal{H}'$ represents the effect of an external field, in our case, that of the interaction between the charges in the molecule as described in terms of their dipole moments, $\Sigma_i e_i r_i$, and the electric field of the light wave, $E_0(e^{2\pi i \nu t} + e^{-2\pi i \nu t})$. We said nothing about the other part of the Hamiltonian, $\mathcal{H}_0$, because this part drops out of the time-dependence problem (8). The physical meaning of this is that the field exerted by the nuclei and electrons on each other in a molecule or atom is generally so much larger than the perturbing field of the light wave, that the perturbation makes only a small change in the charge distribution. The molecule in its excited state looks very much like the molecule in its ground state, and, in fact, the laws of conservation of momentum which give rise to the Franck-Condon principle (6) ensure that this is so. Of course, this refers to the direct result of the excitation process. Once the molecule reaches this excited state, which happens in a very short time compared to vibrational relaxation, it may very well redistribute its electrons and nuclei into quite a new equilibrium configuration; the equilibrium configuration of some molecules even in their lowest excited electronic state, ethylene and formaldehyde being typical examples, is quite different from that in the ground state. But this occurs immediately after the initial excitation process which itself perturbs the charge distribution only very slightly.

To investigate which excitation processes are allowed, we cannot ignore that part of the Hamiltonian operator that describes the field in which the particles of the unperturbed molecule find themselves—or at least, we must find some justification for doing so. The reason for this is that hidden in the transition moment, as it has been written in this text, is the fact that the wavefunctions ψ are state functions and thus describe not only the pure electronic states (i.e., the electronic charge distribution at fixed nuclear and environmental configurations), but also the ro-vibronic states and, for condensed systems, the translational or librational states as well. The rotational and librational states provide some of the width of both isotropic and linearly and circularly polarized absorption bands (Section 3.5), but their specific importance in chiral phenomena is negligible. Rotational fine structure is largely confined to spectra of molecules too small to be optically active, and librational effects due to the solvent interactions are generally continuous and, except insofar as the interactions may affect the molecular conformation, have no effect on the sign (22). However, the situation may be very much different with vibrational effects, and consequently the justification for ignoring them in some cases and for specifically accounting for them in others has to be considered.

In the treatment of isotropic absorption in Section 2.3, for purposes of simplicity specific vibrational effects were ignored by integrating over the entire band, as was the variation of the radiation density with frequency by normalizing it over the band $\int \rho(\nu) d\nu = 1$ (Eq. 3.315). These are perfectly valid and reasonable things to do, and chiral phenomena may be treated similarly if (1) the band under investigation exhibits no vibrational structure under the conditions of the measurement *and* (2) we are uninterested in the particular vibronic states that contribute to the band. If these conditions do not pertain, then it is imperative that a closer accounting be made of vibrational effects. With respect to the second of the two conditions, interest or disinterest will depend on the purpose for which we are examining the chiral activity and, in part, on whether the vibrations of the nuclei have a profound effect on the magnitude or sign of the circular dichroism and optical activity. It is possible to generalize that for strong absorption bands *that have large intrinisic values of the electric dipole transition moment* μ_{ab}, and especially for measurements in solution, the vibrational structure is of little significance in determining the magnitude and especially the sign of the optical activity. The reason for this will become clearer as we examine the vibronic mechanisms in more detail. When the electric dipole transition moment is very small and the isotropic absorption, therefore, weak, the situation is distinctly different. It is for this case that the Herzberg-Teller rules for vibronic contributions to the isotropic absorption become important and for which related extensions by Weigang (24) and others describe the vibronic contributions to chiral absorption and refraction.

We have strayed from our original intent to examine the steady state Hamiltonian, $\mathcal{H}_0$. Making note of the fact that all that was required to see that isotropic absorption and optical activity depend on the electric and/or magnetic transition moment was the perturbation part $\mathcal{H}'$, we may ask why it is necessary to explore the unperturbed $\mathcal{H}_0$. The answer was hinted at a few paragraphs back where it was pointed out that the ψ are state functions specifying the entire array of states and substates of the molecule. We have now seen that for our purposes it is possible to ignore the rotational and librational parts, and under certain conditions delineated in the previous paragraph, the vibrational states as well. Suppose these conditions are not met—what then?

The answer is that is it not possible, then, to assume that either the isotropic absorption or the chiral activity can be described by integrating over the band, suppressing the fact that the transition moment wavefunctions are not pure electronic functions, but are instead ro-vibronic functions $\psi_{e,v,r}$, where the subscripts refer rather obviously to electronic, vibrational, and rotational quantum numbers. We have already determined that the rotational parts may be incorporated without distinction into the vibrational parts; so we need only consider the vibronic function $\psi_{e,v}$. Fortunately, it is not necessary to evaluate the complete vibronic functions because it is usually possible to separate the vibrational parts from the electronic parts by utilizing the Born-Oppenheimer approximation (5):

$$\psi_{e,v} = \psi_e \psi_v. \qquad (3.601)$$

This beautiful simplification requires a very lengthy mathematical justification, which can be found in the original paper, but the intuitive physics of the situation is very similar to that of mechanical resonances, such as the response of the ear or the microphone to musical tones. We are all familiar with the fact that as two resonant frequencies get very close, it not only becomes difficult to distinguish them, but in fact, their exact overlap is marked by such a strong interaction (alternate reinforcement and cancellation) that we hear a beat rather than a constant pure tone. As the frequencies get further apart, the distinction becomes clearer, and our ear or an electronic detector distinguishes them clearly as nearly pure tones. So too with electronic and nuclear wavefunctions, the former being in the frequency range of approximately 25 to 250 kcycles/cm (visible and ultraviolet frequencies) and the latter in the range of 2.5 to 25 kcycles/cm (infrared region), with very little overlap between the two regions except perhaps for the lowest energy electronic transitions of some heavy metal ions or complexes.

Now we can see why we have been alluding to the steady state unperturbed Hamiltonian operator. If the wavefunctions were not separable, then the Schrödinger equations for a many-particle system (even for a single proton and electron) would require that we be able to describe with some accuracy the *inter*particle potential energy in the Hamiltonian, a very difficult thing to do. Furthermore, the solutions of the equations would then span the electronic and nuclear coordinates of an atom or molecule together, so that all solutions would be many-dimensional surfaces in space for the time-independent equation in the space *and* time for the time-dependent equation. For some purposes this might be desirable, for example, for the prediction of reaction coordinates and kinetics; but for most other purposes, it would introduce enormous complexity without adding much to our knowledge.

Fortunately we can almost universally work within the limitations and the freedom of the Born-Oppenheimer approximation.

For a single particle, the Hamiltonian operator consists of a kinetic energy part $(h^2/2m)(\partial^2/\partial q^2)$, which describes the motion along the space coordinate q of a particle of mass m, having the angular moment $h/2\pi i \; \partial/\partial q$, and moving in the charged field having a potential $V(q)$:

$$\mathcal{K} = \frac{h^2}{2m} \frac{\partial^2}{\partial q^2} - V(q). \tag{3.602}$$

For a system of many particles the kinetic energy is just the sum of the kinetic energy of the particles, but the potential energy must account for the interparticle interaction, as well as for the interactions with any external field. Therefore, for a nucleus and an electron of mass M and m, respectively, and coordinates with respect to fixed axes in the molecule, Q and q:

$$\mathcal{K} = \frac{h}{2M} \frac{\partial^2}{\partial Q^2} + \frac{h}{2M} \frac{\partial^2}{\partial q^2} - V(Q,q). \tag{3.603}$$

The specification of $V(Q,q)$ as coulomb interactions is not too difficult for only two particles, although accuracy at close distances is not very good, but for three or more particles, $V(Q,q)$ becomes very complex. It is impractical to use Eq. 3.603 to solve the Schrödinger equation for the complete wavefunction $\psi_{e,v}$ of any but the simplest atoms and di− or triatomic molecules.

Using the Born-Oppenheimer approximation, this problem is avoided; let us examine its significance for the rotational strength. The electric and magnetic dipole operators, μ and $\mathbf{m}$, may each be separated into the sum of an electric and nuclear part (as vectors, the vector sum of the parts is equal to the whole):

$$\mu = \mu_e + \mu_v,$$

$$\mathbf{m} = \mathbf{m}_e + \mathbf{m}_v. \tag{3.604}$$

Then we can write the transition moments μ_{ab} and $\mathbf{m}_{ab}$, using Born-Oppenheimer wavefunctions ψ_e and ψ_v. For convenience in referring to the original work, we adopt Weigang's nomenclature, letting capital letters stand for electronic wavefunctions and lower case letters for vibrational wavefunctions. However, in accordance with the usage in this book, we have retained the designations μ or μ_e for the electric dipole operator and $\mathbf{m}$ and $\mathbf{m}_e$ for the magnetic dipole operator. The original work uses μ_m for the latter. Thus we define $\langle \psi_{e,v} | \equiv \langle Nn |$, which by the Born-Oppenheimer approximation is $\langle N|\langle n|$. Then the electric dipole transition moment, μ_{ab}, is

$$\langle a|\mu|b\rangle = \langle Nn|\mu|Kk\rangle, \tag{3.605}$$

which by Eq. 3.604 is

$$\langle a|\mu|b\rangle = \langle N|\mu_e|K\rangle\langle n|k\rangle + \langle n|\mu_v|k\rangle\langle N|K\rangle. \tag{3.606}$$

The solutions to the Schrödinger equation require that wavefunctions for the *electrons* in two different electronic states be orthogonal (both cannot be in the same space at the same time). This orthogonality is expressed in vector space as vectors at right angles to each other. The second term in Eq. 3.606 is therefore zero because $\langle N|$ and $\langle K|$ are such wavefunctions and the simple (dot) product of their vector representation, $\langle N|K\rangle$, must therefore be zero. The electric transition moment, therefore, reduces to

$$\langle a|\mu|b\rangle = \langle N|\mu_e|K\rangle\langle n|k\rangle. \tag{3.607}$$

Note that $\langle n|$ and $\langle k|$ are the wavefunctions for the *nuclei* in two different *electronic* states and so need not be orthogonal. If they are, this term is also zero and results in the absence of an electric dipole transition between the n and k vibrational levels of the vibronic states $\langle a|$ and $\langle b|$. In fact, in a polyatomic molecule of high symmetry, many vibrations may be precluded from appearing in the manifold of the ultraviolet absorption band for just this reason.

In general, the value of $\langle n|k\rangle$ is governed by the Franck-Condon principle, which requires that little or no change in the vibrational coordinate take place during the electronic transitions, that is, that the nuclei are not displaced during the time the photon and electron interact. In this case, the Born-Oppenheimer oscillator strength and, consequently, the absorption intensity, are proportional to

$$f_{ab} \sim |\langle a|\mu|b\rangle|^2 = |\langle N|\mu_e|K\rangle|^2 |\langle n|k\rangle|^2. \tag{3.608}$$

In exactly the same way the rotational strength is

$$R_{ab} \sim \langle a|\mu|b\rangle \cdot \langle b|m|a\rangle = \langle N|\mu_e|K\rangle \cdot \langle K|m_e|N\rangle |\langle n|k\rangle|^2. \tag{3.609}$$

In the light of the Born-Oppenheimer approximation, it therefore appears that the oscillator strength and the rotational strength are governed by electronic transition moments which are the same for all transitions between allowed vibrational states of the two electronic transitions and by the square of the vibrational overlap functions $\langle n|k\rangle$. If this were true for all transitions, the shape, which is determined by these overlap functions (but not the magnitude, of course) for ordinary achiral absorption and circularly dichroic absorption for the same electronic transition $a \rightarrow b$, would always be identical, and as a matter of fact, for strongly allowed transitions where both $\langle N|\boldsymbol{\mu}_e|K\rangle$ and $\langle K|\mathbf{m}_e|N\rangle$ have large collinear components, this is true. When either or both are highly forbidden by symmetry considerations, this is when the Herzberg-Teller borrowing of intensity from higher energy transitions enters the picture and opportunities arise for the band shapes to be quite different in chiral and achiral absorption. We examine the *result* of these considerations in the manner in which Weigang obtained them (24). Others (25, 26) have produced the same or similar results, but this formulation is most appropriate to that in which our discussion has been cast.

In the absence of an allowed transition between the pure electronic states $\langle N|$ and $\langle K|$, the mechanism by which intensity appears in ordinary absorption is through borrowing from a third state, $\langle S|$. The criteria for such borrowing have been elaborated in exquisite detail in a paper by Gerhard Herzberg and Eugene Teller (7). These same considerations are involved in chiral phenomena. We note first that the state $\langle S|$, from which the intensity is borrowed, need not necessarily be the excited state next in energy to the state $\langle K|$, but usually will be a fairly close one because the degree of borrowing depends in part on the energy separation between the states.

The overall rotation in these terms is given by a somewhat formidable looking expression which becomes even more formidable appearing when the appropriate expressions for the coefficients are substituted (23). However, examining it term by term will reveal its very straightforward meaning.

$$R_{ab} \equiv R_{Nn,Kk} = \mathrm{Im}\left\{ \langle N|\boldsymbol{\mu}_e|K\rangle \cdot \langle K|\mathbf{m}_e|N\rangle |\langle n|k\rangle|^2 \right. \tag{3.610a}$$

$$+ \sum_r \langle N|\boldsymbol{\mu}_e|K\rangle \cdot \mathbf{B}_r \langle n|k\rangle \cdot \langle k|\mathbf{Q}_r|n\rangle \tag{3.610b}$$

$$+ \sum_r \mathbf{C}_r \langle K|\mathbf{m}_e|N\rangle \cdot \langle n|\mathbf{Q}_r|k\rangle \cdot \langle k|n\rangle \tag{3.610c}$$

$$+ \sum_r \sum_s \mathbf{C}_r \mathbf{C}_s \langle n|\mathbf{Q}_r|k\rangle \cdot \langle k|\mathbf{Q}_s|n\rangle. \tag{3.610d}$$

Before examining each term, let us define each of the previously undefined quantities. The vibrational overlap functions $\langle n|k\rangle$ have a maximum value of one and a minimum value of zero, the former representing the fact that in both electronic states the nuclear wavefunctions for a given vibrational mode are identical; that is, the vibrational amplitude is the same function of the space coordinates in both electronic states. Since the potential energy curves, which represent the way in which the energy varies as a function of the coordinates of the electron with respect to the nuclei, are not likely to be identical in two different states, the maximum value is rarely, if ever, realized, but large values do occur.

Integrals of the type $\langle n|Q_i|k \rangle$ are also vibrational but are dependent not only on the wavefunctions, but also on the sign and magnitude of the nuclear (vibrational) displacement normal coordinates Q_i. It may be apparent that these integrals resemble the pure electronic transition moments and are indeed the corresponding vibration transition moments that control the intensity of infrared vibration-rotation spectra (27).

A normal coordinate is parametrically related to the displacement coordinates of the atoms along bonds or with respect to angles between the bonds. It is the displacement of all the atoms involved in a vibration of a given frequency in which the criterion for the magnitude and direction of the displacement is that the vibrating atoms all reach maximum and minimum displacements simultaneously.

The coefficients B_r and C_r give rise to the borrowed intensity. If these are zero, then only the first term in Eq. 3.610 survives, and this is essentially the only term of importance when both the electronic electric and magnetic dipole transitions are allowed, as in a dissymmetric chromophore. The coefficients are defined:

$$C_r = \sum_S \left[\lambda_{SK} \langle N|\boldsymbol{\mu}_e|S \rangle + \lambda_{SN} \langle S|\boldsymbol{\mu}_e|K \rangle \right], \tag{3.611}$$

$$B_r = \sum_S \left[\lambda_{SK} \langle S|\mathbf{m}_e|N \rangle + \lambda_{SN} \langle K|\mathbf{m}_e|S \rangle \right]. \tag{3.612}$$

Since these coefficients contain the pure electronic transition moments for transitions from both the ground state $\langle N|$ and the excited state $\langle K|$ to the third state $\langle S|$, they will both be nonzero only if at least one of the transitions, $N \rightarrow S$ or $K \rightarrow S$, is electrically or magnetically allowed, or if one is electrically and the other magnetically allowed. The magnitude and *sign* of each of these is independently dependent on the geometry of the system through the coordinates r_i, on which the operators $\boldsymbol{\mu}_e$ and $\mathbf{m}_e$ depend (Eqs. 3.307 and 3.407). The weighting coefficients that determine the relative extent from borrowing from the $N \rightarrow S$ and the $K \rightarrow S$ transitions are given by

$$\lambda_{SK} = \frac{\langle S|\mathcal{H}_r|K \rangle}{\mathcal{E}_K - \mathcal{E}_S}, \tag{3.613}$$

where

$$\mathcal{H}_r = \left(\frac{\partial \mathcal{H}_e}{\partial Q_r}\right)_0. \tag{3.614}$$

Note that the weighting coefficients depend on the *difference* $(\mathcal{E}_K - \mathcal{E}_S)$ in energy between the appropriate states and that the borrowing becomes greater, other things being equal, if there is a nearby state to borrow from. Finally, borrowing of vibronic intensity in optical activity can only occur if λ_{SK} or $\lambda_{SN} \neq 0$, and this in turn requires that the operator $\mathcal{H}_r$ be finite. This parameter, which describes the way in which the energy of the system changes as one of the normal coordinates Q_r changes during a vibration from its value when the molecule is in its equilibrium geometry, is identically or nearly zero for totally symmetric vibrations. We will have more to say about the symmetry of vibrations in Chapter 5. However, it is sufficient here to point out that vibrations will be totally symmetric or not depending on whether the nuclear displacements during the vibration do or do not change the dipole moment of the molecule. So the requirement that $(\partial \mathcal{H}_e / \partial Q_r)_0$ be finite means that only antisymmetric vibrations contribute to *vibronic borrowing* mechanisms, but since in a polyatomic molecule there are always some antisymmetric vibrations, this is not a serious restriction. Note that this does not mean that symmetric vibrations do not contribute to optical activity, only that they do not contribute to the mechanism for borrowing this activity from other electronic transitions than the one in question.

Now let us examine the effect of each term in Eq. 3.610 separately.

The first term, Eq. 3.610a: Except for the vibrational overlap integral $\langle n|k \rangle$, this term looks no different from the original expression (the right-hand side of Eq. 3.609); it differs only because of the Born-Oppenheimer separation of the vibronic states $\langle Nn|$ into pure electronic, $\langle N|$, and pure vibrational, $\langle n|$, states. The first term therefore refers simply and directly to allowed electromagnetic transitions for which we have previously ignored the vibrational states by summing over the entire vibronic band envelope, that is, over all vibrational states whose symmetry is allowed for the particular electronic transition. This latter procedure is the one followed by Moffitt and Moscowitz (28) and is a valid and useful one for the strong transitions of dissymmetric chromophores. The effect of the more elaborated expression (Eq. 3.610a) is thus just to show the origin of the shape factor of the band by the dependence of the rotational strength on the overlap integrals between each vibrational mode. When $\langle N|\mu_e|k \rangle$ and $\langle K|m_e|N \rangle$ are associated with strong transitions and have large collinear components, this term dominates the entire expression for the rotational strength. The remaining terms, b, c, and d, are, therefore, important only if the $N \rightarrow K$ transition is electrically or magnetically forbidden or the allowed moments are perpendicular or nearly perpendicular to each other (as in a chromophore with a plane of symmetry).

The third term, Eq. 3.610c: This term is for just such a case, that is, a planar chromophore in which the pure electronic transition moment is forbidden or perpendicular to the allowed magnetic dipole transition whose moment, $\langle K | \mu_m | N \rangle$, it contains as a factor. We shall see in Chapter 7 that the $n \rightarrow \pi^*$ transitions of the ketones fall into this class, and the nearby transitions give rise to the borrowed electric dipole transitions which account for the coefficient C_r being of a reasonable magnitude.

The second term, Eq. 3.610b: The argument for this term is more or less symmetric with that for the third term, except that magnetic borrowing for electrically allowed transitions is not so well known. Also since electrically allowed transitions will absorb radiation (polarized or unpolarized) strongly, it may be difficult to measure the small contributions to optical activity by this mechanism in the presence of strong optical absorption. Cases like this would, however, represent good candidates for measurements, in an absorption band, of the optical activity rather than the circular dichroism, but this would also tend to obscure the vibronic nature of the contributions.

The fourth term, Eq. 3.610d: This term contributes regardless of whether there are allowed pure electric or magnetic transition moments between $\langle N |$ and $\langle K |$, since these do not appear. However, the term is the product of two borrowing mechanisms each of which is likely to be fairly small. The analysis of these contributions is difficult because of the ambiguities due to the large number of sign possibilities coming from the products of the various transition moments.

We cannot leave this discussion of vibrational contributions to electronic chiral phenomena without making a few additional brief remarks, the first and most important being that the discussion has only touched on some of the more subtle or more theoretical points. In particular, we have not examined in detail under what conditions vibrations of various symmetry are more or less likely to make contributions and how this affects the electronic band shape. For example, Moffitt and Moscowitz (28) have shown that in the particular situation of a magnetically allowed electrically forbidden transition (Eq. 3.610c), the circular dichroism band is likely to be displaced to slightly lower energies than the achiral ordinary absorption band through the different contributions of symmetric and antisymmetric vibrations. The relationship between the measuring frequency and the transition frequency in determining the appearance of vibronic contributions (a subject we have totally ignored except by inference) has been covered in a succinct paper by Hameka (25). The question of optical activity of entirely vibrational origin will be discussed in Chapter 10. In discussing the frequency dependence on band shapes, we have ignored exciton or degeneracy effects and so have not really answered one of the questions we put at the very beginning of Section 3.5, namely, are the typical shapes illustrated for optical activity and circular dichroism in Fig. 2.10 the only shapes to be expected. The answer is a simple no, but we defer the discussion to the chapter on polymers where these effects can be discussed in more detail. Finally, we should be alert to the fact that since the shape of the achiral and chiral absorption bands may not be the same because intensity may be borrowed from both electric and magnetic transitions for the latter and, as a practical matter, only from electric dipole transitions for the former, the ratio of the chiral extinction to that of the achiral extinction may be a sensitive diagnostic test for the nature of the optical transition involved in a given chiral absorption band (29). This ratio, which has become known as the Kuhn anisotropy factor (30), is given to a good approximation as $g = 4R_{ab}/D_{ab}$, where D_{ab}, the line or dipole strength, is directly equal to the square of the electronic transition moment; it is measurable from the ratio of the circular dichroism to the ordinary extinction, $\Delta\varepsilon/\varepsilon$, in the same solvent. If the shape of the bands are different, the g will vary with frequency through the frequency range of the bands (29). An early and important example of this was given by Lowry in the first English text devoted to optical activity (31). In addition, Caldwell (32) has shown that vibronic interactions in a single electronic transition can lead to a change in sign in a CD band superimposed on, or in the absence of, changes in sign due to the borrowing mechanism of individual vibrational components discussed above.

3.7 COMPLETELY QUANTAL THEORIES

Completely quantal theories of optical activity differ from those we have been discussing in one important aspect: The radiation field is treated as a quantum field rather than as an electromagnetic wave. When we do this, we lose almost all contact with physical intuition based on the experiences of our senses and our contact with classical theories. Of the quantum theories, perhaps those that use the scattering of photons as the method of approach come closest to being physically intuitive, but even then we have to rely very heavily on some complex mathematical apparatus to distinguish between left- and right-scattering chiral charge distributions. These theories are very important to the continued peeling away of the shrouds that cloud our comprehension of optical chirality. However, in terms of the thrust of this book, which is to promote discourse between theoreticians and experimentalists, these theories add very little. They have the advantage of being able to excise second-order effects (more properly third or fourth order, since the effects we have been discussing are already second-order effects requiring the interaction of two fields or a field and a charge). These are smaller more subtle effects which should provide sensitive experimental tests of the theories. So we bypass their consideration here but not because they are insignificant. In addition to the theories of Hameka and of Chiu, to which we made reference in Chapter 1, several others should be noted, especially those of Stephen (33), who introduced the idea of treating refractive index differences as a scattering phenomenon, of Gō (34), of Atkins and Barron (35), and of Hutchinson and Lawetz (36, 37), who made direct use of the Dirac formulation (38) of the molecular scattering cross section for photons.

3.8 SUMMARY

Because this chapter is concerned with the origin of chiral activity in terms of transitions between stationary states, a diversity of topics has been treated—from the description of the states in terms of electronic *configurations* of atomic or atomic-like orbitals, to the vibronic interactions that control the shape of the optically active absorption and refraction, profoundly in those cases when the intrinsic optical activity is zero or small because of the magnitude of μ_{0a} or m_{0a} or their near perpendicularity, and to a much lesser extent when $\mu_{0a} \cdot m_{0a}$ is large. Perhaps the most important things we should recall are in the sections that lie between these two apposite topics and their relationship to the classical theory in Chapter 2: To explain the classical expressions for the optical activity and circular dichroism in terms of the angle of rotation δ, of plane-polarized light and the ellipticity ψ, and of circularly polarized light when it is passed through the active sample,

$$\delta = \frac{\pi z}{\lambda}(\eta_1 - \eta_r), \qquad \psi = \frac{\pi z}{\lambda}(k_1 - k_r),$$

we called on quantum mechanical perturbation theory to explain isotropic optical absorption and chiral activity. The problem consisted of relating the polarizability and magnetic susceptibility tensors α and β in the classical equations

$$\mu = \alpha E \qquad \text{and} \qquad m = \frac{1}{c}\beta\frac{dE}{dt},$$

for the moments induced in each molecule by the electromagnetic fields of the light, to the quantum mechanical transition moments. Averaged over all orientations in order to describe the behavior in isotropic gases or solutions, the tensors α and β are found to depend on the oscillator and rotational strengths of the electronic transitions with a frequency dependence that is in part intrinsic:

$$\alpha(\nu) = \frac{q^2}{\varepsilon_0 m} \sum_b \frac{f_{ab}}{\nu_{ab}^2 - \nu^2}, \qquad \beta(\nu) = -\frac{c}{3\pi h} \sum_b \frac{R_{ab}}{\nu_{ab}^2 - \nu^2}.$$

If the dissipative effects of the energy are included in an arbitrary fashion, these tensors are given by Eqs. 2.023 and 2.407, the latter after substitution in Eq. 2.319 for the optical rotation.

Quantum mechanical perturbation theory then yields the result that the oscillator and rotational strengths are related to the dipole transition moments:

$$f_{ab} = \frac{8\pi^2 mc}{3e^2 h} \nu |\langle a|\boldsymbol{\mu}|b\rangle|^2,$$

and

$$R_{ab} = -\langle a|\boldsymbol{\mu}|b\rangle \cdot \langle b|\mathbf{m}'|a\rangle,$$

where $\mathbf{m}' = i\mathbf{m}$. So far as chiral phenomena are concerned, the important points are that $\langle a|\boldsymbol{\mu}|b\rangle$ and $\langle b|\mathbf{m}'|a\rangle$ must have parallel components because otherwise, as the dot product of two perpendicular vectors, the rotational strength will be zero, and that the observed rotation or circular dichroism, as measured by the ellipticity, is related to the geometry of the molecule through the molecular parameters $\langle a|$, $\langle b|$, $\boldsymbol{\mu} = e\mathbf{r}_i$, and $\mathbf{m} = (\mathbf{r}_i \times \mathbf{p}_i)e/2mc$.

REFERENCES AND NOTES

1. C. A. Coulson and I. Fischer, *Philos. Mag.*, Ser. 7, **11**, 386 (1949).

2. J. N. Murrell, *The Theory of the Electronic Structure of Organic Molecules*, John Wiley & Sons, New York, 1963; see also J. N. Murrell and A. J. Harget, *Semi-Empirical Self-Consistent-Field Molecular-Orbital Theory of Molecules*, John Wiley & Sons, London, 1972.

3. J. C. Slater, *Quantum Theory of Matter*, 2nd ed., McGraw-Hill Book Co., New York, 1968.

4. P. S. Bagus, B. Liu, A. D. McLean, and M. Yoshimie, in *Energy Structure and Reactivity, Proceedings of the 1972 Boulder Summer Research Conference on Theoretical Chemistry*, O. W. Smith and W. B. McRae (Eds.), John Wiley & Sons, New York, 1973.

5. M. Born and R. Oppenheimer, *Ann. Phys.*, 4th Series, **84**, 457 (1927).

6. J. Franck, *Trans. Faraday Soc.*, **21**, 536 (1925); E. U. Condon, *Phys. Rev.*, **32**, 858 (1928).

7. G. Herzberg and E. Teller, *Z. Phys. Chem.*, **B21**, 410 (1933).

8. E. Schrödinger, *Ann. Phys.*, **79**, 361, 478; *ibid.*, **80**, 437; *ibid.*, **81**, 109 (1926). Discussions are found in a myriad of books on quantum mechanics and quantum chemistry. For the time-dependence problem and in particular, its application to the interaction of electromagnetic radiation with matter, particularly but not exclusively useful is the text by H. Eyring, J. Walter, and G. E. Kimball, *Quantum Chemistry*, John Wiley & Sons, Chapman & Hall, New York and London, 1944.

9. J. C. Slater and N. H. Frank, *Introduction to Theoretical Physics*, McGraw-Hill Book Co., New York, 1933.

10. W. Kauzmann, *Quantum Chemistry*, Academic Press, New York, 1957, p. 615.

11. C. Sandorfy, *Electronic Spectra and Quantum Chemistry*, Prentice-Hall, Englewood Cliffs, N.J., 1964, originally published by Revue D'Optique as *Les Spectres Electroniques en Chimie Theorique*. See especially Chapter 5; also references 3 and 8.

12. J. C. Slater, *Proc. Natl. Acad. Sci. (U.S.A.)*, **13**, 7 (1927).

13. L. J. Oosterhoff, in *Modern Quantum Chemistry*, Vol. 3, O. Sinanoglu (Ed.), Academic Press, New York, London, 1965.

14. E. U. Condon, *Rev. Mod. Phys.*, **9**, 432 (1937).

15. Reference 10, p. 169.

16. M. Wolfsberg, *J. Chem. Phys.*, **23**, 793 (1955).

17. L. Rosenfeld, *Z. Phys.*, **52**, 161 (1928).

18. J. P. Mathieu, *Les Theories Moleculaires*, Centre National de la Recherche Scientifique, Paris, 1946.

19. D. W. Davies, *The Theory of the Electric and Magnetic Properties of Molecules*, John Wiley & Sons, New York, 1967.

20. D. J. Caldwell and H. Eyring, *The Theory of Optical Activity*, Wiley-Interscience, New York, 1971.

21. L. Landau and E. Lifshitz, *The Classical Theory of Fields*, M. Hammermesh (Transl.), Addison-Wesley Press, Cambridge, Mass., 1951, Chapter 6.

22. Strong solvent interactions which result in induced chiroptical effects will be discussed in Chapter 10.

23. In the discussion of Eq. 3.610, in the next section.

24. O. E. Weigang, Jr., *J. Chem. Phys.*, **43**, 71, 3609 (1965); *ibid.*, **55**, 5711 (1971).

25. H. F. Hameka, *J. Chem. Phys.*, **41**, 3612 (1964).

26. R. T. Klinbgiel and H. Eyring, *J. Phys. Chem.*, **74**, 4543 (1970).

27. E. B. Wilson, Jr., J. C. Decius, and P. C. Cross, *Molecular Vibrations*, McGraw-Hill Book Co., 1955.

28. W. Moffitt and A. Moscowitz, *J. Chem. Phys.*, **30**, 648 (1959).

29. S. F. Mason, *Mol. Phys.*, **5**, 343 (1962).

30. W. G. Kuhn and H. Braun, *Z. Phys. Chem.*, **B8**, 281 (1930); also W. G. Kuhn, *Trans. Faraday Soc.*, **46**, 293 (1930).

31. T. M. Lowry, *Optical Rotatory Power*, Longmans Green and Co., London, 1935: reprinted by Dover Publications, New York, 1964.

32. D. J. Caldwell, *J. Chem. Phys.*, **51**, 984 (1969).

33. M. J. Stephan, *Proc. Cambridge Philos. Soc.*, **54**, 81 (1958).

34. N. Gō, *J. Chem. Phys.*, **43**, 1275 (1965); N. Gō, *J. Phys. Soc. J.*, **23**, 88 (1967).

35. P. W. Atkins and L. D. Barron, *Proc. R. Soc.*, **304A**, 303 (1968); P. W. Atkins and L. D. Barron, *Chem. Br.*, **7**, 244 (1971).

36. V. Lawetz and D. A. Hutchinson, *Can. J. Chem.*, **47**, 577 (1969).

37. D. A. Hutchinson, *Can. J. Chem.*, **46**, 599 (1968).

38. P. A. M. Dirac, *The Principles of Quantum Mechanics*, 4th ed., Oxford University Press, London, 1958, pp. 201–206; 239–244.

CHAPTER FOUR

MODELS OF OPTICAL ACTIVITY

4.1 STATEMENT OF THE PROBLEM

In this chapter we finally come to grips with the problem of explaining the optical activity of a nominally planar or centrosymmetric chromophore located in a molecule lacking these elements of symmetry. We will describe two different approaches. The first, to which the bulk of this chapter is devoted, retains the model of the planar or centrosymmetric chromophore buried in an asymmetric environment. We will be referring, therefore, to the activity induced by the asymmetric molecular environment of the chromophore. The vibronic borrowing discussed in the last chapter may sometimes be the pathway by which magnetic or electric dipole intensity is transferred into the particular optical transition of the symmetric chromophore, especially a transition that is only weakly allowed, and it certainly determines the band shape, but it is not involved in the electrostatic perturbations of the symmetric chromophore we will discuss here. Once again, then, by integrating or summing over the band envelope, the individual vibrational contributions are bypassed. What we do discuss here is the fact that our intellectual or computing apparatus, or both, has been insufficient to the task of calculating, with any accuracy, the sign and magnitude of the chiral activity of a large chiral polyatomic molecule in which the major chromophore is achiral in zeroth order, that is, achiral in the absence of the asymmetric environment. Typical examples would be the carbonyl chromophore in methylcyclohexanone, the ethylene chromophore in transcyclooctene, and the benzene chromophore in phenylalanine (Fig. 4.00). For these environments, the epithet "asymmetric chromophore" is used as the descriptive terminology, and the molecules containing them constitute the vast majority of chiral compounds (1). The electrostatic perturbation of the symmetric environment has been analyzed theoretically according to three major schema or mechanisms that depend on the electronic structure of the molecule, and empirically according to

5-methylcyclohexanone   transcyclooctene   phenylalanine

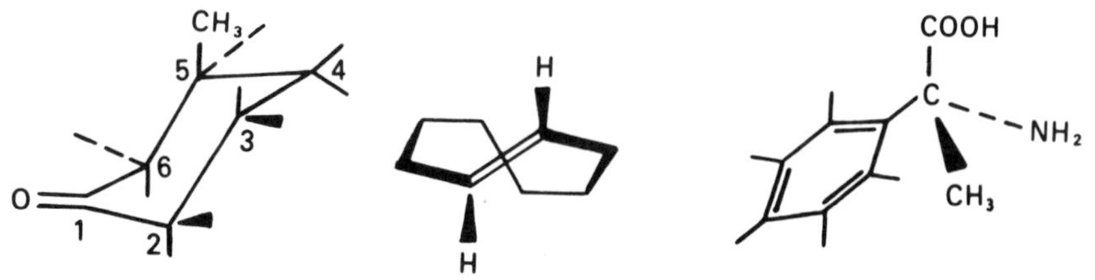

Fig. 4.00 Symmetric chromophores in asymmetric environment: 5-methylcyclohexanone—the oxygen and the carbon atoms at positions 1, 2, and 6 are in a plane of symmetry, the asymmetry is produced by the axial methyl group at position 5 in the ring. Transcyclooctene—if the π electronic systems of the olefinic carbon atoms are considered to be coplanar, the asymmetry is provided by the relationship between this plane and the positions of the hydrogen atoms; the absolute configuration is determined by the sense of twist of the transcyclooctene ring. Phenylalanine—the planar benzyl chromophore is in the asymmetric environment of substituents of the tetrahedral carbon atom.

several more. We will see that the three mechanisms are in fact just different interactions that can occur between charges having some chiral properties, and when the most general perturbation formula for optical activity is examined, it will be found that each of the mechanisms is contained in one or more of its terms. Thus each mechanism consists of some particular part of the interaction between electric and magnetic transition moments of different groups of atoms (chromophores) in the molecule, or of different transitions of the same group. In Chapter 2 we saw that the field of a light wave induces the moments. Using perturbation theory and classical physics, it is possible to express these moments as sums over the states of a molecule and also as sums over the groups of atoms or chromophores in the molecule—in the latter case, only if the assumption is made that there is no electron exchange between groups. Inserting the resulting expressions in the Rosenfeld equation leads to the set of terms that are identified as the mechanisms of optical activity. Indeed the first term will be that of the intrinsically dissymmetric chromophore, which will be treated separately in the last section of this chapter and constitutes the second of the two approaches. At that point, we will be less modest about our intellectual apparatus and attempt to describe the intrinsic dissymmetry of the chromophore in order to calculate its chiral properties directly. In many cases it may be necessary to modify the concept of a chromophore, treating the traditional atomic grouping as part of a more dissymmetric one.

Although the first models or mechanisms for optical activity of asymmetrically perturbed chromophores were developed before this approach was initiated by John Kirkwood (2), we will examine the problem utilizing

Kirkwood's method as subsequently modified by Kirkwood, Moffitt, Fitts, Tinoco, and Schellman. The earlier models, the "one-electron theory" of Condon, Altar, and Eyring (3) and the "coupled oscillator theory" of Kuhn (4) and Born (5), anticipated the later more complete analysis, each accounting for one of the mechanisms. Let us look at the problem in terms of a real molecule as it was perceived, for example, by Kirkwood. Secondary butyl alcohol, a very simple optically active molecule, contains no chromophore that absorbs in the visible or near ultraviolet, but in the conventional wisdom it exhibits "end absorption" at wavelengths below about 220 nm. Of course we know that this means there are strong absorption bands below this wavelength, and it is clear that the optical activity of the *l* or *d* isomer at longer wavelengths must be due to these transitions. The optical rotation of the pure liquid is 13.9 deg/cm at 546 nm, some 35,000 cm^{-1} from its nearest absorption band (6). It was at the time (in 1937), and still would be, difficult to write explicit and accurate wavefunctions from which to calculate electric and magnetic transition moments for this molecule in a dissymmetric conformation. The problem is further complicated by the fact that rotation is somewhat restricted about the bond between the ethyl group and the asymmetric carbon atom, C* (Fig. 4.01), but it is not forbidden. Kirkwood, in fact, was quick to point out that if a mechanism were developed in which the optical activity arose from interaction between the electrons of the ethyl group and those of the asymmetric carbon atom, it would be necessary that the rotation about the bond connecting them be restricted. Otherwise, all the asymmetric interactions would cancel. That the rotation about this bond is restricted is a well-known fact; there are deep minima in the potential energy curve for the staggered forms in which the carbon-carbon bond of the ethyl group makes a dihedral angle of about 60 degrees with the bond of the substituents to the asymmetric carbon atoms, as illustrated by the Newman projections in Fig. 4.01. While these do not accord with the conformations that Kirkwood chose for his original calculation, it is clear that the interaction should be calculable for each conformation and that the contribution of each of the overall optical activities will be weighted by the Boltzmann statistics appropriate to their relative energies.

It is not necessary to perceive of this problem entirely in terms of chromophores which are relatively inaccessible to measurements at the frequencies of their absorption bands. We have referred to the carbonyl chromophore, for example. Its longest singlet absorption band ranges from about 270 nm in formaldehyde to perhaps 400 or 500 nm in a conjugated enone or diketone. But again, we generally assume that the carbonyl group has a plane of symmetry that makes it intrinsically optically inactive. Yet we know from the observation of a Cotton effect at the same wavelength

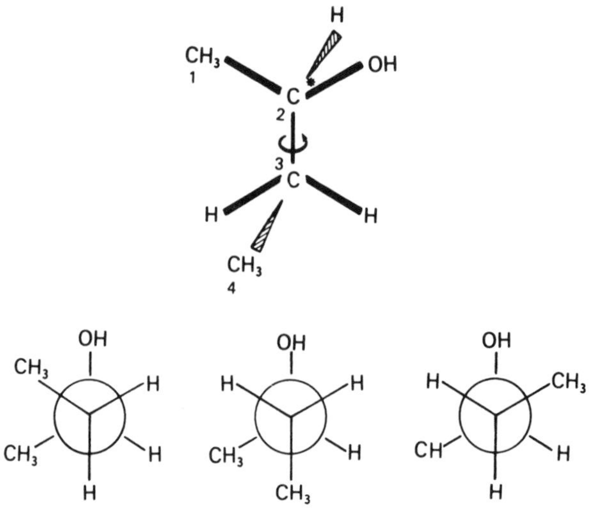

Fig. 4.01 Newman projections of secondary butanol perpendicular to the bond between carbon atoms 2 and 3.

as the isotropic absorption band that in any ketone having an asymmetric center, that is, lacking an overall plane or center of symmetry, the resolved isomers show optical activity specifically originating in the so-called $n \rightarrow \pi^*$ long-wavelength transition of this group. These, then, are typical of the types of problems that have led to the development of the models described in this chapter. Kirkwood's original solution was to consider the interaction as one that arises between the asymmetric polarizabilities of the separate groups, and we will soon see in Section 4.4, for example, how this concept is still contained in the more modern developments.

Let us start by reemphasizing what we are seeking, namely, a theory or basis for understanding the optical activity of a molecule whose electronic structure we cannot describe with exactitude, a molecule that consists of one or more groups each of which, if isolated, would have one or more elements of symmetry that are incompatible with intrinsic chirality. From earlier considerations we concluded that the optical transitions associated with each of these groups separately either have null magnetic or electric transition moments or that these moments are perpendicular to each other. We will see that perturbation theory permits us to expand this picture to induce collinear components of the moments in one of several ways. In one of these, the group whose chirally active optical transition is under consideration will have a strong magnetic transition that interacts with an electric moment on one or more other groups. In another, the chiral interaction will be between strong electric dipole transition moments on both groups. These two are coupled oscillator mechanisms with that particular name applying only to the latter, the former being termed the electric-magnetic coupling mechanism (7). Finally, coupling between electric and magnetic dipole transition moments on the same group through transitions to different states is induced by the static field of the asymmetric environment. This is the "one-electron" mechanism.

4.2   GENERAL THEORY

As described in Chapter 3, perturbation for both achiral and chiral absorption theory required only that the state of the system be different after the interaction then before it. Equation 3.305 described the state of the system at any time as a linear combination of some fraction of the initial and final states. Neither the initial nor the final states, however, were specified as consisting either of unexcited or excited subgroups. There was no need for this because we were describing the optical activity of a single intrinsically chiral group, that is, an intrinsically disymmetric chromophore. We now need to go beyond that. Building on this thought, both Condon and his colleagues in 1937 (3, 8) and Kirkwood, also in 1937 (2) and again in 1957 with Moffitt and Fitts (9), developed separate mechanisms for the induction of optical activity in intrinsically symmetric chromophores. Tinoco (10), in 1960, saw that these were only parts of a more general theory. The Moffitt, Fitts, and Kirkwood papers had already covered much of the ground but were specifically designed to explain the chiral activity of repeating polymers and contained only the interaction between electric dipoles on different groups. Tinoco's theory, which is laid out in explicit detail in his 1962 paper (11), allows the states of the system to consist (1) of a zeroth-order ground state which is the product wavefunction over all the N interacting groups:

$$\psi_0 = \prod_i^N \psi_{i0},\tag{4.200}$$

where $\prod_i^N \psi_{i0}$ stands for $\psi_{10}\psi_{20}\cdots\psi_{N0}$; and (2) of an excited state in which only one of the i groups are excited:

$$\psi_{ja} = \frac{\psi_{ja}}{\psi_{j0}} \prod_i^N \psi_{i0}.\tag{4.201}$$

In order to facilitate comparison with the original literature, we have adopted the convention that the unperturbed ground state is designated by the symbol 0. This is different from the usage in the rest of this monograph (except for Chapter 9) where we beg the question of whether the starting state of the optical transition is the ground state by using the letter a for the starting state, although in all cases, except in the discussion of vibronic borrowing in Section 3.6 and of circularly polarized emission in Section 10.4, it is understood that the transitions we are discussing are generally from the ground state of the molecule. Furthermore, the excited state can be N-fold degenerate if the molecule consists of N identical groups (an N-mer of a repeating polymer). When we use perturbation theory, in which the perturbation arises from the interaction between singly and doubly excited states on different groups, in order to develop first-order wavefunctions of the states of the molecule, the electric and magnetic transition moments are developed as sets of many terms (see reference 11, Eqs. IIIB-18a-f and IIIB-20a-u). Each of these terms are dependent on the physical separation between the groups and on the energy difference between the transitions. In general, as one might expect, the greater the separation between the groups and the larger the energy difference between the transitions, the smaller is the interaction and thus the smaller the individual contribution to the rotational strength. The final expression for the rotational strength (12) associated with the transition to the Ath excited state of a molecule containing N groups is

$$R_{OA} = \sum_i \mathrm{Im}\,\boldsymbol{\mu}_{i0a}\cdot\mathbf{m}_{i0a} \tag{4.202-1}$$

$$+2\sum_{j\neq i}\sum_{b\neq a}\frac{\mathrm{Im}\,V_{i0a,j0b}(\boldsymbol{\mu}_{i0a}\cdot\mathbf{m}_{jb0}\nu_a + \boldsymbol{\mu}_{j0b}\cdot\mathbf{m}_{ia0}\nu_b)}{h(\nu_b^2 - \nu_a^2)} \tag{4.202-2}$$

$$-\sum_{j\neq i}\sum_{b\neq a}\frac{\mathrm{Im}\,V_{iab,j00}(\boldsymbol{\mu}_{i0a}\cdot\mathbf{m}_{ib0} + \boldsymbol{\mu}_{i0b}\cdot\mathbf{m}_{i0a})}{h(\nu_b - \nu_a)} \tag{4.202-3}$$

$$-\sum_{j\neq i}\sum_{b\neq a}\frac{\mathrm{Im}\,V_{i0b,j00}(\boldsymbol{\mu}_{i0a}\cdot\mathbf{m}_{iab} + \boldsymbol{\mu}_{iab}\cdot\mathbf{m}_{ia0})}{h\nu_0} \tag{4.202-4}$$

$$-\sum_{j\neq i}\frac{\mathrm{Im}\,V_{i0a,j00}(\boldsymbol{\mu}_{iaa} - \boldsymbol{\mu}_{i00})\cdot\mathbf{m}_{ia0}}{h\nu_a} \tag{4.202-5}$$

$$-\frac{2\pi}{c}\sum_{j\neq i}\sum_{b\neq a}\frac{V_{i0a,j0b}\nu_a\nu_b(R_j - R_i)\cdot(\boldsymbol{\mu}_{j0b}\times\boldsymbol{\mu}_{i0a})}{h(\nu_b^2 - \nu_a^2)} \tag{4.202-6}$$

where the interaction energies given by

$$V_{i0a,j0b} = \int \psi_{i0}\psi_{ia}\psi_{j0}\psi_{jb}V_{ij}\,d\psi_i\,d\psi_j \tag{4.203}$$

constitute the interaction potential between groups i and j, and $i = 0$ to N.

The first term in Eq. 4.202 is, of course, the now familiar expression of the Rosenfeld equation, except that it is now summed over the i *identical* chromophoric groups in the molecule. It represents the contribution of the $0 \rightarrow a$ transition of these groups and is finite if μ_{i0a} and $\mathbf{m}_{i0a}$ have finite collinear components, that is, if the transition $0 \rightarrow a$ is electrically and magnetically allowed along the same coordinate in space. This is the definition of an intrinsically dissymmetric chromophore (Chapter 7). The major contributions to the rotational strength of asymmetric chromophores comes from the remainder of the terms which involve interactions $V_{i-,j-}$ ($i \neq j$) between *different* groups although the transition moments may be from a single group. We shall ignore the fifth term, which is generally very small because of the large denominator (the frequency ν_a rather than a small frequency difference in the second, third, and the sixth terms) and a small numerator; the numerator contains the product of the difference between the *permanent* dipole moments in the ground and excited states ($\mu_{iaa} - \mu_{i00}$) with the magnetic transition moment $\mathbf{m}_{ia0}$, both of which tend to be small quantities. The fourth term has a similarly large denominator but may not be quite so small if it involves large electric and magnetic transition moments. Consider now the terms that give rise to the various mechanisms.

The second term couples an electric transition moment on one group, μ_{i0a}, with the magnetic dipole transition moment on another group, $\mathbf{m}_{j0a}$, etc. This is the μ-$\mathbf{m}$ mechanism of Schellman.

The third and fourth terms couple electric and magnetic dipole transition moments on the same groups, μ_{i0a} and $\mathbf{m}_{i0a}$, for example, perturbed by the static field of all the other groups. This is the Condon, Altar, and Eyring one-electron theory.

The sixth term couples electric dipole transition moments on different groups, $\mu_{j0b} \times \mu_{i0a}$, in such a way as to give rise to a magnetic moment in the direction of one of the originating transition moments. This is the Kuhn-Kirkwood coupled oscillator theory, which in the Kirkwood formalism replaces the dipole transition moments by the anisotropic polarizabilities to which they give rise (recall Eqs. 2.019 and 2.023).

4.3   THE ONE-ELECTRON THEORY

We have seen that the one-electron mechanism of optical activity results from the interaction between electric and magnetic transition moments in a chromophoric group for which in zeroth order the dot product of these moments vanish. This is one of two distinguishing features of the one-electron theory, the one it shares with the coupled oscillator and the μ-$\mathbf{m}$

mechanism in distinguishing it from the chiral activity of an intrinsically dissymmetric chromophore. Since only the transition moments in a single group are involved, the one-electron mechanism may or may not be a major contributor to the optical activity of a dimer or polymer. A dimer for this purpose is defined as two identical chromophores having some fixed structural arrangement with respect to each other, sufficiently separated in space so that there is substantially no electron exchange between them. Furthermore, if we look at the one-electron terms in parts 3 and 4 Eq. 4.202, we note that each term is itself made up to two terms that involve, in every case, the dot product of electric and magnetic components from different transitions, $0 \to a$, $0 \to b$, or $a \to b$. In ordinary absorption, this type of mixing results in a borrowing of intensity between transitions; similarly, in optical activity. Furthermore all the transitions either start or end on a and b, which requires that they originate or terminate on the same orbital in order to mix. In particular the transitions and the moments that mix them are

Transitions	Moments	Terms
$0 \to a$, $0 \to b$	$\mu_{i0a} \cdot \mathbf{m}_{i0b}$	4.202-3a
	$\mu_{i0b} \cdot \mathbf{m}_{i0a}$	4.202-3b
$0 \to a$, $a \to b$	$\mu_{iab} \cdot \mathbf{m}_{i0a}$	4.202-4a
	$\mu_{iab} \cdot \mathbf{m}_{i0a}$	4.202-4b

Terms 3a, b and 4a, b provide chiral interactions between the transitions on the ith groups, any or all of which may contribute. Because of the relatively large denominator (ν_b compared to $\nu_b - \nu_a$) of the fourth term, subterms 4a and 4b are less likely to make significant contributions than 3a and 3b. Note also the fact that both μ_{i0a} and $\mathbf{m}_{i0a}$ may be finite and contribute to the rotational strength through this mechanism, but this is not contradictory to their failure to contribute to the disymmetric chromophore term since here (in terms 3a, b and 4a, b) they do not appear as the self dot product $\mu_{i0a} \cdot \mathbf{m}_{ia0}$, which is of necessity zero for a symmetric chromophore (the two moments, if finite, being at right angles to each other). Finally, while all the transitions that are involved in this mechanism are of the same group (note that only the ith group transition moments appear in these terms), the perturbation comes from the electrostatic interaction with other groups, $V_{iab,j00}$ (groups i and j) and $V_{i0b,j00}$ (groups i and j), and the summation is over all the j perturbing groups for each chromophoric transition on the group i. The absorption spectra of most compounds, organic compounds in particular, are determined by isolated "chromophoric" groups. To a good approximation, the transition responsible for the absorption is localized on the atoms within the group. In molecular orbital terminology the transition involves the transfer of a single electron from one of the occupied orbitals of the ground state to an unoccupied orbital. It is from this that the one-electron theory takes its name; but, of course, the same is true for dissymmetric chromophores; so that the name is only historically useful, not uniquely descriptive. Because of the requirement that the mixing transitions must originate or terminate on the same orbital, certain, but not very strong, restrictions are put on the types of transitions that can give rise to chiral activity by this mechanism. Consider, for example, the carbonyl chromophores to which the one electron mechanism has been applied. The types of transitions which can occur are

1. Rydberg transitions from the nonbonding orbitals, $n \rightarrow R^*$;
2. valence transitions of the π orbitals, $\pi \rightarrow \pi^*$, $\pi \rightarrow \sigma^*$;
3. valence transitions of the σ orbitals, $\sigma \rightarrow \pi^*$, $\sigma \rightarrow \sigma^*$;
4. valence transitions of the nonbonding orbitals, $n \rightarrow \pi^*$, $n \rightarrow \sigma^*$;
5. Rydberg transitions of the σ and π orbitals, $\sigma \rightarrow R^*$, $\pi \rightarrow R^*$

If the lowest transition is an $n \rightarrow \pi^*$, as is frequently the case, then $n \rightarrow \sigma^*$, $n \rightarrow R^*$, and $\pi \rightarrow \pi^*$ transitions can mix if they can develop moments in the direction of the $n \rightarrow \pi^*$ moments, but not $\sigma \rightarrow \sigma^*$, nor $\pi \rightarrow \sigma^*$, nor $\pi \rightarrow R^*$.

This then is the basis of the "one-electron" mechanism, the perturbation of the transitions of a single chromophore, orbitally plane- or centrosymmetric, by the asymmetric environment. In the treatment by Condon and his colleagues (3, 8), the interacting transitions were not specified in quite so detailed a fashion (12), and the interaction potential was specified in a slightly more arbitrary way than by Eq. 4.202. Nevertheless, it is instructive to look at the original treatment in order to appreciate more intuitively the physics of the situation. Condon pointed out that it is necessary to show that an asymmetric field can give rise to allowed collinear components of the electric and magnetic transition moments. (Note that we are careful to use words rather than symbols for these transition moments because they are no longer the unperturbed moment μ_{i0a} and $\mathbf{m}_{ia0}$ that appear in the terms of Eq. 4.202. Rather they are the moments obtained by the application of the perturbation energies $V_{iab,j00}$ or $V_{ia0,j00}$ to these moments. They can, of course, now be designated as μ'_{i0a} and $\mathbf{m}'_{ia0}$, and the rotational strength associated with the $0 \rightarrow a$ transition is then $R_{0a} = \mathrm{Im} \mu'_{i0a} \cdot \mathbf{m}'_{ia0}$.) The asymmetric field in the Condon treatment was provided by writing the potential energy of the chromophoric electron in the field of the rest of the molecule as an asymmetric function of the coordinates:

$$V = \tfrac{1}{2} k_1 x^2 + \tfrac{1}{2} k_2 y^2 + \tfrac{1}{2} k_3 z^2 + A x y z. \tag{4.300}$$

The term in A produces the dissymmetry. Consider the kind of potential field represented by this expression. Suppose, for example, the $k_1 > k_2 > k_3$, and at first let $A = 0$. Then $\frac{1}{2}k_1 x^2$ is a line along the x coordinate with amplitude proportional to $k_1/2$ and symmetric with respect to the coordinate origin; that is, it has the same value for $x = x'$ and for $x = -x'$. Similarly, for $\frac{1}{2}k_2 y^2$ and $\frac{1}{2}k_3 z^2$. Consequently, with $A = 0$ for any value of x, y, and z (other than zero), the potential V is the surface of a flattened $(k_3 > k_2)$ ellipsoid whose axes are fixed by the values of k_1, k_2, and k_3 and whose principal axes lie along the Cartesian coordinates. Such an ellipsoid would have a center and three planes of symmetry, so that an electron in a potential field of this type would be found with equal probability at a positive or negative value of each of the coordinate axes. In Chapters 6 and 8, we will see how formal symmetry rules lead to the fact that this is, of necessity, an achiral situation; the rotational strength associated with an electron in this field must be identically zero. Recognizing this, the potential $Axyz$ is introduced. Since this is a linear rather than a quadratic function of the three coordinates, it is at once obvious that the value of V will no longer be independent of the sign of the x, y, and z coordinates. With $A = 0$, the sections of the ellipsoid in the plane of some fixed value of the coordinates, say $z = 0$ or $z = z'$, will be ellipses whose principal axes are in the direction of the other two Cartesian coordinates, in this case, x and y. However, with $A > 0$, the equation of the intersection of an equipotential (where the potential has only one value for one of the coordinates, say $z = z'$) is

$$\frac{k_1 x^2}{2} + \frac{k_2 y^2}{2} + Az'xy = \left[V - \frac{k_3(z')^2}{2} \right]. \qquad (4.301)$$

In the words of the original paper (3), "This is the equation of an ellipse whose axes have been twisted out of parallel with the z and y axes in a counter-clockwise sense as viewed from the positive z axis." In other words, the equipotential surface is a kind of twisted ellipsoid something like that one might get if an ordinary ellipsoid of three unequal axes were subject to a torsional stress twisting it around the z axis in the sense of a right-handed screw. The angle of twist in the xy plane is given by

$$\tan 2\theta_{xy} = \frac{Az'}{(k_1 - k_2)/2}. \qquad (4.302)$$

This is an important point to note, because if we permute the axes and k's for twists around the other axes, we get $(z' \rightarrow x' \rightarrow y')$, $(k_1 \rightarrow k_2 \rightarrow k_3)$,

$$\tan 2\theta_{yz} = \frac{Az'}{(k_2 - k_3)/2}, \tag{4.303}$$

$$\tan 2\theta_{zx} = \frac{Ay'}{(k_3 - k_1)/2}. \tag{4.304}$$

The important point here is that the *sign of* $\tan 2\theta_{zx}$ *is different from* the sign of $\tan 2\theta_{yz}$ and $\tan 2\theta_{xy}$ because for $k_1 > k_2 > k_3$, $k_3 - k_1$ is negative, while $k_1 - k_2$ and $k_2 - k_3$ are positive. Similarly for any other combination of relative magnitudes of k_1, k_2, and k_3 (except, of course, $k_1 = k_2 = k_3$), there will always be one of these angles which differs in sign from the other two. This produces equations for the optical activity (terms 3 and 4 of Eq. 4.202) that lead to a dependence of the sign of the rotational strength on the geometric location and, therefore, on the absolute stereochemistry of the perturbing groups in the molecule.

In the Condon, Altar, and Eyring treatment, the asymmetrically perturbed electron is assumed to be a classical harmonic oscillator whose quantized energy levels are given by

$$\mathcal{E}(n_x, n_y, n_z) = \left(n_x + \tfrac{1}{2}\right)h\nu_x + \left(n_y + \tfrac{1}{2}\right)h\nu_y + \left(n_z + \tfrac{1}{2}\right)h\nu_z, \tag{4.305}$$

where $\nu_{x,y,z}$ are the frequencies of the oscillation along the Cartesian axes and n_x, n_y, and n_z are the corresponding quantum numbers designating the state of the system ($\mathcal{E} = \tfrac{1}{2}h\nu$ is the zero-point energy of a classical harmonic oscillator). The wavefunction of the unperturbed electron of the chromophore may be written as the product wavefunction for the electron along the three Cartesian axes.

$$\psi(n_x, n_y, n_z) = \psi_{n_x}\left(\frac{x}{a_x}\right)\psi_{n_y}\left(\frac{y}{a_y}\right)\psi_{n_z}\left(\frac{z}{a_z}\right), \tag{4.306}$$

where the $a_i = x, y, z$ are related to the force constant for the oscillator, that is, the k_x, k_y, k_z in the potential function. By using standard perturbation treatment (13) in which the perturbation energy is given by the asymmetric part of Eq. 4.300, that is, by $Axyz$, a new perturbed wavefunction $\psi'(n_x, n_y, n_z)$ is obtained such that the transition moments $\langle \psi'_a(n_x, n_y, n_z)|\mu|\psi_b(n_x, n_y, n_z)\rangle$ and $\langle \psi_b(n_x, n_y, n_z)|m|\psi'_a(n_x, n_y, n_z)\rangle$ now have finite collinear components. Therefore, the imaginary part of their dot product is finite, and the asymmetrically perturbed system takes on the attributes of a dissymetric system with a finite rotational strength. The details of this treatment are given in the original paper and are repeated in the excellent review by Kauzmann, Walter, and Eyring (14). In Chapter 7, we will see an application of this treatment.

4.4   THE COUPLED OSCILLATOR

The formalism that describes the contribution of two or more groups whose electronic transition dipole moments may couple to give rise to optical activity is given by the second and the last terms of Eq. 4.202. Of these, the second, which arises only when one of the groups has a strong *magnetic* dipole transition, will be described in Section 4.5. We limit ourselves, therefore, in this description of the coupled oscillator to groups having strong absorption bands originating in allowed *electric* dipole transitions.

Attempts to deal with the origin of optical activity in coupled oscillators started with Oseen in 1915 (15) even before quantum mechanics permitted a precise description of transition moments. This, as well as additional attempts (16, 17) shortly after the introduction of the quantum description of the interaction of radiation with matter, were largely unsuccessful but did provide a focus for further development. It was not until Moffitt, Fitts, and Kirkwood (9, 18, 19) developed the application of the theory to molecules containing many groups, polymers for example, that this mechanism began to show some accommodation between theory and experiment.

The theory of coupled oscillators can cover several different situations within the class of groups having strong electric dipole transition moments. There can be just two groups, as in a dimer, or as many as in a polymer, and the groups can be identical or different. If identical, the treatment involves the interaction between transitions that are degenerate in zeroth order, but whose energy splits because of the coupling. When the splitting is large enough, the resulting chiral activity is observed in CD as two adjacent bands of opposite sign or in ORD as a curve with three extrema. We will see examples of this in the strong optical activity of repeating polymers (Chapter 8).

We will consider here the simple case of two chromophoric groups and represent their electric dipole transition moments by arrows *directed along a symmetry axis in each group*. These transitions may be in or out of phase with respect to each other. They are represented pictorially in Fig. 4.02, where the arrows are placed in a solid figure to indicate that they do not lie in the same plane, in which case we would retrieve the optically *inactive*

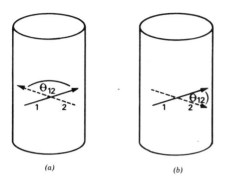

Fig. 4.02 The phase relationship between coupled transition moments of two groups having symmetry axes in the direction of the transition moments (shown in a cylinder to emphasize their noncoplanarity). Unit vectors along these axes (see text) are $\mathbf{b}_1$ and $\mathbf{b}_2$. (a) "In-phase" transitions; (b) "out-of-phase" transitions.

situation. If we arbitrarily designate the transitions in Fig. 4.02a as in phase, then those in Fig. 4.02b are out of phase; the choice is arbitrary (7) since both give rise to rotational strength of equal magnitude but opposite sign and the proper assignment can be made on the basis that the band of lowest energy is associated with a negative coupling energy. If the chromophores are not identical, then the band of lowest energy is usually spectroscopically identifiable, while if they are identical, the high- and low-energy bands are those that arise from exciton splitting and can be described in that framework. Consider how the sixth term of Eq. 4.202 applied to just two groups leads us to these conclusions for the rotation strength contributed by dipole coupling R_D:

$$R_D = -\frac{2\pi}{c} \sum_{b \neq a} \frac{\nu_a \nu_b V_{10a,20b} (\mathbf{R}_2 - \mathbf{R}_1) \cdot (\boldsymbol{\mu}_{20b} \times \boldsymbol{\mu}_{10a})}{h(\nu_b^2 - \nu_a^2)}. \tag{4.400}$$

First, we note that if the two chromophores are coplanar, then the vector cross product $(\boldsymbol{\mu}_{20b} \times \boldsymbol{\mu}_{10a})$ is a vector perpendicular to the vector $\mathbf{R}_{21} = (\mathbf{R}_2 - \mathbf{R}_1)$ connecting the midpoints of the transition moments $\boldsymbol{\mu}_{20b}$ and $\boldsymbol{\mu}_{10a}$. Consequently, the scalar dot product of the optical factor $(\mathbf{R}_2 - \mathbf{R}_1) \cdot (\boldsymbol{\mu}_{20b} \times \boldsymbol{\mu}_{10a})$ is zero. Therefore, a coplanar arrangement of chromophores does not give rise to chiral activity by the coupled oscillator mechanism even if the molecule as a whole does not have a plane of symmetry. A common form of the interaction potential $V_{i0a,j0b}$ between two groups is that of the point dipole approximation:

$$V_{10a,20b} = \frac{\boldsymbol{\mu}_{10a} \cdot \boldsymbol{\mu}_{20b}}{R_{21}^3} - \frac{3(\boldsymbol{\mu}_{10a} \cdot \mathbf{R}_{21})(\boldsymbol{\mu}_{20b} \cdot \mathbf{R}_{21})}{R_{21}^5}. \tag{4.401}$$

From Eq. 4.401 we can infer additional constraints on the activity produced by this mechanism. We have just observed that if the transition moments are coplanar in the groups 1 and 2, no chiral activity is pro-duced; now we see that if they are perpendicular to each other, the first term, $V_{10a, 20b}$, drops out, and if they are both perpendicular and related by a linear rather than a screw translation, the second term drops out as well, because in that case μ_{10a} and μ_{20b} are both perpendicular to R_{21}. This latter situation, which would imply planes of symmetry with respect to the chromophores, may, of course, be rare, but the closer to this configuration the transition moments are, the smaller will be this contribution to the optical activity. The inverse R^3 power of both terms of Eq. 4.401 leads to an inverse dependence of the rotational strengths on R^2; the contribution diminishes rapidly with increasing distance between the chromophoric groups. This is, of course, the origin of the long-known experimental observation that "vicinal" effects decrease with increasing distances be-tween the interacting groups. It is, therefore, apparent that the appearance of the geometric factors (the relative orientation of the interacting groups and their distance apart) implies that *even the strongest absorption bands may contribute little, if anything, to the optical activity by the coupled oscillator mechanism if the chromophoric groups are unfavorably disposed relative to each other*. Weigang has recently developed another generaliza-tion of the geometric requirements for the induction of optical activity by a coupled oscillator mechanism (20). This treatment is notable for the fact that it shows that transitions that are both strongly electric and magnetic dipole allowed, induce geometry-dependent relationships between different groups in a similar but not identical way to the geometry requirements of the static perturbation treatment of the one-electron mechanism for electric-dipole-forbidden, magnetic-dipole-allowed transitions.

The interaction potential and the optical factors of Eqs. 4.400 and 4.401 may alternatively be written in terms of the anisotropic polarizabilities, α_i and β_i, of the group. Values of these polarizability components directed along unit vectors b_1 and b_2 coincident with the chromophore symmetry axes (axial symmetry of the chromophoric groups is assumed; so there are two, not three axes) are often available from Kerr effect and light-scattering measurements or from theoretical calculations. In this form, the rotational strength from the dipole-coupling mechanism is

$$R_D = -\frac{2\pi}{hc} \cdot \frac{\nu_1 \nu_2}{\left(\nu_2^2 - \nu_1^2\right)} \frac{1}{b} \alpha_1 \alpha_2 \beta_1 \beta_2 \left[\frac{b_1 \cdot b_2}{R_{21}^3} - \frac{3(b_1 \cdot R_{21})(b_2 \cdot R_{21})}{R_{21}^5} \right] R_{21} \cdot (b_1 \times b_2).$$

$$(4.402)$$

Aside from the availability of values of α_i and β_i, there is a more intrinsic reason for using this form. Look back at Eqs. 4.400 and 4.401: In Eq. 4.400

the rotational strength arising from dipole coupling between transitions is summed over all possible transitions (i.e., all transitions from the ground state for which the potentials, $V_{10a,20b}$, and the transition moments, μ_{10b} and μ_{20b}, are nonzero; note summed over *all* transitions). Many of these transitions must be of very high energy and appear in the far vacuum ultraviolet where it is difficult if not impossible to detect them, let alone to measure the intensities and polarization directions with any accuracy. The transition moments μ_{10b} and μ_{20b}, required, respectively, for Eq. 4.400 and for the potential energy function that appears in it (Eq. 4.401), may therefore be unobtainable for these high-energy states. As a consequence, if this dipole-coupling contribution is to be calculated, some alternative method must be provided. If we recall from Chapter 2 that each transition makes a contribution to the molecular polarizability (see Eq. 2.023), then it should be clear that some mathematical procedures could be devised to allow the contribution of these higher energy transitions to be accounted for by the polarizability parameters α_i and β_i, because of their contributions to the index of refraction and its anisotropy in the visible or near ultraviolet where it can be experimentally measured. Equations 4.402 and 4.403 do just that. There is an extensive literature on these polarizabilities (21–23). In Eq. 4.402, by replacing the scalar products of the vectors in terms of the angles between them, a tractable expression in terms of experimentally available parameters is obtained:

$$R_D = -\frac{2\pi}{hc} \cdot \frac{\nu_1 \nu_2}{(\nu_2^2 - \nu_1^2)} \cdot \frac{1}{6R_{12}^2} \alpha_1 \alpha_2 \beta_1 \beta_2 (\cos\theta_{12} - 3\cos\phi_1 \cos\phi_2)\cos X_{12} \sin\theta_{12}$$

$$(4.403)$$

where: θ_{12} is the angle between the symmetry axes of group 1 and 2 whose sign is determined by the right-hand rule; X_{12} is the angle between the vector formed from $\mathbf{b}_1 \times \mathbf{b}_2$ and the vector $\mathbf{R}_{12}$; ϕ_1 is the angle between $\mathbf{b}_1$ and $\mathbf{R}_{21}$; ϕ_2 is the angle between $\mathbf{b}_2$ and $\mathbf{R}_{21}$; α_i and β_i are the mean anisotropic polarizabilities and are defined, respectively, by

$$\alpha_i = \tfrac{1}{3}\left(\alpha_{11}^{(i)} + 2\alpha_{22}^{(i)}\right) \qquad (4.404)$$

for a cylindrically symmetric group, where the polarizabilities α_{22} perpendicular to the cylinder axes are equal, and by

$$\beta_i = \frac{\alpha_{11}^{(i)} - \alpha_{22}^{(i)}}{\alpha_i}. \qquad (4.405)$$

Equation 4.403 reveals at once how the sign of the rotational strength of the chiral activity is determined by the phase relationships between the electric dipole transition moments: X_{12} will be $\gtrless 90$ degrees depending on the phase, that is, on the sign of $\mathbf{b}_1$ and $\mathbf{b}_2$ (see Figs. 4.02a,b); the $\cos X_{12}$ changes sign at 90 degrees, while none of the other parameters change sign with the phase, and, in fact, the two phase possibilities shown in Fig. 4.02 correspond to transition moments that generate, respectively, left- and right-handed helices if the moments are translated in a screw sense corresponding to the in- and out-of-phase transition moments. This relationship is related to absolute configuration in the molecule. Thus, true enantiomers of the same molecular groupings will have the same value of all the parameters in Eq. 4.403 except for the sign of $\cos X_{12}$ and will, therefore, give rise to chiral activity of opposite sign.

The expression of the contribution of coupled electric dipole transitions in separate groups in a molecule, in forms designed to elucidate alternately the dipolar coupling or the interaction between anisotropic polarizabilities, is reflected in the names "coupled oscillator" and "polarizability theories" given to the same phenomenon. There have been other related treatments that give rise to higher order and consequently smaller contributions of the same type. Among these treatments are those of de Mallemann (24) and Boys (25) which are higher order in perturbation theory. Also Fitts (26) has pointed out that the one-electron theory which arises from the third and fourth terms in Eq. 4.203 does account for the interactions of neighboring groups with a particular chromophoric group and thus is not completely separable from the polarizability theory unless the latter is limited to the coupled oscillator form of the sixth term of Eq. 4.202.

4.5   THE μ-m MECHANISM

We have now discussed or dismissed all the nonintrinsically dissymmetric terms in Eq. 4.202, except for the second, which involves the interaction between the electric and magnetic dipoles on different groups. Since magnetic dipoles do not "couple" directly with electric dipoles (27), some more subtle mechanism must be operative to produce chiral activity from a term which depends on the interaction of a strong magnetic dipole transition moment in a group whose electric dipole transition moment for the same transition ($\sigma \rightarrow a$) must be null or perpendicular to the magnetic dipole. However, electric moments are produced by the orbital motion of electrons associated with magnetically allowed transitions. Thus, another magnetic transition on a second group can produce an electric dipole quadrupole moment that can interact with the magnetic transition on the first group to produce chiral activity. We have seen how the electric moments can in turn induce magnetic moments. Let us look at a rather more specific example than we have given for the other two mechanisms, in part because it makes it easier to understand, but also because this particular example ties the various mechanisms into a single molecular system, with the operative mechanism depending on the particular geometric arrangement of the chromophoric atoms in the molecule and on their electronic transitions. One of the best known examples is that of the diketopiperazine group where the electric dipole transition moment μ_{i0a} on one amide group (half the diketopiperazine) interacts with the quadrupole moment Q_{j0b}, associated with the magnetic transition $\langle 0|\mathbf{m}|b\rangle_j$, of the other group, to produce the activity. The nature of this interaction is given by

$$V = \frac{\mathbf{R}\,Q_{j0b}\cdot\mu_{i0a}}{R^{-5}} - \frac{5}{2}\frac{\mathbf{R}\,Q_{j0b}\cdot\mu_{i0a}}{R^{-7}}, \qquad (4.500)$$

where Q_{j0b} is the quadrupole tensor of the group j. The quadrupole moment $(Q_{j0b}=Q_{jn\pi^*})$ that is operative here is that associated with the magnetically allowed transition $(\mathbf{m}_{jn\pi^*}>0)$ on the carbonyl group in which the electron is promoted from the nonbonding p_y orbital to the π^* (linear combination of the p_x orbitals of carbon and oxygen). This corresponds to the classical circular motion of charge from a quadrupolar distribution. A quadrupole moment is just what the name implies—the moment of a charge distribution that can be described as though it consists of four-point charges more or less symmetrically displaced. A quadrupole tensor is the description of this charge distribution in terms of its Cartesian components, including its values off the Cartesian axes. Note that quadrupole moments do not give rise to optical activity of disordered systems except through such second-order mixing schemes as described by this μ-m mechanism (see Chapter 9). The dipole moment is the transition dipole moment, $\mu_{i0a}=\mu_{i\pi\pi^*}$, which is the moment of the lowest $\pi\pi^*$ transition to the amides (28), and lies in the NCO plane approximately in the direction along the line connecting the nitrogen and oxygen atoms as shown in Fig. 4.03. The quadrupole moment is a tensor rather than a vector property. A tensor with respect to space is characterized by having values that cannot have its components restricted to three orthogonal directions. Mathematically a tensor for the values of a property in three-dimensional space

Fig. 4.03 The transition moment direction of the lowest energy $\pi\rightarrow\pi^*$ in the amide plane as determined in myristamide by D. L. Petersen and W. Simpson [*J. Am. Chem. Soc.*, **79**, 2375 (1957)].

consists of a 3 by 3 matrix:

$$a_{xx} a_{xy} a_{xz}$$

$$a_{yx} a_{yy} a_{yz}$$

$$a_{zx} a_{zy} a_{zz}$$

each row or column of which may be considered to be a vector. Physically the fact that it has elements like a_{xy} or a_{yz} means that its interaction with a vector field component, say E_x, or in the case of a quadrupole component Q_{xy}, with a dipole moment component μ_z, produces an effect in both the x and y directions corresponding to a circular (or helical in three dimensions) motion of charge. When we remember the concepts of Chapter 2, it becomes apparent that in classical terms this can produce chiral activity. For example, consider the two amide groups of diketopiperazine juxtaposed, but not bonded, so that different geometric arrangements between the groups may occur. We will see that the same atomic groups can give rise to optical activity alternately from one or more of the mechanisms we have been discussing or from some combination. In the coplanar arrangement (Fig. 4.04, left), only the *electromagnetic coupling* mechanism is operative because the cross product of the electric dipole transition moments required by the coupled oscillator term (Eqs. 4.202–4.206) is zero ($\mu_1 \times \mu_2 = 0$). Neither does the one-electron mechanism contribute because it requires substantial asymmetry in the environment of each individual group. Since each group is coplanar in itself and the N atom is only very slightly different from the carbon atom, there is little or no asymmetry in this configuration. If the groups are juxtaposed perpendicular to each other (Fig. 4.04, right), the situation is quite different because now $\mu_1 \times \mu_2$ is

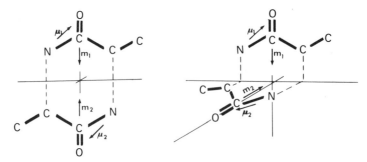

Fig. 4.04 The amide $n \rightarrow \pi^*$ transition electric and magnetic transition moments as they would be juxtaposed in a planar conformation (left) and a nonplanar conformation of diketopiperazine (right).

finite and maximum, so that the *coupled oscillator* contributes heavily. In this case, the one-electron mechanism may also contribute unless there are additional substituents suitably juxtaposed to destroy the asymmetry of the environment of each group in the presence of the other. The electric-magnetic coupling arising from Eq. 4.202-2 is now, however, absent since $\mu_1 \cdot m_2$ and $\mu_2 \cdot m_1$ are null by virtue of their perpendicularity.

One of the most thorough attempts (28) to do a total interpretation of this kind gave reasonable agreement between calculated rotational strengths and observations for the long wavelength $n\pi^*$ transition of molecules containing two peptide groups, but the results for the lowest $\pi\pi^*$ transition are completely inadequate, presumably because delocalization of the π and π^* orbitals makes a simple description inadequate.

4.6 DIRECT QUANTUM CALCULATIONS: THE DISSYMMETRIC CHROMOPHORE

The first discussion of the direct quantum calculation of the optical activity of intrinsically dissymmetric chromophores appeared almost 30 years after Rosenfeld formulated the quantum mechanical expression for the rotatory strength parameter. In this treatment, the necessity for in-troducing an asymmetric perturbation was eliminated (29). The impetus came from the optical resolution and measurement of a beautiful helical chromophore by Newman, Lednicer, and Tsai (30), namely, the hexaheli-cene molecule. With a nuclear framework that contains neither a center nor a plane of symmetry, and with a system of conjugated valence π electrons, which appeared to be well separated energetically from the remaining core electrons, it seemed reasonable to attempt to treat this molecule by calculating the electric and magnetic dipole moments for the low-energy $\pi \rightarrow \pi^*$ transition(s) directly. The nonplanar geometry was accommodated in the specification of the dipole operators. In its initial form, the method is based on the fact that the transitions may be treated in terms of one-electron molecular orbital wavefunctions and operators rather than as transitions of the state wavefunctions of the many electron mole-cules in which they occur (31). If we define the transition moments μ_{i0a} and m_{i0a} of Eq. 4.202 in terms of the overall state wavefunctions $\langle a|$ and $\langle b|$, rather than in terms of the wavefunctions of the planar or centrosym-metric chromophore, then the first term, Eq. 4.202-1, encompasses the entire origin of the rotatory strength. No induction or perturbation is required. All the other terms are null, or more correctly, are inapplicable. (For consistency with the literature and the discussion of ketones in Chapter 7, we designate the states $\langle a|$ and $\langle b|$ by the total configuration wavefunctions $\langle\psi_0|$ and $\langle\psi_N|$, where 0 indexes the ground state and $N = 1, 2, 3, \ldots$ indexes the first and successively higher energy states.) Using Rosenfeld's rotatory parameter β to define the electromagnetic transition moments between just two states (see Eq. 2.402), the rotational strength associated with the $0 \to N$ transition in the chiral molecule is given by

$$R_{0N} = \frac{he^2}{2mc\mathcal{E}_{0N}} \langle\psi_0|\nabla|\psi_N\rangle \cdot \langle\psi_N|\mathbf{r}\times\nabla|\psi_0\rangle. \tag{4.600}$$

The $\mathbf{r}$ and ∇ operators have been defined in Eqs. 3.409 and 3.410. In the one-electron orthogonal orbital approximation, expansion of the state wavefunctions by the method of Slater determinants (32) leaves only those valence molecular orbitals $\langle\psi_\xi|$ and $\langle\psi_{\eta'}|$ involved in the electronic promotion, and these are spin independent; the Greek letters index the molecular orbital, and the prime indicates which is the electronically excited orbital. With this approximation, the energy $\mathcal{E}_{0N}$ between the states is taken as either the calculated energy difference between the molecular orbitals or as the empirical energy difference obtained from the observed spectrum. In either case, to clearly designate this, we replace $\mathcal{E}_{0N}$ by $h\nu_{0N}$. The rotational strength is then given by

$$R_{0N} = \frac{he^2}{mch\nu_{0N}} \langle\psi_\xi|\nabla|\psi_{\eta'}\rangle \cdot \langle\psi_{\eta'}|\mathbf{r}\times\nabla|\psi_\xi\rangle, \tag{4.601}$$

where the factor of 2 has disappeared because in the Slater determinental expansion, a factor of the $\sqrt{2}$ enters for each of the transition moments. When Hückel molecular orbitals are used, the computation of the transition moments $\langle\psi_\xi|\nabla|\psi_{\eta'}\rangle$ and $\langle\psi_{\eta'}|\mathbf{r}\times\nabla|\psi_\xi\rangle$ is particularly simple. The ψ's consist of linear combinations of atomic orbitals (LCAOs). For example, for ethylene the bonding orbital of highest energy is

$$\psi_\xi = \frac{1}{\sqrt{2}} \left[(2p_x)C_1 + (2p_x)C_2\right],$$

and the lowest energy antibonding orbital available for electron promotion is

$$\psi_{\eta'} = \frac{1}{\sqrt{2}} \left[(2p_x)C_1 - (2p_x)C_2\right].$$

In the simplest approximation, the $2p_x$ orbitals are taken as Slater-type orbitals (STOs) (see section 2.4). The magnitudes of the STOs have been tabulated over a range of values of the geometric parameters and of the screening parameter Z, adjusted for the appropriate atomic element (33).

Aside from applying symmetry considerations to eliminate calculating any unnecessary Cartesian components, the rest is arithmetic, complex arithmetic if the number of atomic orbitals is large. The science derives not from mathematics but from decisions that must be made intuitively or on the basis of informed guesses as to: (1) the number of atoms and therefore of atomic orbitals to include in the molecular orbital of the chromophore; (2) the level of approximation at which the molecular orbitals are to be calculated (from STOs in the Hückel approximation, SCF-MOs of various parameterizations in the Intermediate Neglect of Differential Overlap method (INDO), and its variants, or ab initio Hartree-Fock treatments) (34); (3) the use of single or multiple excitations; (4) the (questionable) use of configuration interaction—see Section 3.2 and Chapter 3, references 1–4; (5) where more than one conformer is possible, what basis to calculate the effects of conformational equilibria; and (6) if structural parameters are not available from x-ray or microwave determination, what values to assume for bond distances and bond angles.

4.7 SUMMARY

Chromophores having the symmetry elements of a plane or center of reflection and which are as a consequence nominally chirally inactive may nevertheless exhibit considerable optical activity and circular dichroism when their molecular environment is less symmetric. A formalism originating in the work of John Kirkwood and essentially culminating in the work of Ignacio Tinoco, Jr., has been developed to explain (in the absence of our ability to write exact quantum descriptions of the molecule) how the asymmetric environment induces this activity. This formalism (Eq. 4.202) contains not only the Kirkwood-Kuhn mechanism, but also the special mechanism developed by Condon and his colleagues, namely, the one-electron theory and the Schellman electromagnetic coupling mechanism. By examining Equation 4.202 in some detail, we have seen that the applicable mechanism by which the chiral activity is induced depends on the juxtaposition of groups in the molecule and on the nature of the interactions between the optical transitions of these groups. These methods have been used in part because they are physically intuitive and in part to avoid the computational and conceptual difficulties of trying to solve the Schrödinger equation directly for the wavefunctions required to compute the electromagnetic transition moments. The increasing availability of sophisticated computational facilities and advances in quantum mechanical methods have made the direct quantum mechanical approaches more tractable and more attractive. Examples of applications of both approaches will be discussed in Chapter 7.

REFERENCES AND NOTES

1. But note that both methylcyclohexanone, transcyclooctene, and several other systems that will be discussed in Chapter 7 have been the subject of attempts to calculate the optical activity directly without the intervention of electrostatic perturbations.

2. J. G. Kirkwood, *J. Chem. Phys.*, **5**, 479 (1937).

3. E. U. Condon, W. Altar, and H. Eyring, *J. Chem. Phys.*, **5**, 753 (1937).

4. W. Kuhn, *Z. Phys. Chem.*, **84**, 14 (1929); W. Kuhn and H. Freudenberg, in *Hand-und Jahrbuch der Chemichen Physik*, Vol. 8, P. 3, K. L. Wolf (Ed.), Akadem. Verlag., Leipzig, Germany, 1936.

5. M. Born, *Proc. R. Soc. (London)*, **A150**, 84 (1935).

6. T. M. Lowry, *Optical Rotatory Power*, Longmans, Green and Co., London, 1935, p. 364.

7. J. A. Schellman, *Acc. Chem. Res.*, **1**, 144 (1968); note that a less complete but earlier suggestion of the chiral effect of electromagnetic coupling on different groups was made by L. L. Jones and H. Eyring, *Tetrahedron*, **13**, 235 (1961).

8. E. U. Condon, *Rev. Mod. Phys.*, **9**, 432 (1937).

9. W. Moffitt, R. D. Fitts, and J. G. Kirkwood, *Proc. Natl. Acad. Sci. (U.S.A.)*, **43**, 723, 1046 (1957).

10. I. Tinoco, Jr., *J. Chem. Phys.*, **33**, 1332 (1960).

11. I. Tinoco, Jr., *Adv. Chem. Phys.*, **4**, 113 (1962).

12. Note that even this rather formidable formula leaves out a number of small terms. See D. V. Sears and S. Beychock, in *Physical Principles and Techniques of Protein Chemistry*, P. C., S. Leach (Ed.), Academic Press, New York and London, 1973.

13. H. Eyring, J. Walter, and G. E. Kimball, *Quantum Chemistry*, John Wiley & Sons, Chapman & Hall, New York and London, 1944, Chapter 8.

14. W. J. Kauzmann, J. E. Walter, and H. Eyring, *Chem. Rev.*, **26**, 339 (1940).

15. C. W. Oseen, *Ann. Phys.*, **48**, (1915).

16. M. Born, *Optik*, Springer, Berlin, 1933.

17. W. Kuhn, *Hand-und Jahrbuch der Chemichen Physik*, Vol. 8, Pt. 3, K. L. Wolf (Ed.), Akadem. Verlag., Leipzig, Germany 1932, p. 47.

18. W. Moffitt, *J. Chem. Phys.*, **25**, 467 (1966); *Proc. Natl. Acad. Sci. (U.S.A.)*, **42**, 736 (1965).

19. D. D. Fitts and J. G. Kirkwood, *Proc. Natl. Acad. Sci. (U.S.A.)*, **42**, 33 (1956); *ibid.*, **43**, 1046 (1957).

20. W. E. Weigang, Jr., in press.

21. K. G. Denbeigh, *Trans. Faraday Soc.*, **36**, 936 (1940).

22. H. A. Stuart, *Die Struktur Des Frein Moleküls*, Springer-Verlag, Berlin, 1952.

23. C. G. LeFevre and R. J. W. LeFevre, *Rev. Pure Appl. Chem.*, **5**, 261 (1955).

24. P. de Mallemann, *Ann. Phys.*, **2**, 5 (1924); P. de Malleman, C.-R., **181**, 298 (1925); *ibid.*, **184**, 1241, 1374 (1927); *ibid.*, **185**, 350 (1927); *ibid.*, **186**, 1046 (1928); *ibid.*, **188**, 705 (1930); P. de Mallemann, *Trans. Faraday Soc.*, **26**, 281 (1930).

25. S. F. Boys, *Proc. R. Soc. (London)*, **A144**, 655, 675 (1934).

26. D. D. Fitts, in *Physical Methods of Organic Chemistry*, Pt. 3, A. Weinberger (Ed.), Chapter 23.

27. When we talk about coupling in this context, we do not refer to the formalism that says that the rotational strength is the scalar (dot) product of nonperpendicular electric and

magnetic transition moment components but rather to an interaction that proceeds via the induction of such components.

28. P. M. Bailey, E. R. Nielsen, and J. A. Schellman, *J. Phys. Chem.*, **73**, 228 (1969).

29. W. Moffitt and A. Moscowitz, *J. Chem. Phys.*, **30**, 648 (1959).

30. The optical activity was measured by L. Tsai who communicated the result to A. Moscowitz. The resolution was reported by M. S. Newman and D. L. Lednicer, *J. Am. Chem. Soc.*, **78**, 4765 (1956).

31. A. Moscowitz, Ph.D. Thesis, Harvard University, 1957.

32. J. C. Slater, *Phys. Rev.*, **34**, 1293 (1929).

33. J. Miller, J. M. Gerhauser, and F. A. Matsen, *Quantum Chemistry Integrals and Tables*, University of Texas Press, 1959.

34. There are many sources for descriptions of these quantum mechanical methods. See, for example, J. N. Murrell and A. J. Harget, *Semi-Empirical Self-Consistent-Field Molecular Orbital Theory*, Wiley-Interscience, 1972.

CHAPTER FIVE

GROUP THEORY AND MOLECULAR SYMMETRY

5.1 INTRODUCTION

The rules that govern the induced absorption and emission of electromagnetic radiation by matter are called "selection rules." Of the various selection rules that operate for polyatomic molecules, three types are especially significant for determining absorption and optical activity (1). These are the rules controlling the existence of finite values for the electric and magnetic transition moments and depend on the symmetry of the ground and excited states, the spin parity of these states, and the orthogonality of the vibrational overlap functions. We have already seen that the spin selection rules are such that both absorption and optical activity between states of different spin is largely forbidden. The vibrational contributions were discussed in detail in Chapter 3. In this chapter we will be concerned only with the theory and properties of the symmetry rules. These rules will be taken to apply to transitions between molecular states. In the case of *allowed* electronic transitions, the symmetry that controls is the orbital symmetry of these states, while in the case of *forbidden* electronic transitions, the Herzberg-Teller considerations discussed in Section 3.6 require that both the electronic (or orbital) symmetry and the vibrational symmetry be important. In the latter case, the excited states involved are, of course, vibronic, and we may think of the vibronic symmetry of these states.

Group theory is the tool of choice in determining symmetry rules. The application of group theory to molecular problems originates with E. Wigner (2). Group theory is a very powerful tool for this purpose, and fortunately so, because the Schrödinger equation can be solved exactly in only a few simple cases. The approximations required for solution in the not so simple cases are frequently either too inexact or too complex to give rapid answers to the qualitative results we desire concerning the existence and sign of the optical activity to be expected from particular electronic transitions in molecules of fixed configuration and conformation.

Our goal is to examine the symmetry selection rules governing optical activity. We will do this by setting down in a quasi-logical fashion, but without proof, the group-theoretical theorems, pausing first to examine the way in which group theory is related to the problem. We will see that the classifications of the states of molecules by the symmetry of the atomic or molecular orbitals involved in electronic transitions permit a rapid determination of the allowedness of these transitions. These classifications are systematized in the group character tables found in many books on symmetry and spectroscopy. For convenience we tabulate a few of these here, and the remainder, as well, in Appendix I, but in a rather different order than is usual in order to delineate clearly those groups that give rise to intrinsic optical activity from those that do not. These rules apply basically to dissymmetric chromophores. We have already seen how one-electron theory deals with the problem of asymmetric chromophores. The application of symmetry to this problem is also examined here. As formulated by Schellman, this question is concerned with the symmetry of the perturbing potential which induces optical activity in a chromophore that would be optically inactive in zeroth order because it possesses a plane or center of symmetry. It will be found that just as optical activity itself has the properties of a pseudoscalar, so must the perturbing potential be a pseudoscalar in the symmetry group of the unperturbed chromophore.

5.2 SYMMETRY AND QUANTUM MECHANICS

To the mathematician and the quantum theorist, the beauty of symmetry does not lie in its aesthetic appeal to the eye, but rather to the mind (and sometimes perhaps to the heart). The chemist is, however, in a position intermediate between those who follow mathematics and quantum physics and the biologist and the nonphysical scientist to whom symmetry is a joy to behold because the visual or spatial properties of both living systems and inanimate molecules are so intimately related to the formal representations of their symmetry. Thus the beauty of a spiral staircase, its relationship to the symmetry of a helical molecule, and the mathematical or logical connection of their structure with the absorption and rotation of light by helical molecules are all capable of evoking connected thoughts and ideas and with them the satisfaction of an appeal both to the senses and the intellect. How does this come about?

Because the postulates of quantum mechanics are formulated on such an ad hoc basis, there is no simple answer. If we accept these, however, then we can demonstrate the manner in which symmetry controls the absorption and rotation of polarized electromagnetic radiation. Quantum mechanics provides the most direct, but the most complex, route. In Chapter 3, we saw that the requirement for absorption and rotation are, respectively, that the electric dipole transition moment, represented by spatial dipole moment integrals over the states between which the transition can occur, be nonzero, and that the corresponding magnetic dipole transition moments be nonzero as well for optical activity to occur. Since these moments have vector properties, it is at once obvious that their dot or scalar product must have components in the same direction. The moments cannot be orthogonal because in that case the angle θ between the vectors μ and $\mathbf{m}$ is 90 degrees and $\mu_{ab} \cdot \mathbf{m}_{ab} = |\mu_{ab} m_{ab}| \cos \theta = 0$. This is thus another requirement for the existence of optical activity, a requirement that appears in the form of Rosenfeld's equation as a result of the vector properties of the transition moments. Both the finite value of the individual moments and their orthogonality are controlled by symmetry.

Our object in this chapter is to show the relationship between *group theory and molecular symmetry* so that we can use group theory as a tool for understanding the relationship between *molecular symmetry* and *optical activity*. In those cases in which we are interested only qualitatively in the magnitudes of chiroptical effects and in the relative signs for different optical transitions, this will enable us to bypass quantum mechanical calculation on the path between molecular symmetry and optical relationships. We can establish that a formalism involving thought operations on physical systems can be brought to bear on the quantum description of these systems. In the discussion that follows, the time-*in*dependent Schrödinger equation is used to demonstrate the requirement that a symmetry operation have no effect on the physical properties of a molecule in a stationary state. This is different from Chapter 3 where the time-dependent Schrödinger equation was used to describe the passage of a molecular system between stationary states under the influence of an electromagnetic field. The quantum mechanical Hamiltonian operator of the Schrödinger equation is the sum of the potential and kinetic energies of the system. Intuitively, there is no difficulty in comprehending that, if we perform particular operations on the particles of a system; for example, if we interchange identical nuclei of a molecule, the properties (the sum of the kinetic and potential energies for one) of the system will remain unchanged. Analytically, if the molecule has symmetry elements, this operation can be performed by reflecting the coordinates through a plane of symmetry located in the molecule or by rotation of the coordinates of the atoms about a twofold axis fixed in the molecule. *As in a real system, these analytic operations must leave the sum of the potential and kinetic energies of* *the system unaffected.* We can symbolize this in the following way: if

$$\mathcal{H}\psi_i = \mathcal{E}_i\psi_i \qquad (5.201)$$

is the Schrödinger equation for a stationary state of the system, then for any operator R, since $\mathcal{E}_i$ is a constant so that R commutes with $\mathcal{E}_i$,

$$R\,\mathcal{H}\psi_i = \mathcal{E}_i R\psi_i. \qquad (5.202)$$

If, in addition, R is an operation that leaves $\mathcal{H}$, the Hamiltonian operator, invariant, a requirement that our intuition tells us is necessary in order that the properties of the system described by ψ_i be unaffected by our thought experiment, then

$$R\,\mathcal{H} = \mathcal{H}R \qquad (5.203)$$

and

$$\mathcal{H}(R\psi_i) = \mathcal{E}_i(R\psi_i). \qquad (5.204)$$

Therefore, $R\psi_i$ is an eigenfunction of the system with the energy $\mathcal{E}_i$. But by Eq. 5.201, ψ_i is also an eigenfunction of the systems with the energy $\mathcal{E}_i$ (if for the moment we ignore degeneracies). Since the Hamiltonian form of the Schrödinger equation (5.204) is a differential equation, this can only be true if R is a constant. If, in addition, we impose the constraint required by the condition that the probability of finding the electronic charge ψ_i somewhere in space is given by $\int \psi_i^*\psi_i\,d\tau = 1$, then not only must R be a constant, but its magnitude can only be ± 1, since if R is a constant,

$$\int R^*\psi_i^* R\psi_i\,d\tau = R^2\int \psi_i^*\psi_i\,d\tau = R^2, \qquad (5.205)$$

which can be true (3) if and only if $R = \pm 1$.

With this symbolic definition, we learn that we can represent the effect of a *symmetry operation R* (for nondegenerate systems) on the quantum or wave-mechanical description of the way in which the energy $\mathcal{E}$ of the system varies with its state ψ by the numerals ± 1. For degenerate systems no new concepts are required, but the description is more complex (4).

Symmetry operations are, therefore, the particular operations R, which obey the requirement of Eq. 5.203, namely, that their effect on the Hamiltonian is nil. Stated in a more familiar way, the Hamiltonian is invariant under a symmetry operation.

The operations and the symbols of the corresponding *symmetry operators*, which have the appropriate properties, are as follows:

1. inversion through a center of symmetry, i;
2. reflection through a plane of symmetry, σ;

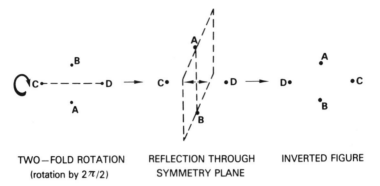

| TWO—FOLD ROTATION | REFLECTION THROUGH | INVERTED FIGURE |
| (rotation by $2\pi/2$) | SYMMETRY PLANE | |

Fig. 5.00 The symmetry operation S_2: Equivalence of the S_2 operation and the inversion operation i.

3. *n*-fold rotation by $2\pi/n$ about an axis of symmetry C_n;
4. *n*-fold rotation by $2\pi/n$ about an axis of symmetry followed by reflection through a plane perpendicular to the axis S_n;
5. any operation that leaves all the points of the system unchanged I.

Some of these operations require no further explanation, but the rotation reflection operation encompasses the symmetry rules for the phenomenon of optical activity. For example, for $n=2$, the operation S_2 consists of a rotation by $2\pi/2$, followed by reflection through a plane perpendicular to the twofold rotation axis. A geometric figure (see Fig. 5.00) on which such an operation can be performed without changing its properties can be generated from four points in space. The *relationship* between the coordinates of the points in the figure, which determines the symmetry, is the same as that of the potential energy of interaction between the electrons or the nuclei of a molecule with its atoms so disposed—as the simplest possible example, if $x_A = -x_B$, then $V_{AB} = -V_{BA}$. It is at once apparent that the operation $S_{n=2}$ is equivalent to the operation of inversion through a center of symmetry. We already know that molecules with a center of inversion are optically inactive. We will see that group theory leads to the conclusion that a molecule possessing a rotation-reflection axis of any order must be optically inactive (5). Consequently, S_2 is a group representation for an optically inactive system.

5.3   QUANTUM MECHANICS AND GROUP THEORY

We have postulated a relationship between symmetry and quantum mechanics defined by the fact that a symmetry operation on the Schrödinger equation must not perturb the energy of the system described by the equation. To see how group theory permits us to determine whether any given system will give rise to the particular observables of the absorption and rotation or scattering of light, we will attempt to establish the *connection between the quantum mechanical observables and group theory*.

The conviction that if we apply the law of nature represented by $\mathcal{H}\psi = \mathcal{E}\psi$ to a physical system, we would obtain a description (the allowed energies and spatial distribution of the electronic charge) of the system is an accepted postulate, and we seek solutions, that is, wavefunctions for which the equality is valid for allowed values of the energy $\mathcal{E}$ and from which we can calculate or predict physical observables. In many cases where these solutions are difficult to achieve, it would, nevertheless, be useful to be able to delineate or qualitatively determine certain properties. In the course of examining how this may be accomplished, we will establish the relationship between group theory and quantum mechanics.

Recall the orthonormal properties of the wavefunctions,

$$\int \psi_i \psi_j \, d\tau = \delta_{ij}, \qquad \delta_{ij} = 0, i \neq j; \quad 1, i = j. \tag{5.301}$$

The scalar product of unit vectors have the same properties:

$$\rho_i \cdot \rho_j = \delta_{ij}. \tag{5.302}$$

By analogy we can set up a correspondence between operations with vectors in geometrical space and operations with wavefunctions in function space. We will use this correspondence later. If the physical system we are attempting to describe has a certain number of identical particles, we can perform a series of operations on the Hamiltonian operator with respect to these particles without changing its value;

$$P \mathcal{H} P^{-1} = \mathcal{H}, \tag{5.303}$$

where P^{-1} is the reciprocal or inverse operation of that represented by P. If there are N particles, there may be $N!$ operations. If by the application of the Schrödinger equation, any of these operations yield functions, ψ_i, which are invariant to further applications of the operations, some special significance may be suspected, since the random selection of ψ functions cannot be expected to exhibit this behavior. To illustrate the consequences of the operations which result in the invariance of the Hamiltonian with respect to the operations, let us examine what properties are required of such operations. We will see that these properties define a mathematical "group."

Suppose that there are a set of functions: $\psi_1, \psi_2, \psi_3, \ldots, \psi_N$ to which we apply an operation P. The ψ_i can be arbitrary mathematical functions or, in particular, wavefunctions describing the charge distribution of the states of an atom or molecule. There are three possible results:

1. The function is invariant: $P\psi_i = \psi_i$.
2. The function is transferred into a linear combination of other members of the set: $P\psi_i = a\psi_j + b\psi_k + \cdots$.
3. The function is transferred into a new function which is not a member of the set: $P\psi_i = \psi_{N+M}$.

If we apply a different operation, P', to the same set of ordered functions, again we may have the three alternatives. Of these, the third alternative is trivial, since we could add the new function to the original set simply by including additional zeros in the appropriate places in the matrix that performs the original operation P.* Consequently, we can consider as distinct operations those which produce results 1 and 2, whether or not they produce result 3. We could also show that result 2 is a special case which can be handled by the same group-theoretical formalism as result 1. We will consider, therefore, a set of manipulations involving operations of the type 1 and from this define the properties of a "group." Later in Section 5.4, we will note that sets of *symmetry operations* defined in Section 5.2 have these properties and consequently constitute "groups."

---

*A familiarity with only the simplest ideas and operations of matrix algebra are required for the discussions in this book. For the development and proof of many of the group-theoretical statements, which are not given here, considerable additional familiarity is required. Four defining statements for a matrix and matrix operations follow. Where needed, additional statements will be given in the text.

1. A matrix is an array of symbols (which may, of course, be numerals) having i rows and j columns which obey the rules of matrix operations. A matrix that consists of only one row or one column is termed a row or column matrix and, in general, will exhibit the properties of a vector.
2. Matrix multiplication is associative but not communitative:

$$M(NL) = (MN)L, \text{ but } M(NL) \neq L(MN).$$

3. As a corollary of 1, the operation of matrix multiplication is carried out from right to left. In the multiplication MNL, the matrix N operates on L to give a matrix Q. The matrix M *then* operates on Q to give the matrix P. Thus, $MNL = MQ = P$.
4. The operation of ordinary matrix multiplication of the matrix M with the matrix N to give the matrix R involves summing the multiplicands of the elements of the ith row of M and the jth column of N to give the ijth element of R. Thus, for the two-by-two matrices,

$$\begin{pmatrix} M_{11} M_{12} \\ M_{21} M_{22} \end{pmatrix} \begin{pmatrix} N_{11} N_{12} \\ N_{21} N_{22} \end{pmatrix} = \begin{pmatrix} M_{11}N_{11} + M_{12}N_{21} & M_{11}N_{12} + M_{12}N_{22} \\ M_{21}N_{11} + M_{22}N_{21} & M_{21}N_{12} + M_{22}N_{22} \end{pmatrix}.$$

5.4 "MANIPULATIONS THAT DEFINE GROUP PROPERTIES"

1. Consider what happens if we apply successively P and P', the latter to the functions generated by the application of the original operator P, to an ordered set of functions. The permutation P' can yield a new set of functions, or it can leave the set of functions unchanged. In the latter case there must be an operator P'', such that its effect is the same as the successive operations P and P'. For convenience in later considerations, we designate P, P', and P'' by P_i, P_j, and P_k and symbolize the equality between the operation P_k and the *successive* application of the operation P_i and P_j by

$$P_i P_j = P_k. \tag{5.304}$$

2. Suppose that one (or more) of the operations P is a permutation such that the successive application of this permutation to the ordered set of functions results in the original set. For this to occur *and* for condition 1 to be obeyed, as well, by the same permutation, it follows that there must be a permutation P_0 which does not change the set. That is, there must be an operation P_i^{-1}, which is the inverse of P_i, such that

$$P_i P_i^{-1} = P_o. \tag{5.305}$$

Note that P^{-1} is not to be read in the mathematical sense of a reciprocal, although its effect is in one sense similar; that is, its product with P_i in the *proper order* produces a permutation P_o, which does not affect any reordering of the set of ordered functions and thus may be considered to have the effect of using the numeral 1 as an operator. However, the clue to the difference lies in the words *proper order* because it is not necessarily true that the P_i and P_i^{-1} commute. Those that do commute do not produce another member of the set.

3. If the permutation P_o as defined in 2 does affect the ordered set in any way, then it must also be true that this permutation may be preceded or followed by any other permutation obeying rule 1 and 2 without changing the latter. Symbolically,

$$P_o P_i = P_i. \tag{5.306}$$

4. Finally, we can require that the associative law holds for the operations with these permutations:

$$P_i(P_j P_k) = (P_i P_j) P_k. \tag{5.307}$$

To set having these four properties, the name "group" is given. Suppose that the ordered set of functions permuted by the operators P are a set of atomic or molecular wavefunctions. Then, not only must the "group" of permutations obey the four laws we have specified above, but also *each of the permutations must leave the Hamiltonian operator unchanged* (Eq. 5.303). We have already seen that there are a set of symmetry operations that do precisely this, that is, leave the Hamiltonian operator unchanged. We could quickly show that various sets of these operations also obey the four laws associated with a group (6). These sets are those associated with the various point groups given in Appendix I.

In addition to the requirement that symmetry operations leave the Hamiltonian operator unchanged, the symmetry of the Hamiltonian itself imposes a symmetry classification on the wavefunctions which obey the Schrödinger equation. In fact, the wavefunctions will possess the same symmetry as the Hamiltonian. Consequently, if we can classify the symmetry of the Hamiltonian, the wavefunctions can also be classified according to the same symmetry.

The methods of solving the Schrödinger equation to obtain the energies and wavefunctions of the states of a system is important in the detailed analysis of the chiral effects associated with optical transitions between these states, but the details of these methods are outside the province of this book. However, the *chiral effects are observables of the systems and the usefulness of the symmetry classification of the wavefunctions arises from the mathematical integrals that express the values of these observables.* Thus, we have already seen in Chapter 3 that the absorption intensity for electromagnetic radiation is given by the square of the electric dipole transition moment integral, μ_{ab}, between the states a and b.

$$\mu_{ab} = \int \psi_b^* \mu \psi_a \, d\tau. \tag{5.308}$$

There is a theorem of calculus that states that this integral will vanish (have zero value) if the integrand is not invariant under the symmetry operations appropriate to the space involved; symmetry operations are defined for this theorem with respect to the properties of an abstract mathematical group that has the four group properties discussed above. Conversely, the integral will have a finite value other than zero if the integral *is* invariant. Therefore, group theory determines the *existence* of a finite value for a physical observable, in this case, an optically induced transition between two electronic states of a molecule. In this way, we have made the connection between *quantum mechanics and group theory* and indirectly between these and (molecular) symmetry. Let us now examine a bit more closely the relationship between group theory and molecular symmetry from which the rules and procedures will be derived, by means of which symmetry can be brought to bear on problems related to optical activity.

5.5   GROUP OPERATIONS AND MOLECULAR SYMMETRY

Certain combinations of symmetry operations may be classified as *groups*. Groups have the very precise mathematical definition given by the properties specified in the previous section, but most important, geometrical representations of the vibrations and of the electronic orbitals of real molecules can be classified as *belonging* to a particular group according to the symmetry operations that may be applied to their geometrical representations. We will illustrate this by an example of the geometrical representations of the normal modes of vibration of a water molecule. For the purpose of this example, consider only one of the $3N - 6$ ($N =$ number of atoms in the molecule) vibrational modes of H_2O, namely, the symmetric stretching mode that gives rise to the band in the infrared and Raman spectra at 3655 cm^{-1}. This mode is represented in Figure 5.01a by the arrows attached to the hydrogen and oxygen atoms. The symmetry elements that may be associated with these geometrical representations of the vibration in which the hydrogen atoms vibrate in phase relative to the oxygen atom are few in number: A twofold rotation C_2 (in the notation of Section 5.2), two reflection planes, one σ_v in the plane of the three atoms, and one $\sigma_{v'}$ perpendicular to the plane. Both planes contain the rotation axis. Each of these operations may be performed on the arrow representations without changing the representation in any way. Except for the identity operation I which leaves the arrows unchanged, these are the only *symmetry* operations applicable to this geometric representation. These four operations form the basis for a group. In the Schoenflies (7) notation, this particular group is given the symbol C_{2v}. Furthermore, the use of this notation and the *character table* associated with it permits an unusually concise representation of the symmetry to be associated with the symmetric vibration of the water molecule. One representation is the set of arrows on the figure, but mathematical manipulations are difficult to carry out with geometric representations. Another representation is the word description given above, again not a particularly good representation for manipulation. We could also give a mathematical formalism by placing the molecule in a set of Cartesian (or other) axes and describe the vibration by the change in the values of the coordinates of the positions of the atoms. This would require three position designations for each atom and thus nine parameters or a 9×9 matrix representation to describe the motion. Of course, this would have the advantage of specifying the magnitude, as well as the direction of the motion; but while it would specify the symmetry of the motion, it would not provide a simple scheme for classifying the symmetry. Finally, we can use the group-theoretical classification, that is, the C_{2v} *character table*:

C_{2v}	I	C_2	σ_v	$\sigma_{v'}$
A_1	1	1	1	1
A_2	1	1	-1	-1
B_1	1	-1	1	-1
B_2	1	-1	-1	1

At the head of the table is the row specifying the symmetry operations associated with this particular group. Recalling their definitions, we see that rotation C_2 about the twofold rotation axis would bring the arrows into exact congruence with each other as would, of course, the identity operation I. Thus, the direction or sign of the arrows are unchanged by these operations, equivalent to multiplying each by $+1$. For these operations (C_2 and I), the vibration is thus represented by either of the first two rows of the table. Similarly, reflection in the plane of the molecule does not change the sign, so that now only the first row remains to represent this

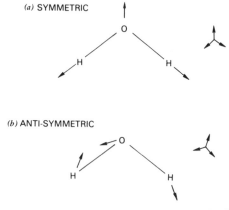

Fig. 5.01 The stretching vibrations of the water molecule. Note that in both the symmetric and antisymmetric vibrations, the sum of the vector displacements of the atoms is zero. This necessary condition for the representations of the displacements is required for the invariance of the Hamiltonian under the definition of a *symmetry operation*. Otherwise, the molecule would move in space, and its translational or rotational energy would change, and so also the Hamiltonian. (*a*) symmetric; (*b*) antisymmetric.

vibration. A rapid glance will show that reflection $\sigma_{v'}$ in the plane perpendicular to the plane of the molecule through the twofold axis again brings the arrows into congruence, and we have a simple representation A_1 for the symmetry of this vibration. Similarly, an examination of the antisymmetric vibration, Fig. 5.01b, will very quickly show that B_1 is the representation for this vibration.

For communication purposes, it is necessary, as will become more apparent later, that a systematic set of rules for defining the operations with respect to a particular coordinate system be chosen. There is now general agreement on such, and for the past decade or two, all such operations have been defined with respect to a right-handed Cartesian coordinate system and applied in a uniform systematic way. The group theory and symmetry operations are, however, internally consistent, and one could, at the risk of causing a crisis of communication, use any systematic classification one chooses.

The choice of the group and, therefore, of the character table to be associated with a particular function is not always quite so simple as in the case of that required for the representation of the intramolecular dynamics of the nuclei of the water molecule. In general, however, the space coordinates of the nuclei, either of the molecule as a whole (in the case of vibrations) or of a chromophoric grouping (in the case of electronic wavefunctions), will provide at least the starting point for an analysis of the symmetry characteristics. *The reason that this is so is that these space coordinates, in general, define the symmetry of the potential function that describes the way in which the energy of the systems varies with the values of the coordinates.*

Just as the vibrations of an H_2O molecule have been used to illustrate the classification of the vibrations according to symmetry representation (A_1 and B_1 for the symmetric and antisymmetric stretching vibration), so can an example of the wavefunctions associated with the valence electrons of a molecule or chromophore be used to illustrate the classification of wavefunctions according to symmetry representations. A detailed description of the transformation properties of atomic and molecular orbitals is outside the scope of this book (9). However, a single example will suffice to describe the procedure.

Consider the π electron system of ethylene. By this designation, we are, of course, already assuming considerable knowledge of the molecular orbitals of the double-bond system. In the first approximation, this system consists of a linear combination of the $2p$ orbitals on the carbon atoms. $2p$ orbitals are axially symmetric and have a single node centered at the nucleus of the atom. The two $2p$ orbitals, one on each of the carbon atoms, can be combined linearly as either a plus or minus combination (neglecting

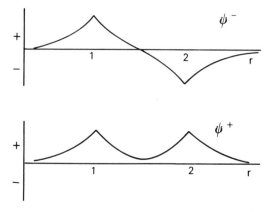

Fig. 5.02 Symmetry of the molecular wavefunctions of ethylene. ψ^+ is symmetric with respect to the twofold rotation axis in the plane of the paper, and ψ^- antisymmetric with respect to the rotation about this same axis. The axis r represents the distance along the line connecting the carbon nuclei at positions 1 and 2, and the vertical axis represents the amplitudes and sign of the wavefunctions.

electron spin):

$$\psi^+ = \frac{(2p_1 + 2p_2)}{\sqrt{2}},$$

$$\psi^- = \frac{(2p_1 - 2p_2)}{\sqrt{2}}.$$

These two combinations may be plotted in terms of the radial parts of the atomic $2p$ functions (Fig. 5.02), the squares of which represent the maximum in the electron density, or alternately, they may be represented by the squares of their angular parts as shown in Fig. 5.03. In either case the following symmetry considerations apply to these orbitals in ethylene considered as a planar molecule in which the nodes of the $2p$ orbitals are in the plane and their symmetry axes perpendicular to the plane:

1. Reflection in the plane of the molecule (yz) inverts the signs of ψ^- and ψ^+.

2. Reflection in the plane (xy) perpendicular to both the plane of the molecule and the plane containing the carbon-carbon bond again inverts the sign of ψ^- but leaves the sign of ψ^+ unchanged.

3. Reflection in the plane (xz) orthogonal to the other two, that is, the plane containing the symmetry axes of both $2p$ orbitals, leaves the sign of both ψ^- and ψ^+ unchanged.

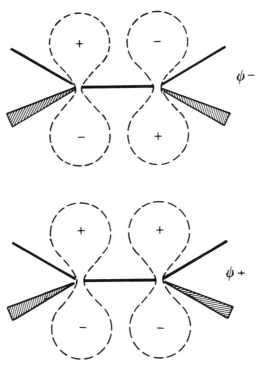

Fig. 5.03 Pictorial representation of the $2p$ atomic orbitals from which the symmetric and antisymmetric molecular orbital functions of Fig. 5.02 are constructed.

The remaining symmetry elements appropriate to ethylene in its planar configuration are three axes of symmetry that lie in the respective planes discussed above and parallel or perpendicular to the C–C bond, a center of inversion at the midpoint of the C–C bond, and, of course, the identity element. This establishes that the character table appropriate to the symmetry of planar ethylene is D_{2h}.

D_{2h}	I	$C_2(z)$	$C_2(y)$	$C_2(x)$	i	$\sigma(xy)$	$\sigma(xz)$	$\sigma(yz)$
A_g	1	1	1	1	1	1	1	1
B_{1g}	1	1	−1	−1	1	1	−1	−1
B_{2g}	1	−1	1	−1	1	−1	1	−1
B_{3g}	1	−1	−1	1	1	−1	−1	1
A_u	1	1	1	1	−1	−1	−1	−1
B_{1u}	1	1	−1	−1	−1	−1	1	1
B_{2u}	1	−1	1	−1	−1	1	−1	1
B_{3u}	1	−1	−1	1	−1	1	1	−1

The application of the remaining symmetry operations for planar ethylene may be summarized as:

$$
\begin{array}{llll}
I & \psi^+ \rightarrow \psi^+, & I & \psi^- \rightarrow \psi^-, \\
C_2(z) & \psi^+ \rightarrow -\psi^+, & C_2(z) & \psi^- \rightarrow -\psi^-, \\
C_2(y) & \psi^+ \rightarrow -\psi^+, & C_2(y) & \psi^- \rightarrow \psi^-, \\
C_2(x) & \psi^+ \rightarrow -\psi^+, & C_2(x) & \psi^- \rightarrow -\psi^-, \\
i & \psi^+ \rightarrow -\psi^+, & i & \psi^- \rightarrow \psi^-.
\end{array}
$$

Matching the appropriate signs of all the symmetry operations with the entries in the D_{2h} character table, we note that ψ^+ belongs to B_{3u} and ψ^- to B_{2g}. In group-theoretical vocabulary, ψ^+ and ψ^- are said to be a *basis* for their respective representations or, alternately, to *belong* to these representations.

This is an appropriate point at which to consider several subtleties of group-theoretical classifications of this kind. It should be apparent that the foregoing analysis of the symmetry classifications of the linear combination of $2p$ orbitals of ethylene really did not carry with it any specification of the nature of the atom to which the $2p$ orbitals belong. Consequently, this analysis should apply, for example, to any pair of atoms bonded by a π bond, except that the symmetry group to which the molecule belongs may not be D_{2h}. For example, the carbonyl group in acetone or formaldehyde has a π bond that is indistinguishable, so far as the foregoing analysis is concerned, from that of the carbon-carbon double bond in ethylene, except that the appropriate symmetry is C_{2v} rather than D_{2h}. There is, in fact, an exact correlation between the symmetry of the π orbitals in these systems; from the correlation tables (10) it can be ascertained that the corresponding orbitals to ψ^+ and ψ^- both belong to the B_1 representation in the C_{2v} group. There will, in fact, be many similarities between the behavioral properties associated with these orbitals predicted from symmetry considerations and observable spectroscopically, for example, the polarization direction of corresponding spectroscopic states.

Still another feature worth some comment is that frequently substitutions that change the point symmetry of the molecule affect only very slightly the physical behavior of the system, which may thus be analyzed—so far as symmetry-controlled properties are concerned—in terms of the parent symmetry. Thus, the spectroscopic behavior of electronic transitions of conjugated systems (the polarization directions of the electric dipole transition moments) are only very slightly altered by the substitution of a deuterium atom or an alkyl group, even though these may have fairly profound effects on the energy of the transition. The sign of the chiral activity, however, may be strongly dependent on the asymmetry of the substitution, but usually more so for weak transitions than for strong ones: see the discussion of vibronic interactions in Section 3.6 and the discussion of the activity of skewed dienes and ketones in Sections 7.31 and 7.21. Finally, we should recall (1) the effect of Herzberg-Teller interactions discussed in Chapter 3 and note that these are ammenable to analysis by group-theoretical considerations, and (2) the fact that the electronically excited states of a molecule may have a quite different geometry and, therefore, symmetry from the ground state, which can have a profound effect on the circular and linear polarization properties. When this occurs, the controlling symmetry classification is one that contains only the symmetry elements common to the geometry of the initial and final states involved in the electronic transition in question.

5.6   REPRESENTATION IN GROUP THEORY

Before proceeding to the next chapter where we finally make the connection between group theory and chiral optical effects, thereby completing our objective of establishing a formal relation of the latter to molecular symmetry, we need to discuss in a little more detail the subject of representations in group theory. As before, for a full understanding the reader is referred to the standard texts. Our object here is to demonstrate the technique of manipulation with representations, which in essence involves only a simple set of operational rules. The statement of these rules is insufficient to give any physical insight, and for this reason we wish to discuss them somewhat more fully without, however, enlarging on the group-theoretical origin. In doing so, we must understand that this is a very superficial treatment designed to provide the tools for application to zeroth-order treatments of symmetry questions and perhaps to intrigue us into going to sources of more extensive treatments for a fuller understanding.

We have already used the word *representation* to designate the Schonflies symbol for the row of a character table that describes the operations of the point group of the molecule or chromophore on the function (vector directions in space, wavefunction, etc.), which forms the basis for the description of a physical property. In group-theoretical terms, representation is more precisely defined as *a set of matrices* (which in the simplest case of unit one-dimensional matrices may be the number one with its sign) *assigned to the elements of a group such that if (P) is the matrix for the element P, and (Q) is the matrix for the element Q, then $(P)(Q) = (R)$, where (R) is a matrix representing the element R of the group.* Thus the matrices (P), (Q), (R), and the matrix (I) of the identity element I, would be a representation of the group if the group consisted only of the elements P, Q, R, and I. As an example, the matrix representation of the C_{2v} group can easily be seen to have this property.

		P	Q			(P)	(Q)		(R)
C_{2v}	I	C_2	σ_v	$\sigma_{v'}$		C_2	σ_v	=	$\sigma_{v'}$
Γ_1	(1)	(1)	(1)	(1)		(1)	(1)	=	(1)
Γ_2	(1)	(1)	(−1)	(−1)		(1)	(−1)	=	(−1)
Γ_3	(1)	(−1)	(1)	(−1)		(−1)	(1)	=	(−1)
Γ_4	(1)	(−1)	(−1)	(1)		(−1)	(−1)	=	(1)

The numbers have all been put in parentheses to emphasize that they are matrices. The table at the right illustrates the property by which we defined the representation. The Greek letter Γ is used to generalize the notation of a representation. The corresponding symmetry symbols in Schonflies's notation may be found by comparing this table to the C_{2v} table on page 132. For example, $\Gamma_3 = B_1$.

Another description of this property which emphasizes its physical significance is given by the statement: Each of the matrices, which form the representation of a group, is the transformation matrix for the operation of a symmetry element on a vector property associated with the molecule, and the simple product of these transformation matrices represents faithfully the combination of symmetry operations that they represent.

Not all of the possible representations of a given symmetry group are this simple. For example, each of the matrices assigned to a given symmetry operation for the vectors attached to a group of points in space can be *at least* as large as $3n \times 3n$, where n is the number of points. So for a

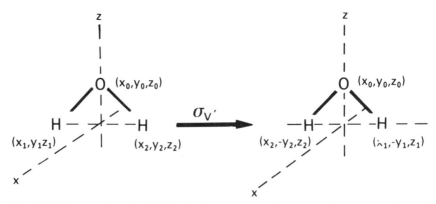

Fig. 5.04 The operation of reflection in the symmetry plane perpendicular to the plane of the nuclei of the H_2O molecule. Note that because of the choice of location of the coordinate system with respect to the nuclear positions, in this case $-Y_0 = Y_0 = 0$ (see the matrix transformation of coordinates in the text).

three-atom molecule, like water, as noted in Section 5.4, we can write a 9×9 transformation matrix that describes the behavior of each of the coordinates of each atom under a given symmetry operation. For example, the operation of $\sigma_{v'}$ on the Cartesian coordinates of the mass points of water (Fig. 5.04) may be written in terms of these coordinates:

$$
\sigma_{v'}
\begin{bmatrix}
X_{H_1} \\
Y_{H_1} \\
Z_{H_1} \\
X_{H_2} \\
Y_{H_2} \\
Z_{H_2} \\
X_O \\
Y_O \\
Z_O
\end{bmatrix}
=
\begin{bmatrix}
0 & 0 & 0 & 1 & 0 & 0 & 0 & 0 & 0 \\
0 & 0 & 0 & 0 & -1 & 0 & 0 & 0 & 0 \\
0 & 0 & 0 & 0 & 0 & 1 & 0 & 0 & 0 \\
1 & 0 & 0 & 0 & 0 & 0 & 0 & 0 & 0 \\
0 & -1 & 0 & 0 & 0 & 0 & 0 & 0 & 0 \\
0 & 0 & 1 & 0 & 0 & 0 & 0 & 0 & 0 \\
0 & 0 & 0 & 0 & 0 & 0 & 1 & 0 & 0 \\
0 & 0 & 0 & 0 & 0 & 0 & 0 & 1 & 0 \\
0 & 0 & 0 & 0 & 0 & 0 & 0 & 0 & 1
\end{bmatrix}
\begin{bmatrix}
X_{H_1} \\
Y_{H_1} \\
Z_{H_1} \\
X_{H_2} \\
Y_{H_2} \\
Z_{H_2} \\
X_O \\
Y_O \\
Z_O
\end{bmatrix}
=
\begin{bmatrix}
X_{H_1} \\
-Y_{H_1} \\
Z_{H_1} \\
X_{H_2} \\
-Y_{H_2} \\
Z_{H_2} \\
X_O \\
Y_O \\
Z_O
\end{bmatrix}
$$

Obviously, all of the symmetry operations, not just $\sigma_{v'}$, were carried out using the matrix representations of the C_{2v} table, that is, with an array of much smaller dimensions. Using the Cartesian matrix representation, we would need four 9×9 matrices and thus $4 \times 81 = 324$ entries rather than the 16 in the C_{2v} character table. Furthermore, by choosing something other than the Cartesian coordinates as a basis set, still other representations can be found. In general, there are an infinite number of representations.

Group theory specifies that for each symmetry transformation within a group, all such representations, except one, are reducible, and the unique one is, therefore, termed an *irreducible representation*. The irreducible representations that can be associated with the arrangements of all the possible crystallographic arrangements of points in space have been worked out and are given by the group theoretical character tables.

It is these *irreducible representations* which have properties of interest for the interaction of radiation with matter. Without proof we will state these properties:

1. An irreducible representation of a group consists of a set of matrices (R), each having pq components where p is the number of rows and q the number of columns in the matrix. [A rapid glance at the character tables will show that many of the matrices (R) are simply the matrix (1) or (-1), and thus, having only one row and one column, have only one element in the matrix.]

2. If $\Gamma_i(R)$ designates the matrix corresponding to the symmetry operation R of the ith irreducible representation Γ_i, and $\Gamma_i(R)_{mn}$ is the mn component of the matrix, then

$$\sum_R \Gamma_i(R)_{mn} \Gamma_i(R)_{mn} = \frac{h}{l_i}, \tag{5.501}$$

where h, the *order of the group*, is the number of symmetry elements in the group and l_i is the dimension of the ith representation.

The dimension of the representation is the product of p and q for the matrices in that representation. Thus, in the C_{3v} group (Appendix I), the representation of A_1 and A_2 symmetry have dimension $l_i = 1$, while the representations of I symmetry have dimension $l_i = 2$. In general, the Schonflies notations of the symmetry of the representation immediately designates the dimension, symmetries A and B having $l_i = 1$, and symmetries E and T having $l_i = 2$ and $l_i = 3$, respectively.

3. Irreducible representations behave like the components of orthogonal vectors:

$$\sum_R \Gamma_i(R)_{mn} \Gamma_j(R)_{m'n'} = 0, \qquad \text{when } i \neq j, \tag{5.502}$$

$$\sum_R \Gamma_i(R)_{mn} \Gamma_i(R)_{m'n'} = 0, \qquad \text{when } m \neq m' \text{ and/or } n \neq n'. \tag{5.503}$$

The three relations that define the latter two properties can be combined into a general orthogonality relation:

$$\sum_R \Gamma_i(R)_{mn} \Gamma_j(R)_{m'n'} = \frac{h}{l_i} \delta_{ij} \delta_{mm'} \delta_{nn'} \tag{5.504}$$

where the δ_{kl} are the Kronecker deltas: $\delta_{kl} = 0$ for $k \neq l$, $\delta_{kl} = 1$ for $k = l$.

It is instructive to consider an example: Consider the group D_{3h}. Let the ith representation Γ_i be A_2 and the jth representation Γ_j be E'. In this case $i \neq j$, and we should find that the expression on the left of Eq. 5.504 is equal to zero. All the elements in the two representations given in the character table have only one entry, so that $m = n = 1$ and $m' = n' = 1$. The symmetry elements R are I, two C_3's, three C_2's, σ_H's, and three σ_v's. Thus, the elements of the representation are

$$\Gamma_i(I)_{mn} = A_1'(I) = 1, \qquad \Gamma_j(I)_{m'm'} = I'(I) = 1,$$

$$\Gamma_i(C_3)_{mn} = A_1'(C_3) = 1, \qquad \Gamma_j(C_3)_{m'n'} = I'(C_3) = 1,$$

$$\Gamma_i(C_2)_{mn} = A_1'(C_2) = -1, \quad \text{etc.}$$

$$\sum_R \Gamma_i(R)_{mn} \Gamma_j(R)_{m'n'} = (1 \cdot 2) + 2(1 \cdot -1) + 3(-1 \cdot 0) + (1 \cdot 2) + 2(1 \cdot -1) +$$

$3(-1 \cdot 0) = 0$, in accord with the equation, since $\delta_{ij} = 0$ for $k \neq j$. For the case $i = j$, consider the representation A_1': $\Sigma_R \Gamma_i(R)_{mn} \Gamma_i(R)_{mn} = (1 \cdot 1) + 2(1 \cdot 1) + 3(1 \cdot 1) + (-1 \cdot -1) + 2(-1 \cdot -1) + 3(-1 \cdot -1) = 12$—again in accord with the equation, since there are 12 symmetry elements so that the order of the group is h and $\delta_{ii} \delta_{mn} \delta_{m'n'} = 1$.

5.7   THE CHARACTER

No discussion of group theory and its applications can be complete (even in the abbreviated fashion treated here) without defining a character and demonstrating that the *direct product of the character of group representations is the character of the representation formed from the direct product of group representations*. For the sake of completeness, we refer to the discussion or proof of this theorem in every book on group theory. Note that the proof of the italicized statement and accompanying discussion is, in fact, quite short. The purpose, then, of excluding its proof here is to emphasize that its significance is in its application and not in its proof. Suffice it to say that the elements of the character tables in this book (as well as all the others) *are* the characters of the *irreducible representations* of the groups. As a consequence, all theorems and operations that hold for irreducible representations—and especially the operations of taking *direct products* which are the determining factors in specifying the allowedness of a chiral observable—may be performed on these representations.

We need, therefore, only to define the *direct product* of representations. The properties of the direct product lead directly from the orthogonality relations, which define the direct product of the characters of matrices and which are different from the ordinary product of matrices. In the ordinary matrix product, the *mn* element of the product is the *sum* of the products of the elements of the *m*th row of one matrix with the elements of the *n*th column of the second matrix. The product matrix, therefore, is of the same dimension; it has the same number of rows and columns as there are in the matrices from which the product was formed. By comparison, for the *direct product* of matrices, the *mn* element is the *product* of the corresponding elements of the *m*th row and *n*th column of the two matrices; thus, the direct product matrix is larger in size than the matrices from which it is formed (the *direct product* matrix of two 2×2 matrices being a 4×4 matrix, and the direct product of two 3×3 matrices being being a 9×9 matrix); thus, the size of the direct product matrix goes as the *square* of the matrices from which it is formed. However, in irreducible matrices, only the character, which is the trace (diagonal elements of the matrix), is needed as a representation of the symmetry in group theory. The *direct product* of the characters, and therefore of the representations, is formed from the diagonal elements of the irreducible matrices. In most cases, for the point groups with which we are concerned, the matrices are simply a single element as is evident from the entries in the character tables. These direct products may be worked out for representations of any symmetry, and we will do so for one as an example. However, they have all been worked out, and, in fact, a simple schema developed using the Mulliken (11) notation of the symmetry representations to determine the symmetry of the direct product of representations.

As our example, let us consider the *direct product* of three different representations of the C_{6v} group, say of the representations for the symmetry species A_1, B_1, and E_1.

C_{6v}	I	$2C_6$	$2C_3$	C_2	3	3
A_1	1	1	1	1	1	1
A_2	1	1	1	1	-1	-1
B_1	1	-1	1	-1	1	-1
B_2	1	-1	1	-1	-1	1
E_1	2	1	-1	-2	0	0
E_2	2	-1	-1	2	0	0

We want the *direct product* $A_1 \times B_1 \times E_1 = ?$ We take the direct product of each of the characters of each representation:

Symmetry Element	Species A_1		Species B_1		Species E_1	=	Species ?
I	1	×	1	×	2	=	2
C_6	1	×	-1	×	1	=	-1
C_3	1	×	1	×	-1	=	-1
C_2	1	×	-1	×	-2	=	2
σ_v	1	×	1	×	0	=	0
σ_d	1	×	-1	×	0	=	0

A glance at the character tables shows immediately that the direct product is the representation of the symmetry species E_2, and an equally rapid glance at the rules that follow shows that this result is obtainable without any reference to the character table itself. We are thus prepared to obtain the symmetry species representations of quantum mechanical parameters to which the rules of group theory apply.

Rules for direct products (12):

1. Groups having a center of inversion i, or a reflection plane σ_h perpendicular to a C_2 axis:

The g, u, or $'$, $''$ in these groups satisfy $g \times g = u \times u = g$, $g \times u = u$, $' \times ' = ''$, and $'' \times '' = ''$.

2. Products of the form $A \times A$, $B \times B$, $A \times B$:
   (a) For all groups:
       Letter symbols: $A \times A = A$, $B \times B = A$, $A \times B = B$.
       Subscripts: $1 \times 1 = 1$, $2 \times 2 = 1$, $1 \times 2 = 2$
   (b) Except for the B representations of D_2 and D_{2h}, where $B \times B = B$, $1 \times 2 = 3$, $2 \times 3 = 1$, $3 \times 1 = 2$.

3. Products of the form $A \times E$, $B \times E$:
   (a) For all groups: $A \times E_k = E_k$, irrespective of the suffix on A.
   (b) For all groups, except D_{6d}, D_{4d}: $B \times E_1 = E_2$, $B \times E_2 = E_1$, irrespective of the suffix on B (if the group has only one E representation, put $E_1 = E_2 = E$).
   (c) For D_{6d}:
       $B \times E_1 = E_5$, $B \times E_2 = E_4$, $B \times E_3 = E_3$, $B \times E_4 = E_2$, $B \times E_5 = E_1$, irrespective of the suffix on B.
   (d) For D_{4d}, S_8:
       $B \times E_1 = E_3$, $B \times E_2 = E_2$, $B \times E_3 = E_1$, irrespective of the suffix on B.

4. Products of the form $E \times E$: (For groups which have A, B, or E symbols without suffices, put $A_1 = A_2 = A$, etc., in the equation below.)
   (a) For O_h, O, T_d, D_{6h}, D_6, C_{6v}, C_{6h}, S_6, D_{3d}, D_{3h}, D_3, C_{3v}, C_{3h}, C_3
       $E_1 \times E_1 = E_2 \times E_2 = A_1 + A_2 + E_2$, $E_1 \times E_2 = B_1 + B_2 + E_1$.
   (b) For D_{4h}, D_4, C_{4h}, C_4, S_4, D_{2d}:
       $E \times E = A_1 + B_1 + B_2$.
   (c) For D_{6d}:
       $E_1 \times E_1 = E_5 \times E_5 = A_1 + A_2 + E_2$, $E_2 \times E_2 = E_4 \times E_4 = A_1 + A_2 + E_4$,
       $E_3 \times E_3 = A_1 + A_2 + B_1 + B_2$;
       $E_1 \times E_2 = E_4 \times E_5 = E_1 + E_3$, $E_1 \times E_3 = E_3 \times E_5 = E_2 + E_4$,
       $E_1 \times E_4 = E_2 \times E_5 = E_3 + E_5$, $E_2 \times E_3 = E_3 \times E_4 = E_1 + E_5$,
       $E_1 \times E_5 = B_1 + B_2 + E_4$, $E_2 \times E_4 = B_1 + B_2 + E_2$.
   (d) For D_{5d}, D_{5h}, D_5, C_{5v}, C_{5h}, C_5:
       $E_1 \times E_1 = A_1 + E_2$, $E_2 \times E_2 = A_1 + A_2 + E_1$, $E_1 \times E_2 = E_1 + E_2$.
   (e) For D_{4d}, S_8:
       $E_1 \times E_1 = E_3 \times E_3 = A_1 + A_2 + E_2$, $E_2 \times E_2 = A_1 + A_2 + B_1 + B_2$,
       $E_1 \times E_2 = E_2 \times E_3 = E_1 + E_3$, $E_1 \times E_3 = B_1 + B_2 + E_2$.

5. Products involving the T (or F) representations of O_h, O, and T_d:
   $A_1 \times T_1 = T_1$, $A_1 \times T_2 = T_2$, $A_2 \times T_1 = T_2$, $A_2 \times T_2 = T_1$; $E \times T_1 = E \times T_2 = T_1 + T_2$;
   $T_1 \times T_1 = T_2 \times T_2 = A_1 + E + T_1 + T_2$, $T_1 \times T_2 = A_2 + E + T_1 + T_2$.

5.8　SUMMARY

In the effort to demonstrate that the symmetry of molecular charge distributions is classifiable according to the methods of group theory, we have shown that symmetry operations that have *group properties* may be used to describe the symmetry of these charge distributions in a way that leaves the Hamiltonian energy operator invariant. This invariance is used as a criterion for the validity of the irreducible representations of any given set of applicable symmetry operations. In preparation for the next chapter, where we examine the utility of the irreducible representations in providing criteria (*selection rules*) for isotropic absorption and for chiroptical phenomena, we have defined a property of the representations called the *direct product*. We will now see what can be done with the direct product of the applicable representations.

REFERENCES AND NOTES

1. In the discussion that follows, the implicit assumption is made that we are concerned only with electronic transitions, although, of course, these may range in frequency from the near infrared for low-lying electronic states to the far vacuum ultraviolet.

2. E. Wigner, *Gruppentheorie und ihre Anwendung auf die Quantenmechanik der Atomspecktren*, Friedr. Vieweg & Sohn Akt.-Ges., Braunschweig, 1931; English language ed., J. J. Griffen (Transl.), Academic Press, New York, 1959.

3. Recall the orthogonality requirement of atomic or molecular wavefunctions is $\int \psi_i^* \psi_j \, d\tau$ $= 0$ or 1, respectively, if $i \neq j$ or $i = j$. Also note that strictly speaking if ψ is a complex wavefunction, then $|R|^2 = 1$ implies that $R = e^{\pm i\delta}$ where $i = \sqrt{-1}$. Since we are always dealing here with real wavefunctions the only allowed values of R are $R = \pm 1$.

4. W. Kauzmann, *Quantum Chemistry*, Academic Press, New York, 1957, Chapters 4 and 10; M. Tinkham, *Group Theory and Quantum Mechanics*, McGraw-Hill Book Co., New York, 1964, Chapter 3.

5. One might more properly say "a charge distribution possessing a rotation-reflection axis," so that in zeroth order the same statement can be made with respect to the contribution of any particular chromophoric grouping to the optical activity of an asymmetric molecule.

6. This is done in so many texts on group theory and its applications to quantum mechanics and various aspects of spectroscopy that it would be folly to do it here again. However, readers unfamiliar with group-theoretical operations would do well to go through an example or two.

7. A. Schonflies, *Kristallsystems and Kristallstructur*, B. G. Teubner, Leipzig, Germany, 1891.

8. See, for example, but not exclusively, L. H. Hall, *Group Theory and Symmetry in Chemistry*, McGraw-Hill Book Co., New York, 1969, Chapter 4; R. L. Flurry, Jr., *Molecular Orbital Theories of Bonding in Organic Molecules*, Marcell Dekker, New York and Basel, 1968, Chapter 6.

9. See, for example, H. H. Jaffé and M. Orchin, *Symmetry in Chemistry*, John Wiley & Sons, New York, 1965, or L. H. Hall, *Group Theory and Symmetry in Chemistry*, McGraw-Hill Book Co., New York, 1969; also, compilations of pictorial representations of molecular orbitals are becoming available, for example: W. L. Jorgensen and L. Salem, *The Organic Chemist's Book of Orbitals*, Academic Press, New York, 1973.

10. E. B. Wilson, J. C. Decius, and P. Cross, *Molecular Vibrations*, McGraw-Hill Book Co., New York, 1955.

11. R. S. Mulliken, *J. Chem. Phys.*, **1**, 20 (1939).

12. D. S. Schonland, *Molecular Symmetry*, D. V. Nostrand Co., Princeton, N.J., 1965.

CHAPTER SIX

THE APPLICATION OF GROUP REPRESENTATIONS TO CHIRAL ACTIVITY

6.1 INTRODUCTION

In the previous chapter we examined the question of the relationship between group theory and molecular symmetry and discussed the way in which the symmetry of certain quantum, mechanical parameters (for example the vibrations of the nuclei of a polyatomic molecule or the charge distribution of a molecular orbital) may be designated by group-theoretical *representations* classified in species according to symmetry. Finally, the method of obtaining the symmetry species of parameters that can be combined by the *direct product* of representations was outlined. In this chapter we will apply this to the determination of the symmetry species of a quantum mechanically predictable observable: the rotational strength. This is not the only use of group theory in chirality although the other uses are subtly related to it. The chapter, therefore, will be divided into two parts; in the first part we will examine the application of the *direct product* of symmetry species to the determinations of the allowedness of the chiral activity of an optical transition between electronic energy levels, and in the second part, the application of group theory to questions of geometric rules for chiral activity. Because of our inability to write exact wavefunctions and the difficulty of solving exact Hamiltonians for large molecules, the geometric rules derived from symmetry considerations appear to be widely applicable and useful. In this model the chromophores are initially considered to have planes of symmetry so that the optical activity arises from one or more of the coupling mechanisms described in Chapter 4. The application of symmetry relationships to these mechanisms originates early in the development of one-electron theory. A number of subsequent relationships, more or less empirical, have resulted in the development of symmetry rules, such as the octant rule of Moffitt and his colleagues (1) and various sector rules (see Chapter 7). A generalization, as well as applications to specific systems, has been made by Schellman and his students (2).

Finally, group theory is directly applicable to the chiral descriptions of molecular crystals, and to the chiral analysis of magnetic ORD and CD, but these subjects are left for separate development in Chapters 9 and 10.

6.2 SYMMETRY AND THE ROTATIONAL STRENGTH OF OPTICAL TRANSITIONS

Since the requirement for invariance of the electric dipole transition moment under symmetry operations appropriate to the applicable point group is established as a criterion for a nonvanishing value of the moment and, therefore, for the appearance of an absorption band corresponding to the energy interval between two molecular states, we are entitled to examine how group-theoretical considerations permit the determination of the variance. We have seen that vector properties can have appropriate symmetry representations (Section 5.4). The transition moment is a vector property. The invariance of this property depends on its behavior under the symmetry operations. The character tables show at once that only certain members of the A species, that is, A, A', A_1, A_g, A_{1g}, can represent a basis vector which is totally invariant under the symmetry operations of the applicable point group. Consequently, we may state a symmetry selection rule for absorption: *Absorption for a pure electric dipole transition may take place only if the irreducible representation of the transition moment belongs to or contains (3) one of the totally symmetric representations of the A species*.

Precisely the analogous selection rule may be written for the magnetic transition moment. However, since rotational strength is a scalar, albeit pseudoscalar with sign but not direction (the dot product of the electric and magnetic dipole transition moments), these selection rules must be applied separately to the magnetic and electric dipole transition moment vectors, that is, to μ_{ab} and m_{ab} rather than to R_{ab}. Thus, *for the rotational strength derived from a transition between two molecular states to be nonzero, the symmetry requirement is that both μ_{ab} and m_{ab} belong to the same totally symmetric representation*, that is, to one of the appropriate invariant A symmetry species.

It is here that the concept of *direct product* of representation enters, because if the symmetry species of the *direct product* of the representations can be obtained from the *direct product* of the representations themselves, then when we use the rules in Section 5.6, the symmetry rules (and therefore the variance or invariance) of the electric and magnetic dipole transition moments can be determined at once from the symmetry species of the representations of the quantities which define transition moments. These quantities are the wavefunctions, and the respective moment operators for the electric dipole transition moment and the magnetic dipole transition moment.

Let us start backward to examine this in slightly greater detail. We will assume that we know the symmetry representations for these parameters for two given states of a molecule that belongs to the C_{4v} symmetry group. This example will suggest a generalization that will then be established by returning to group-theoretical considerations of the *direct product* of characters of representations.

Let the states $\langle a|$ and $\langle b|$ of the C_{4v} molecule have A_1 and B_1 symmetry, respectively, and the electric and magnetic dipole moment operators have A_1 and E symmetry, respectively (refer to the C_{4v} character table in Appendix I). Then the symmetry species of the moments $\mu_{ab} = \mu(\psi_{A_1} \to \psi_{B_1})$ and $m_{ba} = m(\psi_{B_1} \to \psi_{A_1})$ or of their representations are obtainable as follows: By definition,

$$\mu(\psi_{A_1} \to \psi_{B_1}) \equiv \langle a|\mu|b\rangle \qquad (6.200)$$

$$m(\psi_{B_1} \to \psi_{A_1}) \equiv \langle b|m|a\rangle \qquad (6.201)$$

where $\langle a| = \psi_{A_1}$ and $\langle b| = \psi_{B_1}$. The representations Γ_i, for example, $\Gamma\mu(\psi_{A_1} \to \psi_{B_1})$, then have the symmetry species given by the *direct products*: For

$$\Gamma\mu(\psi_{A_1} \to \psi_{B_1}), \qquad A_1 \times A_1 \times B_1 = B_1, \qquad (6.202)$$

and for

$$\Gamma m(\psi_{B_1} \to \psi_{A_1}), \qquad B_1 \times E \times A_1 = E. \qquad (6.203)$$

Neither the species B_1 nor E is invariant to all the operations of the C_{4v} groups; consequently both transition moments are null, and the predicted value of the rotational strength is zero. In fact, in this case, since the symmetry species of the electric dipole transition moment is B_1, the optical transition from the state a to the state b is forbidden, and no achiral absorption band, corresponding to such a transition, should appear either. It is instructive to ask whether there is any state reachable by optical absorption from a state of symmetry A_1 in a C_{4v} molecule and whether such a state could give rise to optical activity. The answer, as we shall see, is given directly in most character tables where *direct product* considerations have already been applied; but for the moment, let us remain sufficiently naive to go through the *direct product* calculation. We wish to see (1) whether there is a state $\langle b|$ for which μ_{ab} is invariant, that is, whether μ_{ab} belongs to the A_1 symmetry species if state $\langle a|$ belongs to the A_1 species, and (2) if there is such a state, whether m_{ab} also belongs to the A_1 species so that the rotational strength for the transition from $\langle a|$ to $\langle b|$, $R_{ab} = \langle a|\mu|b\rangle \cdot \langle b|m|a\rangle$, has a finite value greater than zero.

It is obvious that if the operator μ has an A_1 representation, then by the *direct product* rules, in order for μ_{ab} to belong to A_1, the state b must also belong to A_1. Thus the *direct product* for $\Gamma\mu_{ab}$

$$A_1 \times A_1 \times A_1 = A_1, \qquad (6.204)$$

and optical absorption will be allowed between the states a and b because $\Gamma\mu_{ab}$ belongs to the totally symmetric representation A_1. Consider, however, the magnetic transition moment $\langle b|m|a\rangle$, for which the direct product of the irreducible representation, Γm_{ba}, is

$$A_1 \times E \times A_1 = E. \qquad (6.205)$$

Thus Γm_{ba} does not belong to the totally symmetric species and is, therefore, null. This particular optical transition will consequently appear in ordinary absorption but will be optically inactive.

Although our initial attempt did not produce an optically active transition, we can probe still further. Indeed it is not difficult to see that the *direct prodcts* for both μ and m would belong to the totally symmetric species A_1 if the upper state were an E state and the symmetry species of both moment operators were also E. That is, for $\Gamma\mu_{ab}$,

$$A_1 \times E \times E = A_1, \qquad (6.206)$$

and for Γm_{ba},

$$E \times E \times A_1 = A_1. \qquad (6.207)$$

In this case both transition moment operators belong to the A_1 species and are thus invariant. An optical transition between these states would be allowed in absorption and should be chirally active. However, we will soon see that the chiral activity will not be observable without further perturbation.

It will be apparent by now that this discussion contains the seeds of a generalization. In Section 5, we noted that the *direct product of the characters of a group representation is the character of the direct product of the representation.* Consider then a *triple direct product* of the type shown in Eqs. 6.204–6.207 which are necessary for transition moment representations. Let the characters be designated by $X_i(R)$ for each representation i and symmetry element (R). Then the *triple product* for the characters summed over the symmetry elements, say for example, an electric dipole transition moment is given by

$$\sum_R X_a(R) X_\mu(R) X_b(R),$$

where to avoid confusion from an excess number of symbols, the subscripts a and b are used here and in what follows to designate the states $\langle a|$ and $\langle b|$. The direct product of the representations of the moment μ and the state $\langle b|$ is

$$\sum_R X_\mu(R) X_b(R) = \sum_R X_{\mu b}(R). \tag{6.208}$$

Then the *triple direct product* may be expressed as

$$\sum_R X_a(R) X_\mu(R) X_b(R) = \sum_R X_a(R) X_{\mu b}(R). \tag{6.209}$$

The characters obey the same orthogonality relationships as the representation (Section 5.5), so that

$$\frac{1}{h} \sum_R X_a(R) X_{\mu b}(R) = \delta_{a,\mu b} = \begin{cases} 1, & X_a(R) = X_{\mu b}(R) \\ 0, & X_a(R) \neq X_{\mu b}(R) \end{cases}. \tag{6.210}$$

That is, the *direct product* of the characters is zero *unless* $X_a(R)$ and $X_{\mu b}(R)$ belong to the same symmetry element, in which case it is unity and X_a and $X_{\mu b}$ are characters of the same representation, or of equivalent representations, if the representations are reducible rather than irreducible. It follows that in order for the representation of the dipole transition moment to belong to the totally symmetric species, the species representation of μ and $\langle b|$ must be the same. The generalization of this requirement may be stated in two corollary forms:

1. If the symmetry species of a quantum mechanical observable is given as the *direct product* of the symmetry representation of three vector properties of the system, and if one of the properties (say, $\langle a|$ in this case) is antisymmetric with respect to any symmetry operation of the group, the

representation for *one* but *not both* of the other of the two remaining vector properties must be antisymmetric with respect to the same operations.

2. As in 1, if, however, one of the vector properties is symmetric with respect to all the operations of the group, then the representations of *both* the other properties must also be totally symmetric or *both* must be antisymmetric with respect to the same operations.

With the use of these generalizations, it becomes a trivial matter to determine the allowedness of an optical transition or of chiral activity once the symmetry of the states is known for a molecule or chromophore. It is particularly trivial if the ground state is totally symmetric, as is the case of most organic molecules (but not the case of most inorganic complexes). One need only look at the character tables in Appendix I. The allowed components that is, the polarization directions of both the electric dipole transition and magnetic dipole transition moments, have all been worked out for the 32 crystallographic and 14 additional point groups, and tabulated along with the character tables. These components usually appear, as they do in this volume, as the coordinates x, y, and z for electric dipole components or as the symbols R_x, R_y, R_z for the magnetic dipole components. Thus, returning to the C_{4v} example:

C_{4v}	I	$2C_4$	C_2	2	2	
A_1	1	1	1	1	1	z
A_2	1	1	1	-1	-1	R_z
B_1	1	-1	1	1	-1	
B_2	1	-1	1	-1	1	
E	2	0	-2	0	0	(x,y) (R_x,R_y)

R_i designates the symmetry of those states to which magnetic dipole transitions are allowed, and the Cartesian coordinates (x,y,z) represent the same for electric dipole transition moments. Therefore in this case, ordinary absorption is allowed from a ground A_1 state of A_1 and E symmetry, as determined earlier in this section. From the fact that magnetic transitions are also allowed to states of E symmetry, one might conclude as we did earlier that chirally active transitions of E symmetry are allowed. However, the enclosure of x, y and R_x, R_y separately in brackets symbolizes that in the degenerate E state, one cannot distinguish between the x and y components; so that, in fact, while the transition will be chirally active, the activity will be of the same magnitude and opposite sign for each of two contributions to the degenerate transition. There will thus be no *observable* optical activity from these transitions *unless* the degeneracy is lifted by a perturbing field, in which case chiral activity of equal and opposite sign will be generated but at different wavelengths. As we shall see in Chapter 10, it is just such transitions that give rise to two transitions of equal and opposite sign in magnetically induced CD or ORD (MCD or MORD).

Examination of the character tables shows the symmetry groups that represent molecules or chromophores that are (1) Optically active in the absence of perturbations: $C_1, C_2, C_3, C_4, C_5, C_6, C_7, C_8, D_2, D_3, D_4, D_5, D_6$; (2) optically active only if degenerate transitions are split: $C_{3v}, C_{4v}, C_{5v}, C_{6v}, D_{2d}, S_4, S_8, S_{12}, T, O, C_{\infty v}$; and (3) optically inactive: $C_s, C_i, C_{2v}, C_{2h}, C_{3h}, C_{4h}, C_{5h}, C_{6h}, D_{2h}, D_{3h}, D_{4h}, D_{5h}, D_{6h}, D_{3d}, D_{4d}, D_{5d}, D_{6d}, S_6, S_{10}, T_d, O_h, D_{\infty h}$. Not *all* transitions of molecules with chromophoric systems belonging to any of the groups of set (1) are necessarily optically active. Only for systems with C_1, C_2, and D_2 symmetry do *all* transitions give rise to chiral activity, that is, all transitions from a totally symmetric ground state. For systems of $C_n (n = 3$ to $8)$ and $D_n (n = 3$ to $6)$, symmetry transitions to states of A and A_2 give rise to natural optical activity. Transitions to degenerate states of E or E_1 symmetry will be active only in the presence of a sufficiently perturbing field, as in the case for all transitions in set 2.

The character tables reveal that with one exception, all of the completely optically inactive groups of set 3 have either a center of inversion or both a reflection plane *and* a twofold axis of symmetry, the sole exception being the C_s group, whose only symmetry element is a reflection plane. Molecules whose mirror images are superimposable on each other characteristically have these elements of symmetry. This, then, is the relationship between the optically active point groups and Pasteur's requirement for optical activity, because these groups that belong to set 1 do not contain symmetry elements characteristic of molecules whose mirror images are superimposable.

The groups of set 2 are, of course, intermediate between the other two classes. In the absence of a perturbing field sufficient to destroy the energy degeneracy of their E or T states, they are optically inactive for the reason discussed earlier. However, they offer a clue to the most general statement one can make relating symmetry elements to optical activity. For example, consider the spiropyrollidine molecule, illustrated in Fig. 6.00, which has neither a symmetry plane nor a center of inversion and yet is optically inactive (4). The symmetry elements appropriate to this molecule are a C_2 axis, which bisects both CNC angles, and a fourfold improper rotation axis S_4, which consists of 90-degree rotation about the C_2 axis, followed by a reflection in a plane in which the C_2 axis lies. Since rotation by 270 degrees followed by a reflection perpendicular to the C_2 axis will also produce a molecule indistinguishable from the starting molecule, the S_4 rotation axis

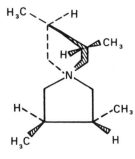

Fig. 6.00 An optically inactive molecule with neither a center of inversion nor a symmetry plane: Its achirality results from the presence of elements of S_4 symmetry which superimpose the molecule on itself by 90- or 270-degree rotation about its C_2 axis, followed by reflection in a plane containing that axis.

is also an appropriate element of symmetry. Containing only these symmetry elements and, of course, I, this molecule belongs to the S_4 point group, which is a member of set 2 and therefore optically inactive, unless the degeneracy of its E states are lifted by an external electric or magnetic field.

Set 1 may thus be set aside from sets 2 and 3 by a more general statement: If S_n is the symbol for an n-fold improper rotation axis for which $n = 1$ is a simple reflection σ and $n = 2$ an inversion i through a center (recall from Chapter 5 that inversion symmetry can always be accomplished by a twofold rotation followed by a reflection in a plane perpendicular to the twofold axis), then the requirement for *natural* chiral activity is that the molecule does not have an improper rotation symmetry element S_n of any order. By this statement, we restrict the definition of *natural* chiral activity to the activity present, even in the presence of a perturbing external field (other than that of the electric dipole radiation being used to measure it).

6.3 CARTESIAN COMPONENTS OF OPTICAL ACTIVITY

Thus far we have ignored the implications of the fact that the activity is developed along different molecule-fixed axes for different states. To a large extent this inattention has been appropriate, since most measurements of optical activity are made in solution where the disorientation of molecules and the consequent inability to distinguish the components is of no importance. However, the polarization of transitions is an important diagnostic tool in assigning absorption bands, in following conformational or configuration changes, and in determination of the structure of macromolecules. There are at least two significant experimental methods for establishing this polarization: linear dichroism of molecules oriented in crystal or in solution by an external field, and chiroptical measurements on oriented systems. The latter method has not been extensively developed, although some theoretical treatments and experimental work has been reported (5–8). There is at least one treatment by Disch and Sverdlich (9, 10) of the experimental problem arising from attempts to measure chiral activity on oriented systems. Such systems of optically active molecules also exhibit linear birefringence and linear dichroism—effects that can interfere severely with attempts to measure circular dichroism or circular birefringence. Experimental problems of this type could be avoided with measurements made with radiation propagating in the direction of the uniaxis of a uniaxial system, such as along the unique axis of a monoclinic crystal or parallel to the orienting field in a solution. A more complete discussion of the effects of orienting fields and chiral activity of crystals is given in Chapter 9. In Chapter 4 we saw how the application of an arbitrary asymmetric potential in the one-electron treatment leads to the development of chiral activity which is dependent on the absolute geometric relationships or configuration of the sources of the asymmetric potential. This same development produces a chiral dependence of the sign of the activity on the direction of propagation of the radiation with respect to the molecular Cartesian axes and, therefore, on the direction of propagation of the radiation if the molecules in the system are oriented. We noted that the following equations,

$$\tan 2\theta_{xy} = \frac{Az'}{(k_1 - k_2)/2}, \tag{6.300}$$

$$\tan 2\theta_{yz} = \frac{Ax'}{(k_2 - k_3)/2}, \tag{6.301}$$

$$\tan 2\theta_{zx} = \frac{Ay'}{(k_3 - k_1)/2}, \tag{6.302}$$

describe the twist angles θ_{xy}, θ_{yz}, and θ_{zx}, with respect to the magnitudes k_1, k_2, k_3 of the potentials along the Cartesian coordinates and the locations x', y', z' of the sources of the potential. This establishes for the one-electron mechanism, at least, the sign of the activity (and incidentally, the allowedness because, of course, x', y', z' may have zero values), since the sign of the coordinates x', y', or z' are dependent on the location relative to the origin of the axes of the perturbing atom or atoms. In addition, it fixes the fact that if the radiation propagates along an axis perpendicular to one of the planes (xy, yz, or zx), as in an oriented system, no contribution from asymmetry along the axis of propagation will contribute to the measured activity. Since we have already seen that the sign of one of these three contributions must be opposite to that of the other two, it is at once apparent that this mechanism provides for the possibility of measuring chiral activity of *different sign* along different axes of a system of molecules of a given absolute configuration. For an oriented system, therefore, the sign of the activity has meaning only if the relative orientation of the molecule's (and crystal's, if appropriate) axes and the direction of light propagation are specified.

We will see in Chapter 7, where the dissymmetric chromophore model is applied to diene-containing systems, that in a more direct quantum mechanical calculation, the *sign* of the Cartesian components of a chiral system are similarly determined by the coordinates of the atoms in the system and give rise, as in the one-electron treatment, to components of different sign along different axes. This results from the nature of the vector cross product, which appears in the magnetic moment operator $\mathbf{r} \times \mathbf{P}$ or $\mathbf{r} \times \nabla$. The phenomenon has only recently been observed in a system of noninteracting small molecules (11). We will see in Chapter 9 that the phenomenon has also been observed in oriented polymers. In a crystal, molecular interactions may produce still more complex effects. We will ignore the sign and examine only the existence of symmetry selection rules that define the Cartesian directions along which chiral activity may develop. This is a straightforward corollary of the analysis of Section 6.2 of the states to which chirally active transitions may take place. In this case we choose the chirally active state and ask along which of the Cartesian components in the axes fixed in the molecule does the activity develop. For example, we noted in the C_{4v} group that transitions to the degenerate E states are chirally active and although not observable in the absence of a perturbing field, could be observed in the presence of one. By referring to C_{4v} character tables again, we find that this activity would develop for light polarized along the x and y axes but not along the z axis.

6.4 SYMMETRY AND ASYMMETRY

Except for the section on Cartesian components of optical activity, the first part of this chapter applied principally to the activity of chromophores that are intrinsically dissymmetric. We have agreed to accept the fact that despite the real dissymmetry of all optically active chromophores, for practical reasons we must frequently accept first-order descriptions of many chromophoric systems as being asymmetrically perturbed systems of elements of planar or centrosymmetry; we can therefore proceed to investigate how group-theoretical considerations can be incorporated into the descriptions of the perturbing asymmetry. This Schellman has done elegantly, and the only excuse for paraphrasing his work here is to provide a background in the framework of the presentation of this volume.

The molecular model of chiral activity to which this approach applies is that of a statically perturbed chromophore, which basically is the model proposed by Condon, Altar, and Eyring (12) and discussed as the "one-electron theory" in Chapter 4. The rotatory strength arises from an asymmetric perturbation of the electronic wavefunctions which in zeroth order have nodal planes or centers of symmetry. Where the resulting rotatory strength for a single transition is finite, it is given by

$$R_{\beta O} = -i\langle O|\boldsymbol{\mu}|\beta\rangle\cdot\langle\beta|\mathbf{m}|O\rangle \tag{6.400}$$

where the perturbed wavefunctions $|\beta\rangle$ and $|O\rangle$ are given to first order by

$$|\beta\rangle = |\beta) - \sum_{\rho}\frac{(\rho|V|\beta)|\rho)}{(\mathcal{E}_{\rho}-\mathcal{E}_{\beta})} \tag{6.401}$$

$$|O\rangle = |O) - \sum_{\sigma}\frac{(\sigma|V|O)|\sigma)}{(\mathcal{E}_{\sigma}-\mathcal{E}_{O})} \tag{6.402}$$

and $|\beta)$ and $|O)$ are the unperturbed wavefunctions whose symmetry would forbid any contribution to the rotatory strength. Since this treatment accounts for the chiral activity induced in otherwise achiral chromophores, we can assume that it will not be an important factor in transitions that are both magnetically and electrically allowed. Therefore, we can treat two cases, that of electrically forbidden, magnetically allowed transitions, and that of electrically allowed, magnetically forbidden transitions. If we take the latter case first, the perturbed functions in the electric dipole transition moment of the rotational strength can be approximated by the allowed functions:

$$\langle O|\boldsymbol{\mu}|\beta\rangle = (O|\boldsymbol{\mu}|\beta). \tag{6.403}$$

The basis for this approximation is that the first-order correction terms to the electric dipole transition moment would be much smaller than those of the allowed, and therefore, intrinsically large, moment. A similar situation would hold for the magnetically allowed transition. The rotatory strengths induced by the perturbing potential V of the surrounding asymmetric environment for the electric dipole and magnetic dipole allowed transitions, respectively, are given by

$$R_{\beta O} = i(O|\mu|\beta) \cdot \left[\sum_\alpha \frac{(\alpha|\mathbf{m}|O)(\beta|V|\alpha)}{(\mathscr{E}_\alpha - \mathscr{E}_\beta)} \right.$$

$$\left. + \sum_\rho \frac{(\beta|\mathbf{m}|\rho)(\rho|V|O)}{(\mathscr{E}_\rho - \mathscr{E}_O)} \right], \qquad (6.404)$$

$$R_{\alpha O} = i \left[\sum_\beta \frac{(O|\mu|\beta)(\beta|V|\alpha)}{(\mathscr{E}_b - \mathscr{E}_\alpha)} \right.$$

$$\left. + \sum_\rho \frac{(\rho|\mu|\alpha)(\alpha|V|O)}{(\mathscr{E}_\rho - \mathscr{E}_O)} \right] \cdot (\alpha|\mathbf{m}|O) \qquad (6.405)$$

In these expressions, the distinction between the $|\alpha)$ and $|\beta)$ functions are that the former represent states that have nonvanishing magnetic transition moments and the latter correspondingly have allowed electric dipole transition moments. A glance at the character tables will show at once that in groups containing reflections or inversions, these functions are segregated from each other. C_{2h} is a typical example:

C_{2h}	I	C_2	i	σ		
A_g	1	1	1	1		R_z
B_g	1	-1	1	-1		R_x, R_y
A_u	1	1	-1	-1	z	
B_u	1	-1	-1	1	x,y	

Here, the $|\alpha)$ functions belong to A_g and B_g representations, while the $|\beta)$ functions belong to A_u and B_u. Consider the consequences of this fact for the rotational strength as given by Eq. 6.404 and 6.405; the second integral of the first term is $(\beta|V|\alpha)$. Since $|\alpha)$ must belong to A_g or B_g for the integral to be finite in the z direction, $|\beta)$ must belong to A_u. In order for the integral to be finite, it must be totally symmetric; for the z component, V must therefore belong to A_u, since $(\beta|V|\alpha)_z$ is $A_u \times A_u \times A_g = A_g$. A similar situation holds for the x and y components where $|\alpha)$ belongs to B_g and $|\beta)$ belongs to B_u: $(\beta|V|\alpha)_{x,y}$ is $B_u \times A_u \times B_g = A_g$. Any other representation than A_u for V would give direct products other than A_g, and thus zero values for this integral and for the rotational strength. A_u is antisymmetric (representation is -1) with respect to inversion and reflection. So the perturbation V must be one that is antisymmetric with respect to a vector coordinate relative to the origin of the axis system; otherwise the first term in both Eqs. 6.404 and 6.405 will be zero. That is, it must belong to a pseudoscalar representation, one that changes sign under the symmetry operations of reflection or inversion.

Since the second terms of Eqs. 6.404 and 6.405 also contribute to the rotational strength, we may inquire whether these also require that V have a pseudoscalar representation, for if it does, then we have a very strong selection rule for the asymmetry conditions that will induce optical activity in a symmetric chromophore by a static perturbation. And, indeed, a very similar argument can be made for the second terms. This can be shown in several ways. For example, the wavefunctions are orthonormal: $|\rho) \cdot (\rho| = 1$; consequently in Eq. 6.404, $(\beta|\mathbf{m}|\rho) \cdot (\rho|V|O) = (\beta|\mathbf{m}|V|O)$. If we are treating totally symmetric ground states, as we usually are (the argument may be further generalized), in this case A_g, the representation A_g for $|O)$ requires that V have the same representation as $(\beta|\mathbf{m})$. (Recall the rules that $A \times A = A, B \times B = A, g \times g = g, u \times u = g$.) At this point in the argument, we can proceed in one of two ways: from the general properties of vectors or from the specific requirements of Eq. 6.404 for an electronically allowed transition. (The electronic transition moment $(O|\boldsymbol{\mu}|\beta)$ must be finite for $R_{\beta O} > 0$). Taking the latter way first, the symmetry representation of $|\beta)$ has to be either A_u or B_u to be finite respectively, for transitions in the z and x, y directions. The magnetic moment operator $\mathbf{m}$ must have the corresponding rotational representation, A_g and B_g, for the z and x, y components. Therefore the representations for $(\beta|\mathbf{m}|$ can only be $A_u \times A_g = A_u$ or $B_u \times B_g = A_u$, which requires that the representation for V be A_u in order that the representation for $(\beta|\mathbf{m}|V|O)$ be totally symmetric—and the same result in the other way, from the general properties of vectors, which require that $(\beta|\mathbf{m})$ for the same Cartesian component be a pseudoscalar; so the V must also be a pseudoscalar and therefore change sign on inversion and reflection through a symmetry plane. The pseudoscalar representation in the C_{2h} group is A_u.

Summing up we are left with the selection rule: *The symmetry requirement for the static perturbation which induces chiral activity in a symmetric chromophore is to be antisymmetric with respect to inversion or reflection along a vector coordinate in the symmetric chromophore.* From this, the form of the perturbing potential required in cases of different symmetry may be generated. We will see in Table 6.1 that the simplest example of a perturbing potential required to induce optical activity in a chromophore of C_{2h} symmetry is a dipole pseudoscalar function, which is a function antisymmetric with respect to a single vector coordinate and thus changes sign under reflection or inversion.

In a real molecule, the perturbing asymmetric potential of the atoms surrounding a symmetric chromophore will not usually belong to any representation of the symmetry group of the chromophore. In order to apply symmetry considerations to this problem, which is the whole purpose of this treatment, the potential must be expressible in terms that can be represented by the symmetry group appropriate to the achiral chromophore. Wigner (13) has shown that it is possible to do this by expressing the potential as a sum of potential functions V_i^v: $V = \sum_{v,i} V_i^v$, which are functions belonging to the ith row of the vth representation of the chromophore symmetry group.

The particular example given above can be generalized by noting that in order for the Cartesian components $(\beta_i|V|\alpha_i)$, $i = x,y,z$, of the perturbation coefficients to be finite in Eqs. 6.403 and 6.404, V must belong to the same representation as that obtained from direct product of the representations for $\beta_i\alpha_i$ and therefore to the same representations as $\mathbf{m}_i \cdot \boldsymbol{\mu}_i$. $\mathbf{m}$ is an axial vector and $\boldsymbol{\mu}$ is a polar vector. The product of axial and polar vectors are pseudoscalars; that is, they have sign but not direction. The symmetry representation of V must therefore be a pseudoscalar representation (or in the case of degenerate representations "contain" a part that is pseudoscalar). Thus the representation of V must change sign under reflection or inversion, as we found in the particular case of C_{2h}. These arguments also apply to the other terms in Eqs. 6.404 and 6.405.

This conclusion applies to all electrically allowed, magnetically forbidden transitions of chromophores located in an achiral framework. *It makes no assumption about the nature of the states involved or about the origin of the perturbing potential other than its symmetry.* The forms of the pseudoscalar functions appropriate to the more common symmetry groups of optically inactive chromophores have been worked out by Schellman and are presented in Table 6.1. In addition to the general form of the functions, the simplest form, which in fact usually plays a predominant role in the induced optical activity, is given in the table. The general functions are expressed in terms of cylindrical coordinates, ρ, z, ϕ, and the simplest forms in terms of Cartesian coordinates. Note that the simplest functions progressively become higher order and that for chromophores with axial or higher symmetry, $C_{\infty v}$ and $D_{\infty h}$, no suitable functions of the electronic coordinates could be found. For chromophores with this symmetry, some mechanism other than a static perturbing field must be found in order to induce optical activity associated with their electronic transitions.

The schematic representations taken from Schellman's 1966 paper of the pseudoscalar functions for some of the more common point groups are given in Fig. 6.01. The planes separating the light and dark regions in these representations are nodal planes. One may think of the chromophore as

Table 6.1 Pseudoscalar functions for the common point groups

Group	Representation	General Form Pseudoscalar	Simplest Example
C_s	A''	$A^0(z)$	Z
C_i	A_u	Function odd in x, y, or $z\, A^0(x), A^0(y)$, or $A^0(z)$	X, Y, or Z
C_{2v}	A_2	$\sum_n A_n(z)\sin 2n\varphi$	XY
C_{2h}	A_u	$\sum_{n=0,1,} A_n^0(z)(\sin 2n\varphi + \cos 2n\varphi)$	Z
D_{2h}	A_u	$\sum_{n=1,2,} A_n^0(z)(\sin 2n\varphi)$	XYZ
D_{2d}	B_1	$\sum_{2,4,} A_n^0(z)\sin 2n\varphi + \sum_{1,3,5,} A_n^6(z)\cos 2n\varphi$	$(X^2 - Y^2)$
C_{3h}	A''	$\sum_{n=0,1,} A_n^0(z)(\sin 3n\varphi + \cos 3n\varphi)$	Z
C_{3v}	A_2	$\sum_{n=1,2,} A_n(z)\sin 3n\varphi$	$Y(3X^2 - Y^2)$
D_{3h}	A_1''	$\sum_{n=1,2,} A_n^0(z)\sin 3n\varphi$	$YZ(3X^2 - Y^2)$
C_{4v}	A_2	$\sum_n A_n(z)\sin 4n\varphi$	$XY(X^2 - Y^2)$
C_{4h}	A_u	$\sum_{n=0,1,} A_n^0(z)(\sin 4n\varphi + \cos 4n\varphi)$	Z

D_{4h}	A_{1u}	$\sum_n A_n^0(z)\sin 4n\varphi$	$XYZ(X^2-Y^2)$
S_4	B	$\sum_{1,3,} A_n^0(z)(\sin 2n\varphi + \cos 2n\varphi)$ $+ \sum_{0,2,4,} A_n^0(z)(\sin 2n\varphi + \cos 2n\varphi)$	Z, XY, X^2-Y^2
D_{5h}	A_1''	$\sum_n A_n^0(z)\sin 5n\varphi$	$YZ(Y^4-10X^2Y^2+5X^4)$
C_{5v}	A_2	$\sum_n A_n(z)\sin 5n\varphi$	$Y(Y^4-10X^2Y^2+5X^4)$
D_{6h}	A_{1u}	$\sum_n A_n^0(z)\sin 6n\varphi$	$XYZ(3X^2-Y^2)(3Y^2-X^2)$
T_d	A_2		$(X^2-Y^2)(Y^2-Z^2)(Z^2-X^2)$ $= -X^4(Y^2-Z^2)+Y^4(Z^2-X^2)$ $+Z^4(X^2-Y^2)$
O_h	A_{1u}		$XYZ[X^4(Y^2-Z^2)+Y^4(Z^2-X^2)$ $+Z^4(X^2-Y^2)]$
$C_{\infty v}$	A_2, Σ^-	Magnetic interaction	
$D_{\infty h}$	Σ_u^-	Magnetic interaction with inhomogeneous field, $\dfrac{\partial H_z}{\partial z}\neq 0$	

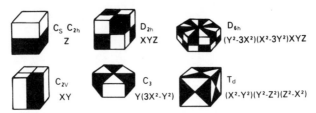

Fig. 6.01 Schematic representation of pseudoscalar potential functions as models for the asymmetric charge distributions that can induce optical activity in the transitions of a symmetric chromophore lying at the origin. (From J. A. Schellman (2)).

located at the origin of the coordinates of these representations. Then a perturbing atom lying in any one of the regions will induce optical activity opposite in sign to that produced by an atom in a corresponding position in a region of different shade. Atoms lying on one of the nodal planes have no influence on the optical activity because the value of the appropriate coordinate of the simplest function in Table 6.1 is zero in these planes.

If the perturbing potential is assumed to be Coulombic (and it need not be), then the simple pseudoscalar functions are given a simple interpretation in terms of a multipole expansion. Thus, a single coordinate, as in the C_s and C_{2h} groups, represent dipole perturbers; the product of two coordinates, as in C_{2v}, represent quadrupole perturbers; the product of three coordinates represent octopole perturbers, etc. We note that as the degree of the perturbing potential increases, the more likely are the perturbing atom or atoms to be near a nodal plane of electron density and the smaller is the contribution of the perturber to the optical activity. Thus as the symmetry of the chromophore increases, the smaller is the optical activity likely to be induced by the static first-order effect. There are other mechanisms, as we have noted, for inducing optical activity, such as Kirkwood-Kuhn dipole-dipole coupling, but the symmetry considerations we have been discussing here do not apply to these mechanisms. In many cases more than one such mechanism can be operative. In such cases, the one-electron perturbation whose symmetry characteristics we have been discussing becomes of decreasing significance as the chromophore symmetry increases. It is instructive to examine the correspondence between this approach through symmetry and the ad hoc approach to the problem by Condon in the early explorations of the one-electron theory. Recall that in Chapter 4, in discussing the one-electron theory, it was pointed out that the introduction of a term proportional to xyz in the description of a charge distribution, which could be represented by a flattened ellipsoid, would suffice to induce optical activity. A flattened ellipsoid, for example, the ethylene molecule for which the charge distribution of its symmetry can be represented by such an ellipsoid, has three reflection planes, three twofold axes of symmetry, and a center of inversion. Examination of the character tables of Appendix I reveals that these symmetry operations belong to the group D_{2h}, an optically inactive point group, of course. If we now go to Table 6.1, we see that the simplest pseudoscalar potential that will induce optical activity in a D_{2h} molecule is xyz, just the potential that Condon, Altar, and Eyring used. Even a single atom bonded to an olefinic double bond, out of the nodal plane of the π electrons, would provide such a potential; we will see examples in the next chapter.

6.5 SUMMARY

The application of group-theoretical considerations to the symmetry of the charge distributions in different molecular states is of two types. In one, particularly applicable to dissymmetric chromophores, the direct product of the characters of the representation of the symmetry properties of the two states and the electric and magnetic dipole moment operators connecting them leads to selection rules for isotropic achiral optical absorption and for chiral activity. In molecules or chromophoric groups with any element of symmetry other than just the identity these rules are distinct along different Cartesian axes related to the symmetry elements. As a consequence, the orientation of the molecule or chromophore with respect to the polarization of light is an important consideration in the appearance of the transition in the absorption spectrum, and in the sign of the chiral activity of oriented molecules.

In the other application of group-theoretical considerations, the symmetry of the potential produced by asymmetrically disposed atoms on otherwise symmetric chromophores is shown to be related not only to the appearance of chiral activity and its relative sign, but also, in fact, to its magnitude, since the higher the symmetry the more likely is a perturbing atom to lie near a nodal plane of the electronic distribution and thus the *less* likely it is to contribute strongly to the induced optical activity.

REFERENCES AND NOTES

1. W. Moffitt, R. B. Woodward, A. Moscowitz, W. Klyne, and C. Djerassi, *J. Am. Chem. Soc.*, **83**, 4013 (1961).
2. J. A. Schellman, *J. Chem. Phys.*, **44**, 55 (1966); V. Madison and J. Schellman, *Biopolym.*, **9**, 569 (1970); T. M. Hooker, Jr., and J. Schellman, *Biopolym.*, **9**, 1319 (1970).

3. The modifier "or contains" is required to include the case where a degenerate species that is formed from the direct product of two or more degenerate species contains as subspecies both degenerate (E or T) and nondegenerate (A or B) species.

4. H. H. Jaffe and M. Orchin, *Symmetry in Chemistry*, John Wiley & Sons, 1965, p. 23.

5. I. Tinoco, Jr., and W. G. Hammerle, *J. Phys. Chem.*, **60**, 1619 (1956).

6. I. Tinoco, Jr., *J. Am. Chem. Soc.*, **81**, 1540 (1959).

7. N. Gō, *J. Chem. Phys.*, **43**, 1275 (1965).

8. N. Gō, *J. Phys. Soc. Jpn.*, **23**, 88 (1967).

9. R. L. Disch and D. I. Sverdlich, *J. Chem. Phys.*, **47**, 2137 (1967).

10. R. L. Disch and D. I. Sverdlich, *Anal. Chem.*, **41**, 82 (1969).

11. In text.

12. E. U. Condon, W. Altar, and H. Eyring, *J. Chem. Phys.*, **5**, 753 (1937).

13. E. Wigner, *Gruppentheorie und ihre Anwendung auf die Quantunmechanik der Atomspecktren*, Friedr. Vieweg & Sohn Akt. -Ges., Braunschweig, 1931; English language ed., J. J. Griffen (Transl.), Academic Press, New York, 1959.

CHAPTER SEVEN

CHROMOPHORIC SYSTEMS

7.1 CHARACTERIZING A CHROMOPHORE

Each transition between two states of a molecule induced by electromagnetic radiation of any wavelength constitutes a chromophoric transition. The particles, electrons, or nuclei in a given molecule that are principally involved in the transition constitute the chromophore as demonstrated experimentally or theoretically by their contribution to the absorption, emission, or rotation of the polarization of the radiation. By this definition we broaden the meaning of the word chromophore from its more usual characterization as an atomic grouping responsible for the absorption of visible or near ultraviolet light. Nevertheless, in most cases what we shall be discussing are, in fact, chromophores that fit the more restricted definition, with only slight modification. The use of the word chromophore (from the Greek *chroma* and *phoros*, color bearer) to characterize a particular atomic group in a molecule is a very convenient fiction (the same fiction, in fact, that distinguishes dissymmetric from asymmetric chiral systems) because very productive approximations lead from it. This is especially true in achiral absorption and emission, which are somewhat less sensitive to the details of other groups in the molecule and their specific structural arrangements. We will see splendid examples of the utility of this approximation *and of its limitations* in the discussions of the circular dichroism of dienes and peptides. Even achiral absorption is by no means, of course, insensitive to the presence or absence of different bonding arrangements or substituents adjacent to any particular chromophore—the well-known Woodward rules (1) representing an important example of such effects. These, in fact, make it clear that even in the limit of the grossest approximation, which defines a chromophore as a small group of atoms, the surrounding atoms must contribute to the chromophore, as, for example, the two carbons of an olefinic bond with respect to the strong absorption band in the 180–210 mm region of the spectrum of olefinic compounds. Furthermore, the same atoms may be only part of still another chromophore. The sigma bonds of the olefinic carbon atoms undoubtedly form part of the chromophore of the ground state to σ^* ($N \to \sigma^*$) valence transition of ethylene and its olefinic analogs, which occurs in the vacuum ultraviolet region of the spectrum, and of the nonvalence Rydberg transitions in these compounds (2).

In this chapter some of the chromophoric systems that have been examined for their chiroptical properties are surveyed. We will be engaged principally in seeing how the theoretical ideas treated in the previous chapters apply to the problem of interpreting the rotational strength, or the lack of it, in particular electronic transitions of these chromophores. Where applicable, we will examine any or all of the three basic approaches we have been studying: the intrinsic chiral strength of a dissymmetric chromophore, the induced chiral strength in an "asymmetric" chromophore, and the symmetry characterization of the chromophore. Some chromophores have received much more attention than others, either because of the range of compounds in which they occur or because of their intrinsic chemical or biological interest. To some extent this will be reflected in the attention given to them here. The divisions into which the chromophoric systems are classified are somewhat arbitrary. Separate consideration is given to conjugated and nonconjugated chromophores containing carbon, oxygen, nitrogen, and sulfur atoms. Some references to chromophoric systems containing these atoms, but not discussed here, are included in the summary.

7.2   NONCONJUGATED CHROMOPHORES

As a class, the nonconjugated chromophores have little in common that may be considered unique. Largely they consist of atoms bonded by sigma bonds, or, by π bonds isolated from any other π bonds that may exist in the system. Each chromophoric group is a separate entity, and while the groups may interact, it is assumed that the charge distribution in the individual groups is only slightly perturbed from what it would be if it were more or less isolated in diatomic or, occasionally, slightly larger polyatomic molecules. These slight perturbations may lead to very significant chiral effects, but the charge distributions as measured, for example, by calculated charge population densities (3) are distinctly localized compared to those of conjugated systems.

7.21   The Ketones

The ketonic group in organic molecules is one of the most versatile in terms of the variety of chemical reactions that may be carried out with it. It is present in a large number of natural products, their synthetic derivatives (some of which are of great biologic interest and medical importance), and the synthetic compounds of the chemical industry. It is easy to understand why there is so much interest in the structure and conformation of such compounds, and there is no shortage of methods of greater or lesser sensitivity for studying these structures. Aside from x-ray crystallography of the solids in crystalline form, almost none are so particularly suited, especially for the natural products and their derivatives of which pure or nearly pure optical isomers may be readily obtained, as are the chiroptical methods of ORD and CD, and also of magnetically induced CD. There is a peculiarly good reason for this; slight variations in the structures produce effects almost undetectable by other methods. We will find that this high sensitivity of chiroptical properties is explicable by the fact that the longest wavelength optical transition of the isolated carbonyl group has a large intrinsic magnetic transition moment and in symmetrically constituted compounds, such as formaldehyde, acetone, and cyclohexanone, a null (forbidden) electric dipole transition moment. The forbidden electric dipole transition moment, however, becomes partially allowed by vibronic interactions (Section 3.6) with higher energy transitions or by a breakdown of symmetry in asymmetrically substituted compounds. This latter effect thus produces large perturbations in a small quantity, the nominally forbidden electric dipole transition moment, and small perturbations in a large quantity, the intrinsic magnetic dipole transition moment; as a result, it creates a sensitive parameter of their product—the rotational strength. In Chapter 3 we made note of the fact that isotropic achiral absorption arising from even fully magnetically allowed transitions is very small. However, we will see now that the associated *chiral* activity can be very large and we refer, therefore, to (in speaking of the rotational strength) the large value of the intrinsic magnetic dipole transition moments.

The simplest ketone analog of formaldehyde is acetone. The three carbon atoms of acetone lie in a plane with the oxygen atom, and while at least two of the methyl hydrogens on each atom have to lie out of the plane, the rotation of the methyl groups about the carbon-carbon bond is almost unrestricted, so that all possible optically active rotamers are present in the pure liquid or in solution simultaneously. The resulting racemate is of course optically inactive. When the carbon atoms are fixed in a ring, for example, cyclopentanone, chiral substitutions become possible. Thus, 3-methylcyclopentanone may exist in one of four optically isomeric forms, depending on whether the carbon of the methyl group is above or below the plane of carbon atoms 2, 3, 4, and 5 (above and below being defined with respect to the relationship of the oxygen atom to that plane) (Fig. 7.00). The introduction of the methyl group at position 3 may also cause the four carbons to skew, thereby destroying the planar symmetry. The ring itself will then be chiral, its chirality determined by the absolute configuration of the methyl substituent. Formaldehyde and acetone (treating the methyl groups as isotropic units because of the free rotation about the carbon–carbon bonds) have two planes of symmetry and a twofold axis through the carbonyl bond and therefore belong to the C_{2v} point group (4). From considerations of Chapter 6, molecules belonging to the C_{2v} group must be optically inactive, but examining the C_{2v} character table (Appendix I), we find also that all electric dipole transitions are forbidden because there are no dipole components of A_1 symmetry. Magnetic transition moments are, however, allowed, the z component of $\mathbf{m}$ having an A_1 representation. Development of optical activity will require a breakdown in symmetry that makes some Cartesian component of both electric and magnetic transition moments to *the same state* allowed. Most, but not all (5), of the significant experiments on ketones have been concerned with their lowest excited state. This state gives rise to a weak absorption band in the near ultraviolet spectrum at wavelengths ranging from about 280 to 320 nm. The fact that it is weak (extinction coefficients below 100) suggests that it comes from a formally forbidden transition and must gain its intensity by one of the perturbative mechanisms we have been discussing. Theoretical calculations of the excited states of ketones have not been too successful in explaining the spectral properties. Robin (2) has reviewed these and claims that except for the $n_1 \to \pi^*$ ($n_0 \to \pi^*$ in Robin's notation) transition, the valence shell spectra of ketones are poorly

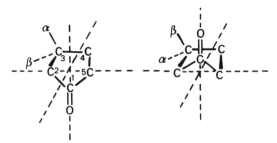

Fig. 7.00 The optical isomers of 3-methyl cyclopentanone. Four isomers are possible by α or β substitution at carbon 3 of either of the two enantiomers with respect to the position of the carbonyl group relative to the symmetry plane of carbon atoms 2, 3, 4, and 5.

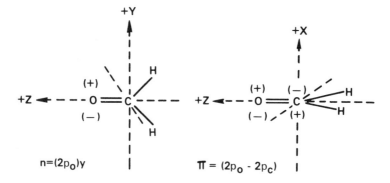

Fig. 7.01 The phase relationship ($\pm$) of the orbitals used in the simplest description of the nonbonding and π^* orbitals of formaldehyde.

understood. The $n_1 \rightarrow \pi^*$ is, however, universally accepted as the weak transition near 300 nm. Note that by using this designation, we are making the common one-electron approximation that the states may be represented by the orbitals from and to which the predominant charge displacement takes place during the electromagnetic transition. In other words, the "configurations" (see Section 3.2) that describe the ground and excited states may, in this case, be loosely approximated for the purpose of describing the transition by just the n_1 and the π^* orbitals, respectively. In Chapter 3 we saw that the simplest description of the n_1 orbital is a 2_p orbital on oxygen. While the current picture of the electron distribution in this orbital is much more complex (6, 7), from the point of view of symmetry, it is a sufficient additional approximation to take this as the $(2p_y)$ orbital on oxygen (Fig. 7.01). Similarly, the antibonding combination, $(2p_O - 2p_C)_x$ of the $(2p_x)$ orbitals on carbon and oxygen of Fig. 7.01, can serve as the π^* for the present symmetry discussion. The symmetry representations of these orbitals in the C_{2v} point group are obtained by examining the effect of the C_{2v} symmetry operations (I, σ_{xz}, σ_{yz}, and C_2) on the orbitals (refer to Section 5.4).
For the n_1 orbital:

		Representation
I	$(2p_O)_y \rightarrow (2p_O)_y$	$+1$
σ_{xz}	$(2p_O)_y \rightarrow -(2p_O)_y$	-1
σ_{yz}	$(2p_O)_y \rightarrow (2p_O)_y$	$+1$
C_2	$(2p_O)_y \rightarrow -(2p_O)_y$	-1

This orbital, therefore, belongs to the b_2 representation (where a lowercase letter is used to distinguish orbital symmetry from state symmetry).

For the π^* orbital:

		Representation
I	$(2p_O-2p_C)_x \rightarrow (2p_O-2p_C)_x$	$+1$
σ_{xz}	$(2p_O-2p_C)_x \rightarrow (2p_O-2p_C)_x$	$+1$
σ_{yz}	$(2p_O-2p_C)_x \rightarrow -(2p_O-2p_C)_x$	-1
C_z	$(2p_O-2p_C)_x \rightarrow -(2p_O-2p_C)_x$	-1

and this orbital, therefore, belongs to the b_1 representation. The electric dipole operators along the x, y, and z axes are characterized, respectively, as b_1, b_2, and a_1. The triple direct products of the symmetry representations of the electric dipole transition moment components of $\langle n_1|\mu|\pi^* \rangle$ are

$$\Gamma \langle n_1|\mu_x|\pi^* \rangle = b_2 \times b_1 \times b_1 = B_2, \tag{7.201}$$

$$\Gamma \langle n_1|\mu_y|\pi^* \rangle = b_2 \times b_2 \times b_1 = B_1, \tag{7.202}$$

$$\Gamma \langle n_1|\mu_z|\pi^* \rangle = b_2 \times b_2 \times a_1 = B_2. \tag{7.203}$$

In the foregoing analysis, and also as we proceed to other systems in this chapter, the approximation has been made that an electronic transition between two states $\langle a|$ and $\langle b|$ may be characterized, for the purpose of computing transition moments, by the one-electron leaving and entering orbitals. Thus the ground state of formaldehyde, which is described by the configuration (see Chapter 4) $\langle \psi_0| = (1s_O)^2(\sigma_1)^2(\sigma_2)^2(n_2)^2(\sigma_{CO})^2(\pi)^2(n_1)^2$, is represented by just the n_1 orbital $\langle n_1|$, from which the electromagnetically excited electron comes, while the excited $n\pi^*$ state, represented above by $\langle \pi^*|$, is the configuration $\langle \psi_1| = (1s_O)^2(\sigma_1)^2(\sigma_2)^2(n_2)^2(\sigma_{CO})^2(\pi)^2(n_1)(\pi^*)$. The rotational strength, which is defined in terms of the state functions $R = [he^2/2m^2c\mathcal{E}_{01}]|\langle \psi_0|\mu|\psi_1 \rangle| \cdot |\langle \psi_1|\mathbf{m}|\psi_0 \rangle|$, is replaced by $R = [he^2/m^2c\mathcal{E}_{01}]|\langle n_1|\mu|\pi^* \rangle| \cdot |\langle \pi^*|\mathbf{m}|n_1 \rangle|$; the factor of 2 disappears because in the expansion of the state functions in terms of the orbital functions, a factor of $\sqrt{2}$ appears in front of each transition moment. The $\langle n_1|$ and $\langle \pi^*|$ orbitals may then be further approximated by any suitable scheme, for example, by a linear combination of atomic orbitals (LCAO). In the example above the $\langle n_1|$ orbital was taken to be the pure $(2p_O)_y$ orbital, but the π^* consists of the antibonding LCAO combination of the $(2p)_x$ orbitals on oxygen and carbon.

It is important to take account of the fact that the symmetry of the excited electronic states is not given by the triple direct products used to determine the symmetry of the transition moments. Thus the Cartesian components of the electric dipole transition moments for the transition to the $n\pi^*$ state were shown to be either B_2 or B_1 in Eqs. 7.201–7.203. But the symmetry of the *state* is given by the direct product of all the orbitals making up the configuration. All doubly filled orbitals are totally symmetric (since the self-direct product of all symmetry representations is always the totally symmetric representation for nondegenerate representations). Therefore, in the C_{2v} point group the symmetry of the filled core orbitals and of the filled π orbital is a_1: $\Gamma(\text{core}) = \Gamma(1s_O)^2(\sigma_1)^2(\sigma_2)^2(n_2)^2(\sigma_{CO})^2(\pi) = a_1$; $\Gamma(\pi)^2 = a_1$. The singlet ground state is $\Gamma\langle\psi_0| = \Gamma(\text{core}) \times \Gamma(\pi)^2 = a_1 \times a_1 = A_1$. As we have seen, the representations for the n_1 and π^* orbitals, each of which contain only electrons in the excited state, are b_1 and b_2, respectively. The symmetry of the $n\pi^*$ *state* $\langle\psi_1|$ is, therefore, $\Gamma\langle\psi_1| = \Gamma(\text{core}) \times \Gamma(n_2) \times \Gamma(\pi^*) = a_1 \times b_1 \times b_2 = A_2$. The pure electronic singlet $n \rightarrow \pi^*$ transition of formaldehyde (or an isolated ketonic chromophore) is thus $^1A_1 \rightarrow {}^1A_2$. The C_{2v} character table shows at once that this is a forbidden electric transition, since there is no Cartesian component for $^1A_2$. Of course, this transition does appear in the spectra of these compounds, weakly in absorption and strongly in the ORD or CD spectra when the compounds are chiral. The theory of how this comes about through vibronic intensity borrowing has been discussed in Chapter 3 and the various inductive mechanisms described in Chapter 4. We will see now how these considerations apply to the keto and other chromophores. Before proceeding, however, let us complete the general symmetry considerations by examining how an A_2 electronic transition can become vibronically allowed. The simplest vibronic mechanism would be for a transition to occur from the ground electronic $^1A_1$ level to the first a_2 vibrational level of the excited $^1A_2$ electronic state; in this case, the symmetry of the resultant state is $^1A_2 \times a_2 = {}^1A_1$, an allowed level. However, by the same token that a $^1A_2$ electronic level is forbidden to a C_{2v} chromophore, so too are vibrational levels of this symmetry forbidden. In formaldehyde, which is a C_{2v} molecule with respect to its nuclear configuration, as well as to its electronic chromophore, the high forbiddenness of the a_2 vibrational level would prevent this mechanism from accounting for much of the intensity borrowing that gives rise to the weak $n\pi^*$ absorption band. However, in a molecule of different nuclear symmetry still containing the C_{2v} carbonyl chromophore (in first approximation), for example, any of the monomethyl cyclopentanones in which an a_2 vibration would be allowed, this vibronic mechanism could be one source of the intensity both for absorption and chiral activity. Alternatively, other vibronic borrowing mechanisms are available. A transition to a level that consists of one quantum each of allowed b_1 and b_2 vibrations, which together have a_2 symmetry, would be one such mechanism.

None of the representations of the electric dipole transition moments (Eqs. 7.201–7.203) are the totally symmetric A_1; this transition is symmetry forbidden and will appear weakly in absorption only because of vibronic or symmetry perturbations, a fact compatible with the conclusion drawn from the identification of the weak long-wavelength absorption band in the ketones with the calculated transition energies of the $n_1 \rightarrow \pi^*$ transition. Other transitions of the ketone chromophore will, of course, be allowed and appear in both the absorption and CD spectra of higher energies (5), but because of the orthogonality of the magnetic and electric transition moments, in no molecule where strict C_{2v} symmetry applies will they give rise to optical activity.

In the decade beginning about 1955, Carl Djerassi and his students and colleagues, and also W. Klyne and his students, began an extensive series of investigations of the optical rotary dispersion of ketones (8–10). They found that a very strong Cotton effect is associated with the $n_1 \rightarrow \pi^*$ transition in the 300-nm region of the spectrum of ketones, especially of those chiral ketones situated in alicyclic steroids, as well as in smaller ring systems. In a series of rapid incisive investigations, they established the high sensitivity of the chiral activity to the location of the ketone in the ring structure, the absolute configuration of adjacent ring junctions, the nature and configuration of nearby substituents, and the conformations of the rings themselves. Dismissed with a single short paragraph in Lowry's 1935 book on optical rotatory power (11), the ketones now appeared to provide an entirely new experimental handle both for investigating the stereochemistry of a multitude of organic compounds and for research into the very nature of optical activity itself. It became clear that a simple chromophore whose electronic structure in the ground state must contain (or nearly so) a plane of symmetry could not only induce strong optical activity, but also activity that is highly sensitive to the precise stereochemistry of the associated molecular architecture. True, this was not the first such observation. The Cotton effects associated with the visible or near ultraviolet chromophores of inorganic complexes and with some nitrogen-containing compounds had already been observed, but the high sensitivity to local structure and the importance of the compounds under investigation quickly aroused the interest of both theoreticians and experimentalists.

The first significant attempt to understand the optical activity of the ketones produced a generalization, the "octant rule," which appeared in a 1961 paper by Moffitt, Woodward, Moscowitz, Klyne and Djerassi (12). Investigations of ketonic optical activity has not ceased since that time; the most recent paper to appear on the theory of the octant rule is a 1976 paper by Bouman and Lightner entitled "The Octant Rule 5. On the Nature of the 'Anti-Octant' and Front Octant Effects by a CNDO/S Study of Rotatory Strength of the Carbonyl $n \rightarrow \pi^*$ Transition" (7). These investigations, from the experimental work of Djerassi and Klyne and the early theoretical work of Moffitt to the self-consistent-field (SCF) molecular orbital calculations of Bouman and Lightner and all the work in between, span the entire range of attempts to put the nature of chiral activity on sound experimental and theoretical grounds (13). For this reason, we devote more space to this chromophore than to any other.

Having established that a symmetry forbidden electronic transition could give rise to large chiral strength when the symmetry of the surrounding molecule is lowered even slightly, it becomes a matter of some importance to determine how this comes about. Interestingly enough, in 1939, long before the experimental work of Djerassi et al., Kauzmann, Walter, and Eyring (14) had chosen 3-methylcyclopentanone as an example with which to demonstrate the effect of a perturbing potential to one-electron theory. By assuming that the cyclopentanone ring is planar, they were able to attribute the asymmetric perturbing potential to the carbon and hydrogen substituents at carbon 3, which are the only atoms out of the plane and not juxtaposed symmetrically to the plane by an identical atom. Choosing to describe the potentials due to the charges on these atoms in terms of Slater orbitals, they found that these led to a final expression for the molecular rotation that contains the product of the three Cartesian coordinates of each of these atoms. Each ith asymmetric atom contributes

$$[\phi]_i = -15.1 \times 10^3 \frac{n^2+2}{3} \cdot \frac{x_i \cdot y_i \cdot z_i}{R_i^3} \cdot D \qquad (7.204)$$

to the rotation at the wavelength of sodium D line. Here x_i, y_i, z_i are the atom coordinates, and D is the radial dependent part of the potential function, dependent only on the distance of the atom from the center of the chromophore. xyz, which we saw in Chapter 4, is just the type of perturbing potential chosen by Condon and his colleagues to illustrate the one-electron theory and is also the simplest pseudoscalar potential, which, as Schellman demonstrated, can lead to optical activity in the D_{2h} point group (Table 6.1). The carbonyl chromophore could be considered D_{2h} only if the assumption is made that *the chromophore is limited to the two atoms of the carbonyl group* and no distinction is made between the electronic structure (essentially the $2p$ orbitals) of the oxygen and carbon atoms. These assumptions are too strong to be really meaningful. Nevertheless, the calculated molecular rotation on this basis was of the right order of magnitude and had the experimental evidence been available at the time, an octant rule might have been proposed, since the product x,y,z, leads directly to an octant rule. Thus the first attempt to explain the optical activity of the symmetric ketone chromophore was by a perturbative mechanism, the one-electron theory. It anticipated by 20 years the next more substantive attempt, initiated by Moffitt and carried on by Moffitt's colleagues after his death. The calculations that gave quantitative and substantive theoretical basis to the octant rule were carried out by Moscowitz and Snyder (15) and published in a most illuminating paper by Moscowitz in 1962 (16). A detailed discussion of the method (both sets of investigators used the one-electron theory) is available in the original papers. We wish only to emphasize here the source of the asymmetry that was chosen in order to give a finite rotational strength. Since the Rosenfeld equation requires that both the electric and magnetic transition moments be finite and have components in the same direction, it was necessary to perturb either or both the ground or the excited state in a way to produce a finite value of the electric dipole transition moment $(n_1|\boldsymbol{\mu}|\pi^*)$, which is null in C_{2v} symmetry. Using the orbital approximation, the choice was to perturb the π^* orbital by mixing $2p$ orbitals on carbon and oxygen with $3d$ orbitals to form a new π^*. In this way a finite electron density is produced where the nodes of the original $2p$ orbitals had zero electron density. A transition from n_1 to the perturbed π^* then gives a circular notion of charge from y to z about the x axis with electric moment components in the direction of the allowed magnetic moment $(n_1|\mathbf{m}|\pi^*)$. To do this, the π^* orbital was approximated as

$$\pi^* = \pi^* + \delta\, 3d_{yz,\text{O}} + \gamma\, 3d_{yz,\text{C}} \qquad (7.205)$$

where

$$\delta = \alpha(2p_{x,\text{O}}|V|3d_{yz,\text{O}})/\Delta\mathcal{E}_\text{O}, \qquad (7.206)$$

$$\gamma = \beta(2p_{x,\text{C}}|V|3d_{yz,\text{C}})/\Delta\mathcal{E}_\text{C}, \qquad (7.207)$$

and α and β are the coefficients of the $2p$ orbitals in the original unperturbed π^* orbitals; the $\Delta\mathcal{E}$'s are related to orbital energy differences between the unperturbed π^* and $3d_{yz}$ orbitals. The potentials V are obtained as described above.

The octant rule, which was established from symmetry considerations and empirical observations, has its origin in the dependence of the molecular rotation on the product of the coordinates of the atoms surrounding the carbonyl group. Cartesian coordinates divide space into eight separate volumes divided by planes formed from each two of the coordinates. In each of these volumes the coordinates exist with a different combination of their signs, their product, of course, being either positive or negative. In this primitive version, then, the octant rule referred to just such a division of space and was stated as follows: "The sign of the contribution [to the chiral activity of the $n_1 \rightarrow \pi^*$ transition of a ketone] which a given atom at point P(x, y, z) makes to anomalous rotatory dispersion will vary as the simple product, $x \cdot y \cdot z$, of its coordinates." (12).

Let us consider one or two applications of this rule and then examine some of the problems and subsequent treatments that were made in order to gain additional insight and predictive accuracy. Before examining the first application, the question arises as to where the origin of the coordinates should be. The two obvious symmetry planes with respect to the

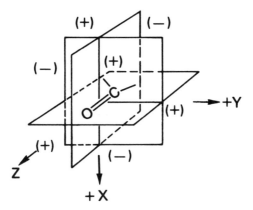

Fig. 7.02   Octant diagram for the contribution of substituents of chirally substituted carbonyl compounds to the rotatory strength of the $n \rightarrow \pi^*$ transition. The carbon and oxygen atoms and the $(sp)_2$ single bonds lie in the zy plane. Front and rear octants are separated by the xy plane, which goes through the center of the carbonyl bond.

plane of the ketone certainly form the basis for two of the coordinates, x and y, but it is not quite so apparent where to place the zero of the z coordinate. Because the unperturbed π^* orbital has a node about halfway between the carbon and oxygen atoms, it seemed to make sense to place the third plane perpendicular to the other two through that point. In the vast majority of organic compounds, to an observer looking along the z axis from the oxygen end, this puts all the perturbing atoms in one of four quadrants behind the carbonyl group. This makes the rule relatively easy to apply through the use of a simple planar diagram (Fig. 7.02). As stated, the octant rule still allows a choice of direction of the carbonyl group along the z axis. This choice was made with respect to optical rotation measurements on molecules of known absolute configuration. The established convention is that the oxygen atom is in the $+z$ direction with respect to the carbon atom, which is correspondingly toward the $-z$ direction. The octant rule projections on the diagrams of Fig. 7.02 are then always made with the observer looking from the $+z$ direction toward the oxygen atom.

If the positions of the atoms are projected on this figure *using the convention* that the oxygen atom is on the z axis in front of the xy plane and the carbonyl carbon behind the xy plane on negative z axis, then the xy, yz, and zx planes separate the octants. For molecules whose atoms all lie behind the carbonyl carbon, only the back octant diagram need be used; substituent atoms whose projections do not fall on or very close to the x or y axes contribute rotational strength with a sign as shown in the back or $-z$ octants. Thus the projection of the chair form of (+) 3-methylcyclohexanone (Fig. 7.03) shows that the ring carbons either lie on nodal octant planes or in the case of atoms 3 and 5 make equal positive and negative contributions, but that the methyl carbon attached to carbon 3 makes a positive contribution not balanced by any corresponding negative one. Furthermore, for the chair conformer illustrated, both the axial and the equatorial configuration of the C-3' methyl give rise to positive rotations, since in either case the methyl carbon lies in the $+x, -y, -z$ octant. The observed rotational strength (17) is positive in accord with the octant rule prediction. The cyclohexanones are conformationally mobile so that comparison of the measurements of the circular dichroism or optical activity of the $n \rightarrow \pi^*$ transition with the octant rule predictions for the possible conformer is a method of studying the conformational equilibrium. Literally hundreds or possibly thousands of examples of the application of the octant rule to problems such as these and others in the stereochemistry of ketones have been published, and several excellent reviews are available (18a–e).

From its very inception the octant rule ran into difficulties because some observations did not accord with the predictions. It will be more instructive here to consider some of the problems rather than the many successful applications that are profusely covered in the monographs and reviews. To

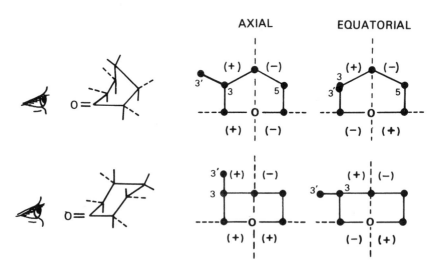

Fig. 7.03 Octant rule projections of the boat (upper) and chair (lower) forms of (+)-3-methylcyclohexanone with the 3-methyl group axially and equatorially substituted.

do this, we will return to an examination of the nature of the $n \rightarrow \pi^*$ carbonyl achiral absorption. This will lead to a more detailed discussion of the mechanism of the induction of chiral activity in the pseudoplanar carbonyl absorber. We have introduced the term "pseudo" to indicate that starting in 1966 with the work of Pao and Santry (6), attempts to understand the chiroptical properties of the carbonyl group (and other nominally planar chromophores) began to take on the aspect of a dissymmetric chromophore treatment. That is, rather than introducing an asymmetric potential of the surrounding chiral system as in the one-electron treatment or the coupled oscillator treatments, the chromophore was extended to include vicinal chiral atoms by directly including their electrons in the chromophoric molecular orbital description: It was now a dissymmetric chromophore.

Consider some of the difficulties that appear in the application of the octant rule. One of the first to be observed was based on the data obtained by Djerassi, Osiecki, Riniker, and Riniker (19) on the rotatory activity of haloketones. In these compounds the substitution of fluorine for another halogen or for a methyl group always results in a contribution to the optical rotation of opposite sign. That the magnitude of the perturbation should be dominated by the electronic nature of the substituent or its size is, of course, not unexpected in view of the polarizability theories of optical activity; but in view of the prevailing ideas of the relation between structure and optical activity, it was indeed somewhat surprising that an inversion of sign should occur. This fact, moreover, did make it quite clear that the failure to account for still another substituent, the hydrogen atoms, in the octant rule projections could raise quantitative difficulties in octant rule applications; the fact that these could be ignored in a qualitative discussion is "probably as much a consequence of widespread automatic cancelling of effects of opposite sign as it is in the result of a small perturbative effect of hydrogen atoms." (12). We will see in the discussion of dienes later in this chapter that even the small size of the perturbation may be open to question.

Another of the complexities in the application of the octant rule arose from the juxtaposition of the observations on the *trans*-decalones and the *trans*-hydroindanones in which the ketone group is, respectively, in a six-membered and a five-membered ring; the latter, it will be recalled, provided the model used in the original application of the one-electron theory. If we ignore for the present problems arising from conformational mobility in these compounds, the observation is that the "amplitude" (20) of the ketone Cotton effect in the *trans*-hydroindanones is very much larger than that of similar *trans*-decalones (Fig. 7.04), despite the fact that the octant projections indicate stronger asymmetric projections in the latter

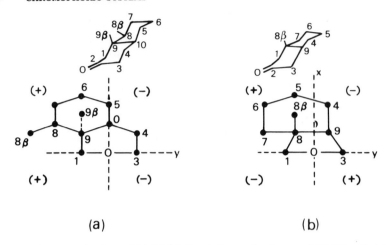

Fig. 7.04 Octant projections of (a) 8β,9β-dimethyl-*trans*-decalone and (b) 8β-methyl-*trans*-hydroindanone.

than in the former. For example, in the projections of Fig. 7.04a, where the decalone conformation is essentially that of a double chair, atoms 6, 7, 8, and the 8β and 9β methyl carbons are all uncompensated in other octants, occupying only the upper left positive back octant. By comparison, in the projection of 8β-methyl-*trans*-hydroindan-2-one, only carbons 5, 6, and 7 are uncompensated; yet the amplitude is +216 (21) as compared to +52 in the decalone (12). In 1961 Klyne recognized that there existed ample evidence (22) that the cyclopentane ring itself was highly skewed and that if the same thing occurred in cyclopentanone, the octant projection of the hydroindanone (Fig. 7.05) would show only three uncompensated carbon atoms in appropriate octants. However, not only is this insufficient to

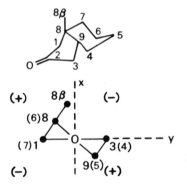

Fig. 7.05 Octant projection of 8β-methyl-*trans*-hydroindan-2-one.

account for the difference between the six- and five-membered ring compounds, but additionally, the observation (23) was made that the absence of the 8β-methyl group in the *trans*-2-one had almost no effect on the magnitude (or sign) of the ketone Cotton effect. This led Klyne to conclude that the primary effect resided in the skew of the cyclopentanone ring independent of the octant projection, although he did rationalize the small size of the difference between the compound with and without the methyl substituent by pointing out that in the skew (or "half-chair") conformation, this methyl carbon was very close to one of the nodal octant planes. Later Ouannes and Jacques in a semiempirical treatment of cyclopentanones also partially rationalized the observations by demonstrating that conformational mobility could account fairly well for the observed amplitude of the Cotton effect of methyl-2(S)-cyclopentanone, but the same considerations were not quite so adequate for some of the other compounds considered (24).

Compounds containing vicinal cyclopropyl rings proved to be in worse conflict with the octant rule because in this case the predicted sign of the $n \to \pi^*$ transition chiroptical activity was demonstrated to be opposite to that experimentally observed (8c). In Fig. 7.06 we see that the octant rule

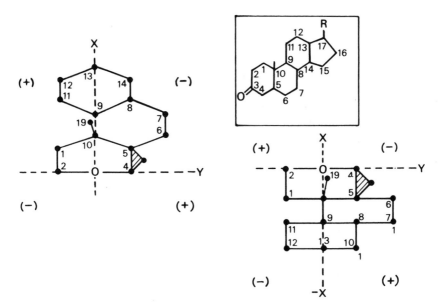

Fig. 7.06 Octant projections of rings A, B, and C and the 4,5-α-cyclopropyl ring of 4,5-α-cyclopropyl-cholestan-3-one with ring A in the boat (upper left) and chair (lower right) conformation. The five-membered ring and the substituent R are not shown in the projections.

predicts a positive sign for the Cotton effect of 4, 5-α-cyclopropyl-cholestan-3-one where ring A is in the conventional chair conformation, while the measured value is negative (25). Molecular models for this compound indicate, however, that the tetrahedral angles at carbons 4 and 5 are more closely preserved if ring A takes on either the half-chair or the boat conformation of cyclohexane. In the latter case the octant rule projection predicts the observed negative optical activity. On this basis it is possible to conclude simply that the octant rule has demonstrated that the conventional chair conformation of ring A is strongly perturbed by the presence of the cyclopropyl group. However, in a series of investigations that followed (26,27), cyclopropyl ketonic compounds for which no such rationalization could be made were found, for the most part, to obey a "reversed" or "antioctant" rule. More recently Lightner and Beavers (28) examined a series of rigid bicyclic ketones containing the cyclopropyl group as a third ring and demonstrated unequivocally that there are many compounds of this type in which antioctant rule behavior is observed. For example, exotricyclo [3.2.2.0^{2,4}] nonan-6-one (Fig. 7.07) is predicted to be only weakly positive on the basis of its octant projection but is observed to have an extremely strong ($R_{n\pi*} = +20.2 \times 10^{-40}$ cgs unit) positive Cotton effect. Its epimer, the endotricyclo[3.2.2.0^{2,4}]nonan-6-one, which is also predicted to have a weak positive effect, is found instead to have a negative rotational strength ($R_{n\pi*} = -3.1 \times 10^{-40}$ cgs unit). Antioctant behavior has also been established in aliphatic ketones containing vicinal cyclopropyl groups (29).

With these complexities in mind, let us return to a more thorough examination of the $n \rightarrow \pi^*$ carbonyl transitions. We have seen that symmetry considerations predict that in an isolated (C_{2v}) carbonyl group no electric dipole transition moments are allowed for the $n \rightarrow \pi^*$ transition, but

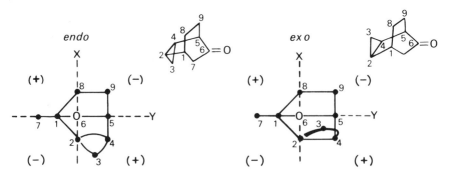

Fig. 7.07   Octant projections of endo- and exotricyclo[3.2 2.0^{2,4}]nonan-6-one.

that there is an allowed magnetic dipole transition moment along the carbonyl axis (in the z direction in our convention). Calculations (1) by Mason (30) of the magnitude of the magnetic transition moment based on experimental observations of the achiral absorption and the optical rotation of a number of ketones (using the relation $g = 4R_{ab}/D_{ab}$); (2) by Emeis and Oosterhoff (31) for *trans-β*-hydroindan-2-one; and (3) an SCF calculation for formaldehyde by Jungen, Labhart, and Wagniere (32) give very large values for this moment. In the case of *trans-β*-hydroindan-2-one, the value 1.19 Bohr Magnetons (BM), is greater than the maximum value of 1.0 BM, which would come from a localized $2p$ nonbonding electron on the oxygen atom. This requires that the nonbonding orbital be delocalized in these compounds. Furthermore, with such a large value of the magnetic transition moment in the z direction, it is not hard to see why any perturbation that allows even a small component of electric dipole transition moment in the z direction would produce substantial optical activity. Consider, for example, the rotational strength expected from even a weak electronic transition if its induced *electric transition moment* is directed parallel to the z direction of the magnetic moment. We require the value of

$$R_{ab} = \mu_{ab} \cdot \mathbf{m}_{ab} = \mu_{ab} \mathbf{m}_{ab} \cos\theta. \qquad (7.208)$$

Since in this case the magnetic and electric moments are parallel, $\theta = 0$ degrees and $\cos\theta = 1$; and since $\mathbf{m}_{ab}$ is the order of 1 BM (0.27×10^{-20} cgs),

$$R_{ab} = 0.927 \times 10^{-20} \mu_{ab}. \qquad (7.209)$$

The magnitude (but not the sign) of the electronic transition moment can be evaluated from its relationship (Eq. 3.316) to the observed band strength $\int (\varepsilon/v)\, dv$:

$$|\mu_{ab}|^2 = 6909 \frac{hc}{8\pi^3 N} \int \frac{\varepsilon}{v}\, dv \qquad (7.210)$$

$$= 9.80 \times 10^{-39} \int \frac{\varepsilon}{v}\, dv \qquad (7.211)$$

It is sufficient to approximate the integral by integrating between finite limits conveniently taken as frequencies above and below the band maximum at which the extinction coefficient becomes less than about 1 percent of the maximum extinction ε_{max}, assuming that the band is Gaussian in shape; for most solution absorption bands this assumption will be accurate to within 5 percent, more than sufficient for the present purpose. In this case

$$\int \frac{\varepsilon}{\nu} \, d\nu = \varepsilon_{max} \frac{\Delta\nu}{\nu_{max}}, \qquad (7.212)$$

where $\Delta\nu$ is the width of the band at the point at which ε is 0.368 the value of ε_{max}. For a typical weak $n \to \pi^*$ band of a ketone at about 33,000 cm^{-1} (300 nm), ε_{max} may have a value of the order 20, and if we assume that half of this derives from an electric dipole transition moment parallel to the z axis, then with a typical half-width of about 6000 cm^{-1}, $\varepsilon_{max}(\Delta\nu/\nu_{max}) = 2$. Substituting in Eq. 7.211, we find $\mu_{ab} = 1.33 \times 10^{-20}$ cgs unit. Thus a weak ketonic $n \to \pi^*$ absorption in a chiral molecule could have a rotational strength of the order of $R_{ab} = 0.927 \times 10^{-20} \times 1.33 \times 10^{-20} \approx 1.23 \times 10^{-40}$ cgs unit ($\sim 1.33 \times 10^{-2}$ DBM) if even half of its electric dipole strength is derived from electric transition moments along the carbonyl axis. Although not as large as some of those observed, for example, the cyclopropyl ketones discussed above, this is nevertheless a very considerable rotational strength, as it corresponds to a circularly dichroic absorption (see Appendix II) of $\Delta\varepsilon = 0.025$ or a molar ellipticity $[\theta]$ of the order of 80 degrees.

We can ask, therefore, whether there is any experimental evidence in the ordinary absorption spectra for the existence of a sufficiently large electronic transition moment along the carbonyl (z) axis in ketones to produce a rotational strength of this magnitude. Furthermore, since this component of the transition moment is electronically forbidden in the C_{2v} symmetry of the carbonyl chromophore, if it does exist at all, where does it come from? Does it arise from a vibronic perturbation, from a static perturbation by an asymmetric environment in the ground state, or by a dissymmetry produced because the excited $n \to \pi^*$ state in its equilibrium geometry is different from the ground state geometry? Each of these can result in a finite electric dipole strength for the nominally forbidden transition and therefore in the appearance of ordinary isotropic absorption regardless of whether the molecule is chiral or not. The second of the three mechanisms is the basis for the one-electron polarizability theory used in the derivation of the octant rule for chiral strength. All three are illustrated diagrammatically in Fig. 7.08. Mechanisms other than these have also been proposed.

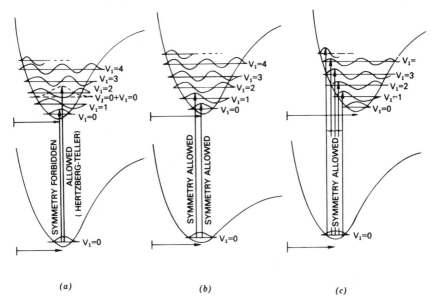

Fig. 7.08 Two-dimensional energy diagram illustrating three mechanistic origins of intensity for $n \to \pi^*$ absorption bands. (*a*) Vibronic allowedness: The pure electronic symmetry in the ground and excited states, C_{2v}, forbids the optical transition and chiral activity. Only vibronic transitions to nontotally symmetric states are symmetry allowed and contribute. (*b*) Asymmetric environmental perturbation: The perturbation produces non-C_{2v} wavefunctions; the resulting rotational strength is calculated by the classical polarizability mechanism. (*c*) Change in geometry between states: The electronic symmetry in the ground state is C_{2v}, but the symmetry of the excited state is reduced to one allowing achiral and chiral activity. Some, but not all, of the intensity may be vibronic, since the transition is electric and magnetic allowed in the symmetry class containing symmetry elements common to both states. This resembles the dissymmetric chromophore, but the molecule is not intrinsically chiral in the ground state. The abscissas in this diagram represent a generalized nuclear coordinate specifying the geometry of the carbonyl group and its two adjacent bonds. This type of diagram is more familiarly presented for a diatomic system in which the generalized nuclear coordinate is just the internuclear distance represented on the abscissas and the potential energy of the system, as in this case, is represented on the ordinate. A more realistic representation of the $>$C=O group requires a three-dimensional surface or more, in which the bending and stretching coordinates of the bonds show individual potential contours.

The higher order interactions of quadrupoles, etc., which result if the perturbation potentials in the one-electron theory are not cut off early in the expansion of the interaction of the perturbing charge distribution with the chromophoric electron, have been examined. A particularly good discussion of this is available (33). We have discussed vibronic perturbation (Section 3.6) and the static perturbation of one-electron theory

(Section 4.3), but we have not yet explicitly treated the third of the mechanisms illustrated in Fig. 7.08c. Referring to the figure, it will be observed that as a result of Franck-Condon considerations, which require that the molecular geometry of the molecule undergoing the transition not change *during* the transition, much of the band intensity may come from molecules thermally excited to higher vibrational states. Since the energy of low-frequency vibrations is of the order of the available thermal energy, kT or greater, at room temperature only a small fraction of the molecules in a sample can contribute to the absorption in this way. Like the other perturbations, this will lead to only weak absorption bands, in keeping with the symmetry-forbidden nature of the undistorted chromophore. We will return later to an additional discussion of this mechanism.

The evidence that these, or any other mechanism, result in a z component of the electric dipole transition moment can come from two sources; one is spectroscopic and the other a quantum mechanical calculation of the transition moments of an asymmetrically substituted ketone. In the quantum calculation the concept of the symmetric chromophore is abandoned entirely. For the present we note only that a treatment of this type using the extended Huckel method (EHT) found that in halogen-substituted acetone held in a chiral conformation, the dominant contribution to the optical activity does come from terms in $\mu_z m_z$ (34). On the other hand, for chiral formaldehyde, neither the extensive experimental investigations of the spectra (35) nor the quantum calculations (36) have turned up any evidence for a z-polarized electric dipole transition moment. In cyclopentanone where the carbonyl group retains its C_{2v} symmetry but in which the ring system is skewed so that the molecule as a whole has only C_2 symmetry, the ordinary spectroscopic intensity of the $n \to \pi^*$ transition is found to be predominantly polarized out of the plane of the carbonyl group and only a very small z polarized moment is observed (37); the vibronic components are found perpendicular to the z axis as in formaldehyde. The calculated electric dipole transition moments in the latter were found to be 0.88×10^{-20} cgs in the molecular plane and perpendicular to the carbonyl axis, and only 0.55×10^{-23} cgs perpendicular to the molecular plane (36). Comparing these to our model calculation of the $n \to \pi^*$ rotational strength for a ketone with $\varepsilon_{max} = 20$, we see that the out-of-plane electric dipole transition moment is three orders of magnitude smaller; and while the in-plane moment is of the same order of magnitude, it is at right angles to the strong magnetic dipole transition and thus cannot be thought of as a major source of optical activity in chiral ketones. Where then does a z component of the electric moment (or a strong y component of the magnetic moment) come from?

Except for referring to the possibility of a geometry change between the ground and the excited $n\rightarrow\pi^*$ states in Fig. 7.08, we have assumed that the C_{2v} symmetry of the chromophore is retained in the excited state. Suppose that this assumption is incorrect, namely, that the equilibrium geometry of the $n\rightarrow\pi^*$ state is such that the carbonyl group and its two-bonded atoms do not retain their planar conformation after excitation. Then, the reduced symmetry may permit electric dipole transition moment components parallel to the magnetic moments. For example, in C_2 symmetry (refer to the character table in Appendix I) both electric and magnetic transitions to ^1A states are allowed along the z axis and to ^1B states along the x and y axes. In 1953, on the basis of simple valence considerations, Walsh (38) predicted that in the $n\pi^*$ state, formaldehyde would be pyramidal with the carbon atom at the peak of the pyramid and the oxygen and two hydrogen atoms forming the base—an almost D_3 geometry in which the chiral moments to $^1A_2$ states would be directed along the z axis. Herzberg-Teller considerations (see Chapter 3, reference 7) require that the symmetry elements common to both the ground and excited states, in this case the twofold (C_2) axis, control the selection rules for the transition moment. The nonplanar geometry of the $n\pi^*$ state has since been confirmed by both experiment (35b,c; 39) and calculation (36b). Similar changes may be expected in the aliphatic ketones, and although it is less likely that in the ring ketones the same pyramidal structure would exist, substantial changes from planarity can occur. A recent comparison of the circularly polarized emission (see Section 10.4) and absorption of ketones has led Dekkers and Closs (40) to propose just such a mechanism as at least a partial explanation of the chiral strength of ketones. They point out that there is a large

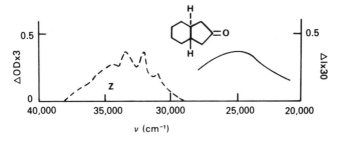

Fig. 7.09 The CD (---) and CPL (——) of β-hydrindanone in n-heptane. Note the difference in shape of the two spectra. The magnitudes are also very different, but other criteria have to be used to establish the relationship because the circularly polarized luminescence is not measured in absolute terms (40). Note vibrational components in the CD spectrum whose origin is discussed in Section 3.6.

discrepancy between the magnitude of the rotational strength in absorption and fluorescence, and in the shapes of the spectra as well (Fig. 7.09). Since absorption and emission spectra are generally almost mirror images of each other if no appreciable change in geometry has occurred between states, these spectra lead to the conclusion that just such changes have occurred. A theoretical analysis of the vibronic mechanisms is in accord with this conclusion but, in addition, predicts that in some cases this difference in the nuclear configuration of excited states can produce bisignate circular dichroism bands such as those that have been observed even in conformationally rigid ketones (Fig. 7.10). The term bisignate is descriptive for a CD curve that has both negative and positive lobes in the spectral region of an apparently single achiral absorption band. Bisignate curves of simple chromophores may have their origin in vibronic borrowing or conformational equilibria, or in the accidental near degeneracy of two electronic absorption bands of different symmetry. In repeating polymers, bisignate curves are usually indicative of exciton interactions between the repeating chromophores. In Section 3.6 we discussed the fact that vibronic borrowing can lead to bisignate CD curves without changes in upper state geometry. Distinguishing between these two origins may not always be simple, but the association of distinct vibrational structure in the ORD or CD spectra with the vibronic analysis of nonpolarized absorption spectra (41) is reasonable evidence of the ordinary vibronic borrowing mechanism. Both may exist, of course, in the same molecular transition.

Let us now finally proceed to the developments that are rapidly turning attention away from the perturbation mechanisms for inducing chiral activity in nominally symmetric chromophores and toward more complete quantum mechanical treatments in which the inclusion of the electrons of

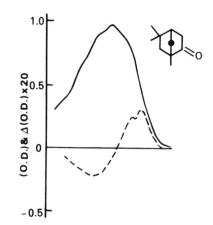

Fig. 7.10 Achiral absorption (——) and circular dichroism (---) of 5-ketobornylacetate (40).

the asymmetrically disposed substituents converts the system into a dissymmetric chromophore. The earliest use of these procedures was for the calculation of the optical activity of hexahelicene (42) and of the conjugated dienes (43) using the simple Hückel molecular orbital method (HMO). Recently, and starting with work of Pao and Santry on ketones (6), a variety of self-consistent field and molecular orbital methods (SCF-MO) have come into use with varying degrees of success. Some of the newest results are highly promising and threaten for the first time to make the statement by Tinoco (with which we started this book) obsolete. The general aspects of the dissymmetric chromophore approach, or direct calculation as it has sometimes been termed (44), are discussed in Section 4.3; in Section 7.31 its application to conjugated chromophores for which it was originally developed is treated in detail. The reader wishing to get a more general view of this approach before examining its application to ketones should turn to these sections first.

In these treatments the basic idea is that the inclusion of the actual dissymmetric geometry of the nuclear framework of the molecule in the calculation of both the molecular orbital wavefunctions and the transition moments will lead to finite collinear components of the electric and magnetic dipole transition moments, and thus to finite rotational strength for a given optical isomer. There is no need to assume or calculate any perturbation potential as required for the mechanisms discussed in Chapter 4. Thus Pao and Santry, starting with the (unstated) assumption that the methyl group in various methyl-substituted cyclohexanones is not free to rotate about the bond to which it is attached to the cyclohexanone ring, created dissymmetric molecules for arbitrary conformations of the methyl rotamer. In the so-called staggered conformation of the methyl group of 3-methyl cyclohexanone (Fig. 7.03), the SCF-MO calculations produced the MO coefficients of the n and π^* orbitals given in Table 7.1 (6).

Comparing these coefficients with those of unsubstituted cyclohexanone, we can see at once the change in the charge distribution, which is especially noticeable in the nonbonding orbital, where, for example, the coefficient on the oxygen atom has changed from -0.61 to -0.71 (an increase in the localization of the nonbonding orbital on oxygen!), and in the asymmetry introduced by the redistribution of that part of the n orbital charge from hydrogen 9 to other parts of the molecule. Note that the π^* orbital has not changed very much and that a node of zero charge density must exist along the z axis between the carbonyl oxygen and carbon 1, since the sign of the coefficients changes between these two atoms. Interestingly, there is also a node in the n orbital between these atoms (the coefficients are $+0.30$ and -0.71). The existence of the nodal point in the π^* orbital had been predicted earlier from a simple Hückel treatment and

Table 7.1   Molecular Orbital Coefficients of Cyclohexanones[a]

	X_1	X_2	X_6	X_O	h_7	h_{15}	h_9	h_{13}
	\multicolumn{8}{c}{n Orbitals of Cyclohexanone and 3-Methylcyclohexanone}							
Cyclohexanone	0.35	−0.32	−0.32	−0.61	0.15	−0.15	−0.21	0.21
3-Methylcyclohexanone	0.30	−0.29	−0.29	−0.71	0.10	−0.11	—	0.17

	Z_1	Z_O	h_8	h_{16}
	\multicolumn{4}{c}{π^* Orbitals of Cyclohexanone and 3-Methylcyclohexanone}			
Cyclohexanone	0.62	−0.64	0.24	0.24
3-Methylcyclohexanone	0.60	−0.64	0.25	0.25

[a]In the n orbitals, X_1 and X_O refer to the $2p_x$ orbitals centered on carbon 1 and on the oxygen atom, respectively; h_7 refers to the $1s$ orbital centered on hydrogen 7. In the numbering convention use in the octant diagram of Fig. 7.11 (used by Pao and Santry), hydrogens 7 and 8 are on carbon 3, hydrogen 13 on carbon 5, and hydrogens 15 and 16 on carbon 6. (In Fig. 7.03 a different numbering system is used.) In the π^* orbital, Z_1 and Z_O refer to the $2p_z$ orbitals in a similar fashion.

formed the basis for the octant plane perpendicular to the $(C)_2C=O$ plane in the ketones; now the Pao and Santry treatment led to the prediction of a similar node in the n orbital consistent with and reinforcing the predictions of the octant rule. The new calculations, however, did indicate that this third octant surface was not a plane, but rather a curved surface running back from the $+z$ direction after it crossed the $C=O$ bond (Fig. 7.11a). Agreement with the experimental observations on several substituted cyclohexanones was excellent, and except for one case, axial 3-methyl-cyclohexanone, it was in agreement with the octant rule. It is clear, however, that the difficulties in the application of the octant rule to the skewed cyclopentanones and to compounds containing vicinal cyclopropyl groups would not yield to this treatment.

Of the several treatments using the direct dissymmetric chromophore approach since that of Pao and Santry, two are especially significant with respect to the problems we have been discussing. The first, by Richardson, Shillady, and Bloor (44), using the INDO variant (45) of the SCF-MO methods, emphasized the effect of the twisted ring in cyclopentanones. More recently Bouman and Lightner (7), using the CNDO/S variant (46), derived a set of nodal surfaces that are different from both the earlier proposed surfaces and that seem to hold considerable promise for accommodating most of the observations to date, including the effect of the vicinal cyclopropyl groups, the antioctant behavior of fluorine, and the

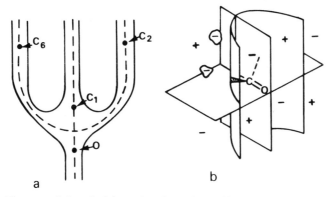

Fig. 7.11 The octant "planes" of the carbonyl $n \rightarrow \pi^*$ transitions as calculated by (*a*) Pao and Santry (6) and (*b*) Bouman and Lightner (7).

orientation effects on aminoketones (47). These new nodal surfaces (Fig. 7.11*b*) result from the nature of the SCF-MOs produced by solution of the Schrödinger equation in the one-electron approximation using the particular semiempirical parameterization that characterizes the CNDO/S method. It is altogether difficult at this stage, and especially without a much more extensive discussion of these methods than is possible here (48), to arrive at a good intuitive understanding of the reason this particular method seems to be more successful than others, but a few pertinent remarks may be made. Each of the atomic orbitals (AOs) that are used in the linear combination of atomic orbitals (LCAO) approximation to create the starting $\langle n|$ and $\langle \pi^*|$ orbitals can give rise to a contribution to the electric dipole transition moment components.

In the SCF method a starting "basis" set of wavefunctions is chosen for each of the orbital electrons, and by successive solutions of the Schrödinger equation, acceptable ones are arrived at—these being orbitals that, when substituted back in the equation for further solution, result in a smaller lowering in the orbital energy or in the total energy than is warranted by the cost or the complexity of an additional calculation.

In particular, as we saw earlier in this section, any components directed along the z axis will contribute to the chiral activity, since the magnetic dipole transition moment is largely directed along this axis in the $n \rightarrow \pi^*$ transition of the ketones. What is observed by Bouman and Lightner is that the hydrogen atom orbitals contribute strongly to the z component of $\langle n|\nabla|\pi^*\rangle$. They point out that it is very likely that the reason that this has not been noticed before "can be ascribed to widespread cancellation of terms (from hydrogen atom AOs) across local symmetry planes" (7) in these molecules. The puzzling antioctant behavior of fluorine in cyclohexanones also yields to these calculations, the sign changes coming primarily from changes in sign of the mixing of $2p_x$ and $2p_y$ AOs on the carbonyl group in going from a methyl to a fluorine substituent, and also from a change in sign of one of the orbital components on the carbon to which the fluorine atom is attached. The precise physical reason for these sign changes is still somewhat obscure, but the effect is completely consistent with the observed chiral activity of equatorial 3-fluorocyclohexanone. Of particular importance are the changes in the nodal surfaces between those of Fig. 7.11b and those of the original octant rule or of the Pao and Santry surfaces (Fig. 7.11a). The new surfaces bring several "antioctant" observations into line with the predictions, for example, that of the β-methyladamantanones (49). The octant projections of these on the original Moffitt octants put the only asymmetric carbons into the same octant, predicting a clear negative sign for the CD of the $n \rightarrow \pi^*$ transitions of both the axial and equatorial compounds; but experimentally, the β-axial compound was found to have almost no, or a small positive, activity (49). In the Bouman and Lightner octants, however, the β-axial methyl projection is just in front of and very close to the curved nodal surface (upper right) and thus in a region where it should make only a very small positive contribution.

One other thing worth noting and remembering about these treatments is that the concept of nodal surfaces is more appropriate to the perturbation or independent systems treatments than to dissymmetric chromophore treatments. That is, these surfaces delineate parts of space relative to axes fixed in a nominally planar carbonyl chromophore, in which perturbing atoms produce chiral transitions. However, these surfaces are produced by not restricting the molecular orbitals to the electrons of the planar chromophore and thus represent a dissymmetric chromophore treatment. Undoubtedly more remains to be understood about the optical activity of the ketones, especially of that of the higher electronic states, but it is clear that the chiral activity of the $n \rightarrow \pi^*$ transition is far better understood than it was even just a few years ago.

Ernstbrunner, Giddings, and Hudec have recently made another interesting application of direct quantum calculations to the chiral activity of ketones (50). Rather than calculate transition moments and rotational strengths for individual molecules, they have used the CNDO/S method (51) to calculate the orientation of the nonbonding orbital axis on the oxygen atom in the ground state configuration of the carbonyl, and of the π^* orbitals (the latter still in the virtual orbital approximation). The dihedral angle of twist between these orbitals was then correlated with observed rotational strengths. The calculated angles and, of course, the actual angles are affected by substituents, which is just another way of saying that the substituents form part of the extended dissymmetric chromophore. In this way the transmission of substituent effects in ketones is correlated in a very systematic way, which brings at least some of the nonoctant (or disignate) behavior of particular substituents into correspondence with predictions. On the basis of these calculations and correlations, the rotational strength is found to be given by $R(n \rightarrow \pi^*) = C(n)\theta(n_0) + C(\pi^*)\theta(\pi^*)$, where the θ's are the twist angles (clockwise twist is assigned a positive value when the ketone is viewed in an octant rule projection) and the coefficients $C(n_d)$ and $C(\pi^*)$ come from calculated or empirical data on rotational strength. Whether a particular substituent exercises an effect of the twists on the n_0 and π^* orbitals appears to be correlated with the existence of a contribution to NMR long-range coupling constant J. The method is not particularly easy to use and has to be correlated separately for different types of compounds, but does appear to be very successful in rationalizing chiral substituent effects in the $n \rightarrow \pi^*$ transitions of ketones. A similar entirely empirical correlation has also recently been made by Kirk (52).

7.22 The Hydroxyl Chromophore and Additivity Relationships

The hydroxyl chromophore was responsible for the major part of the optical activity of the first pure organic compound to be measured (53). The measurement on cane sugar was, of course, made with white light, and it was to be three-quarters of a century before any substantial connection would be made between the molecular structure of the sugars with their numerous hydroxyls and the optical activity of these compounds. Even at this stage in the late nineteenth century, the specific relationship between the optical activity and the hydroxyl chromophore was not recognized; the -OH groups being accepted as simply one of the substituents of a tetrahedral carbon atom that could make the carbon atom optically active. It is interesting to examine briefly how this affected the course of research and of the understanding of chiroptical phenomena for the four or more decades that followed, even up to 1960. A large number of measurements of the optical activity of organic and inorganic compounds were made during this period, including dispersion measurements; the Cotton effect was discovered; and the Rosenfeld equation was developed. But it was only in 1934, for example, that Boys (54) attempted to explain the optical activity of (-)-(R)-2-butanol on the basis of the interaction of isotropically polarizable groups of which the hydroxyl group is one. It was not until 1940 that Mulliken (55), without reference to the implications for chiral behavior, assigned the lowest energy singlet electronic transition in alcohols as an $n \rightarrow \sigma^*$ type. Boys's prediction was the wrong sign [he predicted a negative rotation for the (S)-configuration]. Continued attempts have been made both experimentally and theoretically to treat the optical activity of hydroxyl compounds. Some of these will be discussed below, but first we consider how some of the early work led to the concept that the contributions of atoms or groups of atoms make algebraically additive contributions to the overall optical activity of the molecules containing them. This concept, which arose from measurements on simple sugars and alcohols, dominated much of the interpretation of the observations of optical activity until the late 1940s and even the early 1950s, despite the fact that there was clear experimental evidence, as well as the theoretical developments of the polarizability and one-electron theories, that such additivity was at best only a gross approximation. This additivity of partial contributions to rotation, or the principle of optical superposition as it was designated by Guye and Gautier (56), is expressed (57) as follows: *"The rotatory power of a substance with several asymmetric carbon atoms may be deduced by evaluating the optical effect of each asymmetric carbon atom as if the rest of the molecule were inactive and forming the algebraic sum of these different effects."*

The principle, of course, worked very well for some simple acyclic carbon compounds in which rotation around single bonds is relatively free so that there is little or no steric interaction between adjacent optically active groups, and reasonably well for series of similar molecules in which the same steric interactions recurred. In 1961, Kauzmann, Clough, and Tobias very specifically made this important point:

It is clear that the position of each group relative to other groups in the molecule would have to be known in order to be able to give a sign to its contribution. This means that groups in an optically active molecule cannot be considered to be acting independently in contributing to optical rotation; the contribution by each group is determined by the presence of other groups. In short, no principle of the additivity of independent contributions by groups can serve as a sound basis for a general empirical theory relating structure to optical rotation (58).

Nevertheless, while it no longer dominates interpretation of chiroptical activity, the concept is still in use. For example, in a recent paper (52), D. N. Kirk makes this statement in a discussion of the difficulties encountered in applying the octant rule to cyclopentanones: "Assuming the additivity of effects of substituents, we observe the *difference* (or the algebraic sum) of the separate contributions of the α-axial bromo-substituent and the α-axial hydrogen atom." Kirk acknowledges that the experimental data alone cannot be used to derive the absolute values for the chiroptical effects of substituents using the assumption that the cyclohexane ring maintains an undistorted chair formation, and he is fully aware of the problem—but the concept persists and obviously has some utility. Kauzmann, Clough, and Tobias make their point in a number of ways. For example, they point out that the optical rotations of sugar alcohols in which CHOH groups are replaced by CH_2 groups often tend to be several times as large as those containing the extra CHOH groups; the molar rotation at the sodium D line of 1,6-di-deoxy D-mannitol, $CH_3(CHOH)_4CH_3$, with only four hydroxyl groups is -31 deg dm^{-1} mol^{-1} as compared to the $+4.3$ deg dm^{-1} mol^{-1} rotation of L-idomethylitol, $CH_3(CHOH)_4CH_2OH$, the latter having an additional methyl group and the *same* number of asymmetric carbon atoms. On the basis of considerations like these, a system of pairwise interactions between groups was constructed. Taking into account the effect of conformational equilibria, the method allows the rotations from these interactions to be additive. At best this still restricts the method to groups of similar compounds because, taking a typical example, the interactions between atomic groups across fused rings will be different from interactions between the same groups located in a single ring or in an aliphatic compound. The principle of pairwise interactions has its antecedents in still earlier attempts along similar lines, especially those of Whiffen (59) and of Brewster (60). In Brewster's case the principle revolves around summing the contributions not of atoms or diatomic groups, but of all groups of four nonplanar atoms that thus form the optically active disymmetric units; these units are termed "conformational units" (Fig. 7.12). Brewster's rule (60), designated the "conformational dissymmetry rule," is as follows:

1. "Conformation unit" I is dextrorotatory (at the sodium D line) in the absolute configuration shown in Fig. 7.12.

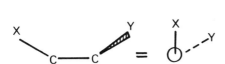

Fig. 7.12 The conformational dissymmetry unit of four atoms and its Newman projection. The four-atom unit will arise again in other contexts. The absolute sense illustrated corresponds to a right-handed screw or helix and has the IUPAC designation M.

2. The magnitude of the rotatory effect is proportional to constants characteristic of the terminal atoms X and Y. (Brewster points out that it seems likely that the constants will be different if the central atoms are other than carbon.)

3. The rotatory power of a full conformation (as II) is the sum of the rotatory contributions of the six contributing constituent conformational units (Fig. 7.13).

Fig. 7.13 Newman projection of the nine constituent conformation units associated with an optically active tetrahedral carbon and its nearest neighbor in a staggered conformation. Note that Brewster's rule does not require that all the tetrahedral or conformational (dihedral) angles be equal, but if they are, as illustrated in this figure, then in addition to the fact that conformational units in which the dihedral angles are 180 degrees ($XCCZ$, $YCCH$, and HCCH) do not contribute because they each contain a plane of symmetry, two of the interactions ($+$HCCZ and $-$HCCZ, designated as $+HZ$ and $-HZ$) algebraically cancel. Therefore, $[\theta]_D = k[XY - HY - HX + HH]$ where the products, kXY, kHY, kHX, and kHH, represent the contribution of each of the conformational units to the molar rotation at the sodium D line.

Again, it is apparent that the rule will only work for rather specific groups of compounds, and even then only imperfectly from the quantitative point of view. As Brewster points out, the system is relative to a standard for a given atom or bond, which may be taken as C–H. It clearly also depends on uniformity of bond angles, a situation not often realized, especially in ring compounds. Brewster also recognized that the rule would not necessarily apply at absorption band maxima; the presence of chromophores that absorb close to the frequency of the sodium D line or even at longer wavelengths can completely vitiate the results. These and other obvious complexities make the principle of optical superposition of very limited value in chiral phenomena. The rules were, of course, instructive and very helpful in pulling together disparate observations. In some cases they provided the only firm basis for stereochemical interpretations, but they perpetuate a fictional concept. Since more sophisticated and probably more rational concepts and methods are also still limited by the complexities of quantum mechanical developments and the cost of large computations, the early rules need not be discarded but only used with considerable caution. There is no need to belabor further the fact that in many cases the additive relationships proved to be inadequate, or that the first application of perturbation theory (58) yielded rotatory strengths two orders of magnitude too small (61). The first clues to the interactions affecting the chiral activity of hydroxyl groups resulted from the dispersion measurements by Mateos and Cram (62) in the 300- to 600-nm range. They demonstrated that in simple acyclic monohydroxy alcohols with two asymmetric centers, one of which has a phenyl substituent, no overall correlations or additivity relationships could be discerned, but that there is a difference between the behavior of threo and erythro compounds (Fig. 7.14). Depending on whether or not the two asymmetric centers were substituted with the same or different alkyl groups, either (but not both) showed the onset of a

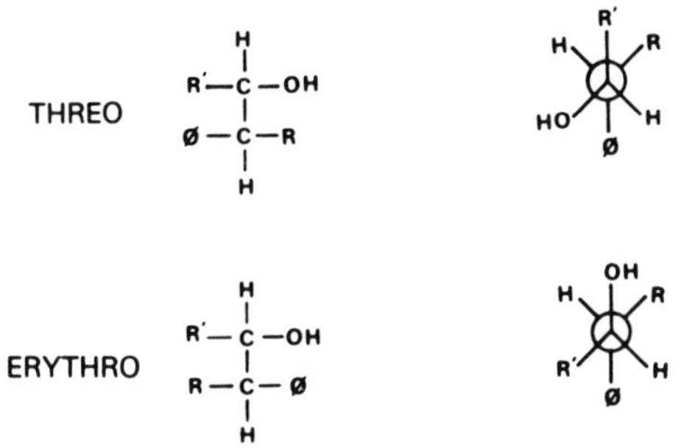

Fig. 7.14 The threo- and erythroisomers of acyclic monohydroxy alcohols with two optical centers in the Fischer convention (left). Note that there are three possible staggered conformations (with respect to rotation about the central carbon-carbon bond) for each isomer.

Cotton effect in the ORD curve just above 300 nm. It is clear that in these compounds the difference resides in the relative steric dispositions of the hydroxyl and phenyl groups; the Cotton effect must manifest itself as a result of some interaction between these groups rather than originate from the superposition of the chiral activity associated with the two asymmetric centers. The exact interpretation of this result is still to be made, but there is no question that it provided a stimulus for additional work on the chiral behavior of the hydroxyl chromophore (as well, of course, as the benzene chromophore).

The hydroxyl group chromophore, like that of the ketone group, is dominated in the ground state by the nonbonding electrons of the oxygen atom, but there the resemblance ends. It will be recalled that the nonbonding oxygen orbital in the ketone is, in zeroth order, a p orbital perpendicular to the nodal plane of the π orbital between the carbon and oxygen atoms, and that even in higher order this symmetry is maintained with other contributions to the nonbonding orbital from adjacent carbon and hydrogen atoms. The lowest energy transition is then an $n \rightarrow \pi^*$, which carries charge in a circular or helical motion to the plane of the π system from the perpendicular plane and thereby produces a magnetic transition moment directed along the carbon-oxygen bond in accord with the simple notions of electromagnetic theory (Chapter 2). In the hydroxyl group this p orbital forms a sigma bond with the hydrogen, while the other p orbital, which is now not required for π bonding with the carbon, contains the nonbonding electron. Furthermore, there is no low-lying π^* orbital into which this nonbonding electron can be promoted. The lowest energy *unoccupied* orbitals are σ^* formed from antibonding combinations of oxygen-hydrogen or oxygen-carbon σ electrons and of Rydberg orbitals formed from higher energy unoccupied s and p orbitals on the respective atoms. The assignment of the lowest energy transition at about 182 nm is not unequivocal (2); there is, however, considerable evidence that it is a valence transition of the $n \rightarrow \sigma^*_{OH}$ type and that the next more intense transition is a Rydberg transition of the $n \rightarrow 3s$ type (63), but the latter may also be $n \rightarrow \sigma^*_{CO}$. The transition to the Rydberg $3s$ state is not accompanied by a change in angular momentum; therefore, the magnetic dipole transition moment is zero. In classical terms this is so because the increase in the orbital radius is exactly compensated by a decrease in the electron velocity, or in quantum terminology there is no change in the orbital (or total) angular momentum quantum number. Still another way of looking at this is that the magnetic transition moment must be invariant to the perturbation of the radiation field in order to be finite; in other words, it would have to have a totally symmetric component. The n orbital changes sign on reflection through its nodal plane and the s orbital does not; therefore, the n orbital has b symmetry (b_1 in this case because it does not change sign on reflection through any other plane), while the s orbital is the totally symmetric a_1. If in the first approximation we consider the hydroxyl chromophore to be an axially symmetric linear chromophore, it belongs to the $C_{\infty v}$ group. Referring to the character table in Appendix I, we see that transition to a_1 states in $C_{\infty v}$ are magnetic dipole forbidden (and electric dipole allowed). Of course, these considerations apply exactly to pure atomic transitions. Because of coupling of the orbital angular momentum in many-electron systems, it is possible that the magnetic transition moment for a Rydberg transition will be greater than zero, but it is not likely to be large. The presence of significant optical activity would therefore be an indication that a valence transition rather than a Rydberg transition is involved (64). There have been at least two investigations of the circular dichroism of alcohols with respect to its electronic origin (65, 66); in both cases, the low-energy transition was observed to have significant optical activity. The strong optically active hydroxyl transition occurs at 182 nm in (+)-S-2-butanol. From the theoretical calculations and the presence and properties of this transition in this and other hydroxyl compounds, the only reasonable conclusion is that this is the valence transition involving the promotion $n \rightarrow \sigma^*_{OH}$.

The chiral activity of (+)-S-2-butanol provides a good example with which to diverge for a brief discussion of the effect of conformational equilibria. It should be obvious from the discussion of the ketones in the previous section and from some of the theoretical considerations of Chapter 4 (for example, of the differences in the origin of the optical activity of a planar and nonplanar diketopiperazine) that the activity of a nonrigid molecule, and especially of one where the vicinal groups whose interactions may contribute strongly to the rotational strength are in close nonrigid relationships to each other, will be strongly dependent on the conformational equilibria. We will see additional examples of this in Chapter 9, where different conformations of polymers give rise to different chiral activity, and in Section 7.25, where the effect of the distribution of rotameric isomers on the quantum mechanically calculated rotational strength associated with a diene chromophore is discussed. The activity of (+)-S-2-butanol is an instructive example with which to introduce the subject.

There are a number of conformation equilibria that can exist in the 2-butanols. The methyl groups, for example, can rotate about the carbon-carbon bond, and the hydroxyl group can rotate about the carbon-oxygen bond. In addition, the carbon backbone or skeleton can rotate about the central carbon-carbon bond. In Fig. 7.15 the Newman projections of the staggered conformations of the latter rotamers are illustrated. For the purpose of this example, we will consider only this and the hydroxyl rotation. The reasonable assumption can also be made that at or near room temperature, energy minima occur only for the illustrated staggered conformations and not for eclipsed conformations. For any fixed conformation of the hydroxyl group, the rotational strength of a given transition, say the $n \rightarrow \sigma^*_{OH}$, will be some Boltzmann-weighted average of the rotation

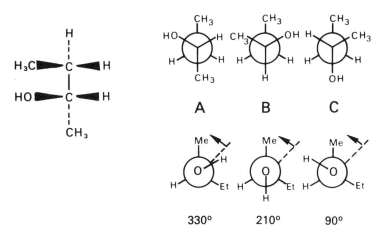

Fig. 7.15 The staggered conformations of (+)-S-2-butanol. Upper row: The conformations A,B,C for rotation about the central carbon-carbon bond. Lower row: The conformations defining the angle of rotation of the hydroxyl bond about the carbon-oxygen bond.

associated with each of the conformations:

$$R(n \to \sigma_{OH}^*) = \frac{\sum_r R_r e^{-\Delta \mathcal{E}_r/kT}}{\sum_r e^{-\Delta \mathcal{E}_r/kT}}, \qquad (7.213)$$

where $\Delta \mathcal{E}_r$ is the difference between the potential energy of the rotational conformer r and the conformer of lowest energy. If, in fact, $[\Delta \mathcal{E}_r/kT] \gg 1$ for only the staggered conformations, then the total rotational strength will be close to the simple average of the rotational strength for each of the three conformations of this type:

$$R(n \to \sigma_{OH}^*) = \frac{\sum_{r=1}^3 R_r}{3}, \qquad (7.214)$$

where $r = A, B, C$ and R_A, R_B, and R_C are the separate rotational strengths associated with the corresponding conformations. In treating this problem Snyder and Johnson (66) made the assumption (by analogy with known data for butane) that conformation B is an energetically unfavorable conformation and therefore calculated the rotational strengths using only the other two backbone conformations. Within this framework, using the perturbation theory approach (Chapter 4 and references 67 and 68), they found that the major contributions to the 182-nm chiral activity comes from terms 4.202-2 and 4.202-6 (la and le of reference 66), each term contributing variably as the hydroxyl bond is rotated about the carbon-oxygen bond. The results of this calculation are illustrated in Fig. 7.16. Note especially how the sign changes as the OH bond is rotated; it is strongly negative for both backbone conformations when the OH rotation angle is 200 ± 50 degrees (the angle is defined in the Newman projection of Fig. 7.15). Except for the near-zero contribution of the backbone conformer C when the hydroxyl rotation angle is near 0 degree, it is elsewhere always positive, as is the observed rotation and circular dichroism of the 182-nm band of (+)-S-2-butanol. The calculated rotational strengths in this region vary from about 0.04 to 0.51 DBM using conformation A alone compared to the experimental value of 0.39 DBM. When conformations A and C are each calculated to contribute half the rotational strength, the values in this region of the hydroxyl rotation vary between -0.05 and $+0.09$ and are again distinctly large and negative in the 200 ± 50 degree region. Using these results and the additional result that the rotational strengths of the next two higher energy transitions (assigned as predominantly $n \to \sigma_{CO}^*$ and $n \to 3s_O$) are calculated to be strongly positive and negative for the same A conformation, Snyder and Johnson conclude that

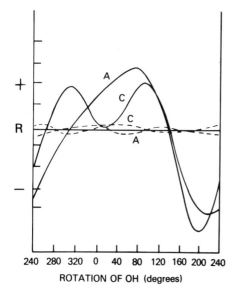

+

R

−

240 280 320   0   40   80   120 160 200 240
ROTATION OF OH (degrees)

Fig. 7.16 The calculated (66) rotatory contribution from the μ-$\underset{\sim}{m}$ mechanism, Eq. 4.202-2 (———), and the dipole-coupling mechanism, Eq. 4.202-6 (---), to the rotational strength of (+)-S-2-butanol as a function of the hydroxyl bond rotation for two different backbone conformations (A and C of Fig. 7.15).

conformation A is the predominant one at room temperature, and that hydroxyl rotation about the carbon-oxygen bond is restricted to the region in the vicinity of 180 to 320 degrees, defined by the Newman projections of Fig. 7.15 (69).

It will not have escaped attention that only a few paragraphs back we pointed out that a transition to an s Rydberg state is not expected to be magnetic dipole allowed, but that Snyder and Johnson assign the third fairly strong optically active transition in 2-butanol as a Rydberg-like $n \rightarrow 3s_O$. This provides us with an opportunity to demonstrate another application of perturbation theory (or the idependent systems approach) to the interpretation of chiroptical activity; this time the situation is quite different from the $n \rightarrow \pi^*$ transition of the ketone, which is magnetic dipole allowed and where the one-electron mechanism was the predominant contributor. We are now dealing with a magnetic-dipole-forbidden transition. There are, of course, other ways of looking at this than perturbation theory. In particular, the system could be treated as a dissymmetric chromophore, in which case we would expect to find that the transition assigned as $n \rightarrow 3s_O$ is a highly coupled transition, with an upper state configuration that is far from a pure atomiclike s orbital. Although it may retain Rydberg-like characteristics, it is in fact, at least in the asymmetric environment of the chiral compound, a partially magnetically allowed transition. But our point here is to demonstrate the application of the perturbation theory approach to this problem. Referring to Chapter 4 and examining the terms 4.202-1 through 4.202-6 of the perturbation theory expression, we see that there are only four terms that do not involve an intrinsically allowed magnetic dipole transition moment $\mathbf{m}_{ia0}$, which is null for the transition in question ($0 \rightarrow a = n \rightarrow 3s_O$). Three of these terms are the first parts of Eqs. 4.202-2, 4.202-3, and 4.202-4, in which there are contributing magnetic moments but of a different transition: $0 \rightarrow b$ or $a \rightarrow b$; the fourth is Eq. 4.202-6, which does not involve any intrinsic magnetic transition moment. The terms in Eqs. 4.202-3 and 4.202-4 are the static interactions $V_{iab, j00}$ or $V_{i0b, j00}$ of the electric dipole moments for the transitions $a \rightarrow b$ and $0 \rightarrow b$ with the permanent charge distributions on another group, the jth group. Since the alcohol has no fixed charges or other groups with large dipole moments, these terms are negligibly small. This leaves only the first term of Eq. 4.202-2 and the term 4.202-6 to consider as possible candidates for inducing chiral activity in the $n \rightarrow 3s_O$ transition. The first term of Eq. 4.202-2 is significantly large only if the electric dipole transition moment μ_{i0a} of the Rydberg $n \rightarrow 3s_O$ is very large, since it contributes as the dot product with an intrinsically small quantity the magnetic transition moment $\mathbf{m}_{jb0}$. (This latter statement can be made of all of the terms in Eqs. 4.202-1 through 4.202-4 and is indeed the reason why they are frequently quite small. But it is also true that in many cases the relevant electric dipole transition moment is large.) The absorption intensity of this Rydberg transition is not particularly large. Note that the optical activity shows that the band near 160 nm consists of two overlapped transitions. As a consequence of the low or moderate intensity and the facts that the coupling magnetic moment is intrinsically small and need not be collinear with the electric moment μ_{i0a} of the $n \rightarrow 3s_O$ transition, only the dipole coupling term, Eq. 4.202-6, is likely to make a significant contribution. Using only this term, Snyder and Johnson did, in fact, calculate rotational strengths for conformer A that are of the same order of magnitude (within half of the experimental value) as the experimentally observed circular dichroism. So for a magnetic-dipole-forbidden transition, perturbation theory does provide a mechanism by which Rydberg's can contribute to chiral activity.

The hydroxyl chromophore is a significant substituent in many thousands of optically active organic compounds other than simple aliphatic alcohols and sugars. Most of the foregoing discussion can be applied to understanding the optical activity of these compounds, especially if they have no chromophores that absorb at longer wavelengths. The extension to polyhydroxyl compounds (polyols) would be particularly interesting. Although the methods are quite different, it is unfortunate that one of the most complete treatments of polyols (70) did not have the benefit of the Snyder and Johnson treatment (the two papers appearing in different journals at nearly the same time). Applequist's (70) calculations were based on his variation of Kirkwood's classical polarizability theory; the sign of the optical activity of a large number of cyclohexanepolyols was successfully predicted, but the magnitudes were generally much larger than experimentally observed or predicted by Whiffen's empirical rules (59).

Before proceeding with the next two sections, where nonconjugated chromophores other than ketones and alcohols are treated, let us summarize briefly what has been covered in the discussion of these two chromophoric systems. We have seen that perturbation theory applied to nominally symmetric chromophores (the C_{2v} system of the ketone and the $C_{\infty v}$ system of the hydroxyl group) can be used to account, in a rather gross but nevertheless informative way, for the chiroptical activity of compounds containing these "chromophores." In the case of the ketones, where we considered only the magnetically allowed $n \rightarrow \pi^*$ transition, the one-electron mechanism sufficed to produce a rule, the octant rule, which, while far from perfect, has been of great help in elucidating the stereochemical configurations of compounds containing this group. When the octant rule failed, nonperturbative direct quantum treatments succeeded in modifying the geometric surfaces that define the space in which perturbing substituents produce chiral activity of a given handedness. Furthermore, these direct dissymmetric chromophore calculations are beginning to produce explanations for more subtle effects—the difference between substituents and types of bonding. We have also seen how symmetry considerations can be of important qualitative help to both the perturbation and nonperturbative direct calculations. Using the hydroxyl chromophore, we examined the difference in the activity arising from different optical transitions of the same atomic grouping; the $n \rightarrow \sigma_{OH}^*$ and $n \rightarrow 3s_O$ transitions of the hydroxyl have their perturbative origin in different mechanisms, the latter because it is a magnetic-dipole-forbidden transition almost entirely in the dipole-coupling mechanism given by the term containing $\mu_{j0b} \times \mu_{i0a}$ in the Kirkwood-Tinoco theory. In addition, we used the chirality of (+)-S-2-butanol to examine the relationship between conformer equilibria and the sign and magnitude of the calculated rotational strengths.

In the discussions of the nonconjugated chromophores that follow this section, we will be less deliberately analytical in keeping with the considerations expressed at the beginning of this chapter, except perhaps for the monoolefins. Nevertheless, an attempt will be made to choose examples that are informative in concept rather than in coverage.

7.23   The Olefin Chromophore

Historically, the olefin chromophore consists of the isolated valence double bond between sp_2 hybridized carbon atoms. We single it out here for two reasons. One is to demonstrate how complex the analysis of a simple chromophoric system can be from both the experimental and theoretical points of view; as a result, empirical correlations assume overwhelming importance. The second is precisely because the theoretical approach to the chiroptical activity of olefins has been continuously different from the approach to the ketones and alcohols, the latter of which have not yet been treated by a molecular orbital method. Antithetically, the olefins have not been formally treated by a perturbation method, although they have been subject to empirical correlations based on perturbation concepts.

The history of the experimental and theoretical analysis of the olefin chromophore is a fascinating one. It is as if the largest available guns were reserved for the smallest available target. The first resolved electronic absorption spectrum of ethylene was reported by C. P. Snow and C. B. Allsopp in 1934(71). In 1940 Price and Tutte (72) published an electron impact spectrum. In 1932 Mulliken (73) had already described the lowest energy singlet transition as a valence transition arising from an electron promotion from the highest filled π orbital to the lowest filled π^* orbital. (In the Mulliken notation, which we record here to facilitate comparison with the literature, the orbital promotion $\pi \rightarrow \pi^*$ is represented as generating the V state from the ground state; i.e., $N \rightarrow V$. The V state is frequently designated as a $\pi\pi^*$ state.) Although some work continued in the interim, especially on the alkyl olefins, it was not until 1955 that a higher resolution spectrum of ethylene and deuteroethylene was published by Wilkinson and Mulliken (74). The higher resolution showed vibrational components in the $N \rightarrow V$ transition that were interpreted as coming principally from a stretching vibration of the carbon-carbon bond. It was another decade before McDiarmid and Charney (75) obtained a still better spectrum of the $\pi \rightarrow \pi^*$ band of ethylene, a broad intense electronic band with a maximum near $60,600$ cm^{-1} (165 nm) in both the proto and deutero forms, and analyzed the prominent 807-cm^{-1} (547 cm^{-1} in the deutero compound) vibrational progression as coming from promotions to an excited state torsional mode in which the hydrogen (or deuterium) atoms at one end of the molecule are twisted with respect to those at the other end. This fit nicely with the predictions made by both Walsh (76) and Mulliken (77) that in the excited V $(\pi\pi^*)$ state the equilibrium configuration of ethylene is perpendicular with respect to the methylene planes. In the planar ground state, ethylene has a center of inversion, three reflection planes, and three twofold rotation axes and therefore belongs to the D_{2h} symmetry group. The perpendicular configuration of the excited $\pi\pi^*$ with only two reflection planes and one twofold axis is C_{2v}. Neither C_{2v} nor D_{2h} chromophores can be optically active, but molecules with intermediate configurations, which have three twofold axes but no reflection planes as symmetry elements, belong to the D_2 symmetry group and thus may be optically active. We will see that this has important implications for the chiroptical properties of olefins, a fact that was early recognized by Moscowitz and Mislow (78) in their attempt to interpret the optical activity of *trans*-cyclooctene. The optical activity of the chiral olefins has been an important part of the attempts to understand their electronic properties. The ability to predict correctly the sign and magnitude of their rotatory strength may be used as one of the criteria of quantum calculations of excited states. With the exception of the work published in the last few years, the history and analysis of these efforts have been reviewed by Robin (2), and except for a few pertinent comments, especially with respect to new work, we will not further review it here.

The aspects most critical to this discussion are those concerned with the interpretation of the origin of certain features in the circular dichroism and ordinary absorption spectra of substituted olefins in the region of the strong $\pi\rightarrow\pi^*$ band. In Fig. 7.17 where the achiral absorption spectra of ethylene, *trans*-butene, and tetramethylethylene are illustrated, it will be observed that a series of very sharp features appear in the vapor spectrum of *trans*-butene, sharper than the vibrational features of the ethylene spectrum rather than more diffuse as might be expected in a molecule with more atoms (more degrees of vibrational freedom) if the features belonged to the same transition in both molecules. That they do not so belong was shown by McDiarmid, who found that the sharp components in the *trans*-butene spectrum are members of a $\pi\rightarrow3s$, that is, a Rydberg transition superimposed on the $\pi\rightarrow\pi^*$ band (79). Keeping in mind that in the previous section, we showed that a Rydberg transition could acquire rotatory strength by a dipole-dipole coupling mechanism, we can proceed first with the understanding that in zeroth order no magnetic dipole moment is associated with an s-type Rydberg transition, which is therefore likely to be only very weakly optically active, if at all. Consequently, it seemed at the time that the first work in this field was being done that significant optical activity, not traceable by approximate molecular orbital theory to the $\pi\rightarrow\pi^*$ transition, would have to originate in still another very weak transition, perhaps unobservable in the ordinary absorption spectrum. In tetramethylethylene (Fig. 7.17) and in cyclic dienes, similar features are present, but the structure is not quite so sharp, and for some time it was thought that these features represented still another transition predominantly valence but with some Rydberg character, the latter arising

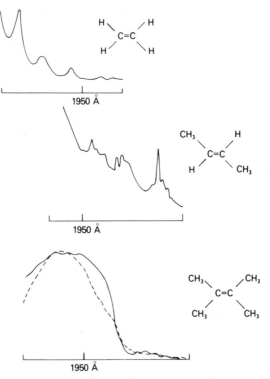

Fig. 7.17 The absorption spectra of ethylene (75), *trans*-butene (79), and tetramethylethylene (2). In the spectrum of ethylene the band structure consists of the vibrational components of the essentially valence transition ($\pi \rightarrow \pi^*$). In *trans*-butene the sharp features of the Rydberg transition ($\pi \rightarrow 3s$ and/or $\pi \rightarrow \sigma^*$) are superimposed on the vibrational components of the valence transition. In the spectrum of tetramethylethylene, the predominantly Rydberg transition (not quite so sharp as in *trans*-butene) has moved to lower energies, as can be seen from its disappearance in condensed media, a characteristic of Rydberg transitions. Solid lines are gas-phase spectra. Dashed line is film spectrum at 77°K. Wavelength increases to the right.

from excited state wavefunctions defined by large mean radii as represented by large values of $\langle \psi^* | \Sigma er^2 | \psi^* \rangle$. As a valence transition it could, if magnetic dipole allowed by symmetry, be associated with large magnitudes of chiral activity in contrast to the expected small magnitudes of the low-lying Rydberg transitions. When, therefore, the chiral activity of *trans*-cyclooctene in this region turned out to have a sign opposite to that expected from the $\pi \rightarrow \pi^*$ transition of a twisted olefin (80), the search was on for possible unknown states of ethylene and substituted ethylene that could contribute to the optical activity in this region. Calculations of varying degrees of complexity on ethylene have continued up to this time

(81). There is still no clear picture of how much valence and how much Rydberg character is to be associated with the $\pi\pi^*$ state, although in accord with experiment, the calculations predict it to be predominantly a valence state. The excited electrons cluster close to the carbon atoms; however, a small, but significant, electron density is contributed from orbitals spread over the rest of the molecule (82). In addition, while again the results are not quantitatively definitive, it is clear that both the Rydberg $\pi\rightarrow 3s$ and two other transitions, designated as $\pi\rightarrow\sigma^*_{CC}$ and $\pi\rightarrow\sigma^*_{CH}$ (83), hover in this spectral region. Substitution of the hydrogens raise or lower the energy of these weak transitions enough to make it somewhat ambiguous as to which transition is responsible for some of the smaller features seen in the spectrum. It is precisely for this reason that the chiral effects have been examined to see to what extent they can be used as criteria for assigning the responsible states and, conversely, to what extent the calculations can be used to understand the chiral activity. The problem can be approached with three questions: (1) What are the lower excited states of monoolefins? (2) Which of these states are responsible for chiroptical effects in this spectral region? (3) To what extent can empirical correlation help answer the problems posed in (1) and (2), or conversely, provide useful tools for investigating the stereochemistry of chiral olefins? We have just briefly reviewed the situation with respect to quantum calculations of the excited states and reached the conclusion that while the exact location of these states in ethylene and more complex monoolefins cannot be precisely specified, there are at least five states to which optical transitions are allowed that may be confidently predicted to appear (at frequencies lower than 80,000 cm^{-1}). These are listed in Table 7.2 for ethylene, together with some of their predicted spectral properties. (At least two other transitions have at various times been predicted to lie in this region, namely, the $\pi\rightarrow 3p$ and $\pi\rightarrow 3d$ Rydbergs (84, 85).) The question of which of these states is responsible for chiroptical activity may be further divided into two parts: What effects may be relegated to the ground state distortion of the molecule? What effects to differences in excited state geometry? These are the same questions that were discussed in Section 7.2 with respect to ketones and to which Fig. 7.08 and the accompanying discussion was addressed. The perturbation treatments used to attack this problem in the ketones have not been used, however, with the olefins. Instead, direct calculations for skewed geometries have been attempted (83, 85–87). We will examine here only the most recent of these (83) and discuss the results with respect to possible assignments of the features in the CD spectrum of *trans*-cyclooctene (Fig. 7.18) (88). The decomposition of this spectrum by Bouman and Hansen (83) yields a weak positive feature at 46,800 cm^{-1}, a strong positive band at 50,800 cm^{-1}, another

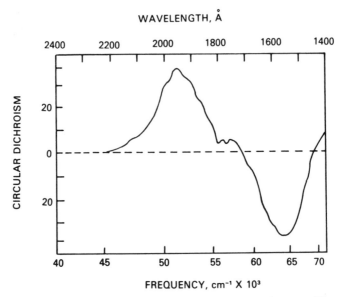

Fig. 7.18 The circular dichroism spectrum of *trans*-cyclooctene (88).

weak positive band at 55,600 cm^{-1}, and a strong negative band at 63,700 cm^{-1}. Because of strong overlapping, this decomposition may not be unique, but it is a reasonable one. In Table 7.2 the calculated optical properties of the transitions assigned by Bouman and Hansen and by McMurchie and Davidson (81) to twisted and planar ethylene, respectively, are placed in juxtaposition with the experimental results for *trans*-cyclooctene for comparison. The accord appears reasonable, and Drake (89), using the CD spectrum of A–B neolup-9-one supports the assignment of the lowest energy feature of the $\pi \rightarrow 3s$ transition, although he has to call on the complex structure of this compound to "stabilize" or "protect" the Rydberg state from an expected blue shift at low temperatures. The single direct calculation (85) of the rotatory strength of the $\pi \rightarrow 3s$ Rydberg in twisted ethylene does give quite a large value, -2.75×10^{-40} cgs (90). However, the calculated energies were substantially high and very pronounced changes from the values given in Table 7.2, which occur in the calculated oscillator and rotational strengths after the introduction of configuration interaction into the calculation. One possible alternative remains. It is clear that a marked correspondence (albeit with a rather low value for the calculated rotational strength) between the calculated rotational strengths for ethylene and the values observed for *trans*-cyclooctene would occur if the $\pi \rightarrow \sigma^*_{CH}$ calculated at 75,250 cm^{-1} in the twisted

Table 7.2 Assignments of the Electronic Transitions of the Olefins

		Ethylene (calculated)		Trans-Cyclooctene (observed)[g]		
Transition[a]	Energy (cm⁻¹)[b]	Oscillator Strength	Rotatory strength[h]	Spectral Energy (cm⁻¹)	Oscillator strength	Rotatory Strength[h]
$\pi \rightarrow 3s$	60,600[c]	0.02	−2.75	46,500	weak	−5
$\pi \rightarrow \pi^*$	60,200[d]	0.026	−68.2	50,600	~0.15	−70
	67,200[e]	0.316	−22.4			
	65,200[c]	0.113	−40.1			
	74,100[d]	0.147				
	51,400[f]					
$\pi \rightarrow \sigma_{CC}^*$	69,800[e]	0.037	−20.3	55,400	weak	−10
	52,700[f]		+24.7			
$\pi \rightarrow \sigma_{CH}^*$	75,200[e]	0.005	−0.4			
$\sigma \rightarrow \pi_{CH}^*$	79,400[e]	0.104	+105.1	63,400	weak	+110

[a]The symbols represent the dominant molecular orbital contribution to the state configuration.
[b]In the references these values are given in absolute electron volts.
[c]Calculated for planar ethylene (81).
[d]Calculated for twisted ethylene (85).
[e]Calculated for twisted ethylene (83).
[f]Calculated for twisted ethylene (87).
[g]Estimated by Bouman and Hansen (83) from the spectrum by Mason and Schnepp (88).
[h]In units of "reduced rotational strength," equal to 10^{-40} cgs.

ethylene moved down in energy to about 46,800 cm^{-1} in *trans*-cyclooctene, replacing the $\pi \rightarrow 3s$ as the origin of the low-energy rotatory strength; in this case no special mechanism would be required to lend magnetic dipole strength to the $\pi \rightarrow 3s$ Rydberg or to explain its unexpected location at such a low energy in this compound. The calculated and observed oscillator and rotational strengths for the first four transitions would then be in one-to-one correspondence, and a $\pi \rightarrow 3s$ contribution, if any, would have to be weak and hidden under the stronger CD bands. For the moment this must remain conjecture, to be settled by additional calculation on the chiral properties of the substituted ethylenes and a calculation of the rotational strength of the $\pi \rightarrow 3s$ Rydberg in twisted ethylene using these later higher order calculations.

Before leaving this discussion of the attempts to predict by direct quantum mechanical calculation the location and rotatory strength of transitions in the monoolefins, we should call attention to two factors that have thus far received only scant attention. One is that the CD band of the $\pi \rightarrow \pi^*$ transition of the olefin rarely deviates in shape from that of the ordinary absorption band. As a consequence, it is reasonable to conclude that the electronic transition and its vibrational components responsible for the electric dipole strength, and therefore for the ordinary achiral absorption, are largely responsible for the rotational activity as well. In ethylene itself, and very probably in exomethylene olefins, a large part of the ultraviolet band's intensity is from promotion to excited torsional states (nonvertical excitation in spectroscopic parlance). In ring compounds this distortion is undoubtedly inhibited by constraints of adjacent bonds. However, to accommodate the required excited charge distribution, some distortion must occur, and this probably means that not only does the double bond undergo some torsional distortion, but it must stretch as well. This has significance for the second factor, which is that almost all calculations are done for a fixed geometry. Since this geometry is not *necessarily* the one that produces the greatest intensity in the CD band, the calculations may be deficient for this reason alone. Until more attention is paid to this problem, even high-order calculations, at least on the olefins, will have to be accepted as indicative, but not necessarily quantitatively accurate.

The second sentence of the most recent and thorough empirical analysis of the circular dichroism of chiral olefins (91) contains the statement that "there is little agreement...concerning the relationship between structural features and the shape of the CD curve" of these compounds. In this *Tetrahedron* paper by J. Hudec and D. H. Kirk, 24 of the first 25 references represent almost all the papers published on the subject since 1965. Some of the relationships are successfully sorted, and reasonable correlations between the signs of the two lowest energy CD bands and the structures are established. Unfortunately, this does not result in a definitive generalization of the chiral effects of the lowest energy transitions of the olefins. The nonchiral D_{2h} symmetry of ethylene itself naturally led to early attempts to consider an octant rule as the perturbative symmetry for inducing chiral activity in olefins according to Schellman's precepts (Chapter 6). Despite the lack of a formal perturbation treatment, empirical attempts from the first treatment by Yogev, Amar, and Mazur (92) to the most recent by Hudec and Kirk, have, therefore, remained centered around an octant rule concept, the first actual designation of octant perturbation being that of Scott and Wrixon (93). Refinements of the perturbation by considering separately the handedness of the formal double bond with respect to allylic substituents at one or both ends of the olefin (94), as well as with the "polarization" of the allylic axial bonds (95), have helped sort out some results; but it is apparent that inordinate complexity is introduced by (1) the several overlapping or near-overlapping electronic transitions and (2) the possibility of varying contributions to the rotational strength arising from the different degrees of torsional distortion of the double bond due to strain in the ground state or from the extent of the transition density to vibrationally distorted upper states. These factors make it almost impossible to construct a generalized rule for all olefins. Hudec's and Kirk's classification of olefins into different types with respect to their structure and spectral or chiroptical behavior should therefore prove very useful despite its still severe limitations. The classification with respect to *chiral effects* depends on obeisance to a right-handed octant rule similar to that of the ketones, but with the viewing point taken as the center of the double bond and each end considered separately; so the "back octants" only are considered. The terms consignate and dissignate are then used to specify whether the CD band in question does or does not

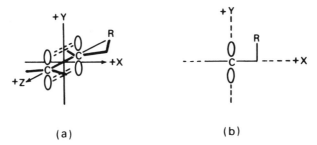

(a)   (b)

Fig. 7.19 (a) Coordinate frame for olefins. R is any "axial-allylic" substituent. (b) Projection of "rear octants" (in the $-z$ direction as viewed from the origin) showing the position of a substituent R in the upper right octant (91).

conform to the right-handed octant designation. In making the octant projection, the coordinate frame chosen is illustrated in Fig. 7.19 (96). The classification with respect to *compounds* proceeds essentially according to the degree and nature (1, 1; 2, 2; cis and trans) of the substituents, other than hydrogen, about the double bond. Four classes of compounds are defined:

Class (A): 1, 1-Disubstituted ethylene (exocyclic methylene compounds)

Class (B): *cis*-1, 2-Disubstituted ethylene (cycloalkenes)

Class (C): Trisubstituted ethylenes

Class (D): Tetrasubstituted ethylenes

The attribution of specific chiroptical effects to substituents in each of these types of compounds is further subdivided according to whether the effects are seen in the stronger or weaker of the two CD bands that can be resolved in the CD spectra of these compounds (see for example Fig. 7.20). Hudec and Kirk assign these two bands to the $\pi \to \pi^*$ and $\pi \to 3s$ Rydberg transitions. It would be valuable to test semiempirical SCF or RPA calculations on molecules of known rigid geometry to see how they accord with these assignments and the observed effects. The empirical correlations are discussed in great detail in the *Tetrahedron* paper. Some of the more significant conclusions are:

1. Characteristically, two bands are resolvable in the CD spectra of most of the compounds at wavelengths longer than 180 nm (frequencies below 55,000 cm^{-1}) as given in Table 7.3. See also reference 93.

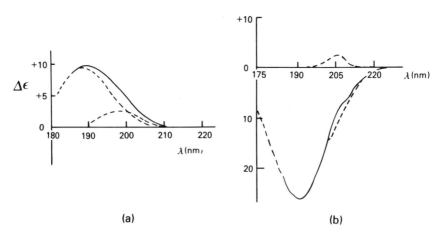

(a)   (b)

Fig. 7.20 Resolved circular dichroism curves of the olefins: (*a*) 3'-methylene-2α,3β-tetramethylene-5α-cholestane and (*b*) 5α-cholest-6-en-3α-ol. (——) experimental curve; (---) resolved component curves of the two longest wavelength transitions (91).

Table 7.3   Characterization of the CD Bands of Monoolefins (91)

Olefin Class	λ_1(nm)	λ_2(nm)
(A) 1,1-Disubstituted ethylenes (exocyclic methylene)	200 ± 2	189 ± 1
(B) 1,2-Disubstituted ethylenes (cycloalkenes)	200 ± 2	Ca. 185
(C) Trisubstituted ethylenes (methyl cyclohexene)	203.5 ± 1.5	192 ± 3
($\Delta^{1(9)}$ -octalin)	210 ± 2	192 ± 3
(D) Tetrasubstituted ethylenes[a]	$195 - 210$	Ca. $185 - 195$

[a]In addition, a weak band appears at 215–225 nm in these compounds.

2. Using the octant projections as described above, the CD of the longest wavelength transitions of classes A, B, C, and D are generally consignate, but exceptions are noted.

3. Some of the exceptions to the octant behavior of the longest wavelength transition may be generalized (more specific exceptions are found in the paper):

   Class A: β-axial methyl substituents can outweigh the octant contributions of carbocyclic rings.

   Class B: Allylic axial C–H bonds, as well as the carbon substituents, produce a consignate contribution.

   Class C: In $\Delta^{1(9)}$-octalins (Fig. 7.21) a large dissignate effect is produced by axial alkyl substitution at the allylic carbon atom trans to the olefinic C–H bond.

   Class D: These compounds show rather weak CD curves as expected from the fact that the olefinic bond tends to be more symmetrically surrounded, and axial-allylic methyl substituents produce dissignate CD bands.

4. The sign of the CD of the second transition is generally the reverse of the lowest energy CD bands, but there are exceptions.

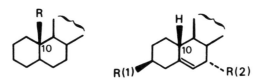

Fig. 7.21 Two of the type of $\Delta^{1(9)}$ octalins that show dissignate CD effects when alkyl substituents are substituted in the axial configuration at the allylic position trans to olefin C–H bond, that is, at C-10.

Some of these conclusions are produced in earlier work, but this large collection of data (the CD data for 228 chiral olefins were analyzed) permits a more thorough classification of the observations than has been available. Perhaps equally important is the fact that these limited generalizations and exceptions provide a target for theoretical treatments.

7.24   The Disulfide Chromophore

The ordinary achiral absorption spectrum of the disulfide chromophore has received less than its share of attention considering its importance in the formation of bridges between polypeptide chains and its presence in a variety of natural and synthetic compounds of biological interest (97). The longest wavelength band is usually observed between 250 and 350 nm, depending on the twist (dihedral) angle of the chromophore; a strong band not quite so sensitive to the angle appears near 240 nm. The first transition is near 250 nm in the open-chain compounds such as cystine, where the dihedral angle is close to 90 degrees. Because several of the low-lying electronic transitions of the chromophore are either or both electric and magnetic dipole allowed when chirally substituted or dissymmetrically twisted, they have inherently large rotational strengths. Electronically, sulfur is in many respects like oxygen; with its 16 electrons, the sulfur atom core configuration has the $1s$, $2s$, $2p$, and $3s$ orbitals and two of the $3p$ orbitals filled. Using $3s$, $3p$ hybridization, the atom forms single bonds with adjacent atoms. In the simplest approximation a disulfide will have, therefore, 4 nonbonding electrons in its highest filled molecular orbital. It is on this basis that Bergson (98) constructed a model to account for the dependence of the position of the longest wavelength absorption band on the dihedral angle between the two planes formed by the two sulfur atoms with each of the adjacent atoms. The treatments of the skewed disulfides as chiral dissymmetric chromophores have largely taken over this model either in its original form or with the addition of d orbitals to the electronic description. The rationale for the inclusion of d orbitals, although it is nowhere explicitly stated, is that these have significant density in the vicinity of the sp_3 orbital of the adjacent sulfur atom; thus despite the fact that in the neutral atomic sulfur atom the $3d$ orbital is not occupied, it may very well contribute either to the ground or excited state electronic configuration of a disulfide. The descriptions have, moreover, emphasized the intrinsic dissymmetry of the disulfide chromophore. In this sense they differ sharply from the early treatments of the ketones and hydroxyl groups, and even to some extent from the olefins, the latter of which, at least initially, had been considered to be an intrinsically symmetric chromophore. The earliest theoretical treatment of the chiral disulfides (99) was of a model twisted H_2S_2 molecule with C_2 symmetry for all angles of twist except of $\phi=0$ or 180 degrees. The chromophore, considered to be coextensive with the molecule, thus has the screw symmetry of a helix and is designated as right- or left-handed and correspondingly by convention as an M or P helix (refer to Fig. 7.12; in H_2S_2 the central atoms are the sulfur atoms; the P helix represents a molecule with skew sense opposite to that shown in the figure). Both from valence considerations and from the calculations, if d orbitals are *not* included in the basis set, the lowest unfilled orbitals in the disulfide system are σ-type antibonding orbitals: between the two sulfur atoms σ_{SS}^*; between the sulfurs and adjacent hydrogens σ_{SH}^*, in H_2S_2; or between adjacent carbons σ_{SC}^*, in alkyl substituted compounds. It is relatively easy to see that the σ_{SS}^* must have b symmetry in the C_2 symmetry group of the four-atom chromophore because it is an antibonding combination of sp_3 hybrids and thus changes sign on rotation about the twofold axis passing through the node. The σ_{SH}^* and σ_{SC}^* orbitals also have nodes between the respective atoms, but rotation about the C_2 axis, which passes through the center of the sulfur-sulfur bond, brings the same signs (orbital phases) into superposition, so their symmetry is a. This requires that the lowest energy transitions be electronic promotions from the ground ^1A state (all doubly filled orbitals) to states containing one electron in either an a or b orbital. As described by Woody (100), the four nonbonding orbitals of the ground state can give rise to four transitions to the σ^* orbital of lowest energy (and, of course, four transitions in the orbital of next higher energy). It is instructive to look at the selection rules for the planar C_{2v} molecule because we can see what induced components must be sought in the skewed C_2 symmetry to give rise to chiral activity. For the transition to the lowest energy σ^* orbital if it has b_1 symmetry:

Orbital Promotion	Expected State Symmetry	*Allowed Transition Moments*			
		$C_{2v}\leftarrow\mu\rightarrow C_2$		$C_{2v}\leftarrow\mathbf{m}\rightarrow C_2$	
$n_1\,(a_2)\rightarrow\sigma^*\,(b_1)$	(Core) $a_1\times a_2\times b_1=B_2$	y	x,y	x	x,y
$n_2\,(b_1)\rightarrow\sigma^*\,(b_1)$	(Core) $a_1\times b_1\times b_1=A_1$	z	z		
$n_3\,(a_1)\rightarrow\sigma^*\,(b_1)$	(Core) $a_1\times a_1\times b_1=B_1$	x	x,y	y	x,y
$n_4\,(b_2)\rightarrow\sigma^*\,(b_1)$	(Core) $a_1\times b_2\times b_1=A_2$	z	z		

The axis system is defined with the four-atom planar C_{2v} disulfide chromophore in the xz plane with the sulfur atoms on the x axis, and n_1, n_2, n_3, and n_4 are the possible linear combinations of the nonbonding sulfur orbitals. In the nonplanar C_2 systems the end atoms move out of the plane. It is clear that although there is no chiral activity in the planar molecule ($\mu\cdot\mathbf{m}=0$ in each case), in the skewed C_2 system collinear components will

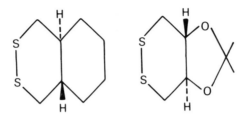

Fig. 7.22 Enantiomeric skewed disulfide chromophores: (9S, 10S)(−)-*trans*-2,3-dithiadecalin (left); (4R,5R)(+)-4,5-isopropylidenedioxy-1,2-dithiane (right).

develop for $n_1 \to \sigma^*$ and $n_3 \to \sigma^*$ even in first order. Furthermore, the magnetic-dipole-allowed $n_4 \to \sigma^*$ can become chirally active by vibronically borrowing electric dipole intensity. In fact, it is this latter transition that is assigned to the second lowest energy band in cyclic disulfides on the basis of theoretical calculations. A comparison of the calculated results for a 60-degree left-handed skewed system with the experimental values (Table 7.4) for (9S, 10S)(−)-*trans*-2,3-dithiadecalin (Fig. 7.22) clearly implicates the ^1B, $n_1 \to \sigma^*$, and ^1A, $n_4 \to \sigma^*$, as the transitions responsible for the CD bands at 290 and 241 nm.

The advantage of this treatment is that it presents a rationale for the high sensitivity of the disulfide chromophore to perturbation by other chromophores in the chiral environment (102), a factor that introduces almost insoluble complexity into the chiral spectra of complex compounds. The n_1 and n_4 orbitals turn out to be π type because the s contributions to the sp_3 hybrids from which they are constructed cancel, while for a similar reason, the n_2 and n_3 are σ type. As a result there are large monopole charges associated with separate contributions of $s \to s$, $s \to p$, $p_x \to p_y$, and $p_y \to p_x$ contributions to the transitions. These in turn give large values to the coupling of transitions in the Kirkwood-Tinoco formalism (see Chapter 4). The 1973 papers by Nagarajan and Woody (100, 102) explore this concept

Table 7.4   Rotational Strength of (9S, 10S)(−)-*trans*-2,3-dithiadecalin[a] (101)

Transition	R(observed) (DBM)	R(calculated) (DBM)
290 nm, $n_1 \to \sigma^*$ (^1B)	− 0.21	− 0.11
241 nm, $n_4 \to \sigma^*$ (^1A)	+ 0.11	+ 0.08

[a]Note that the enantiomorphic (with respect to the disulfide chromophore) (4R, 5R) (+)−4, 5-isopropylidenedioxy-1, 2-dithiane exhibits CD bands of opposite sign at approximately the same wavelength.

and, in addition, describe the dissymmetric chromophore treatment (103) in explicit detail.

The picture of the orbital promotions becomes rather different when the d orbitals are included in the basis set (104). The ground state nonbonding orbitals do not change very much, but the virtual orbitals to which the electronic promotions are made all contain considerable sulfur $3d$ orbital charge. An examination of the skew angle dependence shows a much more complex picture than is seen in the other calculations, with sign changes occurring three times in each of the low-energy transitions (for the same absolute configuration) as the skew angle proceeds from 0 (cis) to 180 degrees (trans). For H_2S_2 these changes occur between 30 and 60, 85 and 90, and 120 and 180 degrees.

An interesting feature of these calculations for H_2S_2 is that for any given absolute configuration, that is, right- or left-handed, the sign of the rotational strength follows exactly the symmetry designations of the transitions. For example, the excited state of the lowest energy transition goes from ^1A to ^1B and back to ^1A as the rotational strength changes sign as indicated above. A very similar observation had been made earlier (105) for dienes and diketones and had been proposed as a general rule by Wagniere and Hug (106). It should be carefully noted, however, that all these observations are made with respect to systems that retain C_2 symmetry throughout. As soon as this restriction is relaxed, as in a chiral molecule containing another obvious optical center, or even in the strictly dissymmetric chromophore in which a vicinal rotamer introduces the possibility of additional chirality, the general rule will not necessarily be obeyed. For example, while in H_2S_2 the sign changes with the transition symmetry, it does not change for $(CH_3)_2S_2$ or for 1,2-dithiolane, $(CH_2)_3S_2$ (Table 7.5). The latter compound is not a strictly C_2 structure even with respect to the nuclear framework of carbon and

Table 7.5 Dependence of Calculated Rotatory Strengths of Right-Handed (M Configuration) Disulfides on the Transition Symmetry (104)

Molecule	Transition	LRSS, SSR' (degrees)	Symmetry	Rotational Strength (DBM)
H_2S_2	$n_1 \rightarrow \sigma_{SS}^*$	60	B	$+2.81$
		90	A	-3.18
	$n_1 \rightarrow \sigma_{SH}^*$	60	A	-1.35
		90	B	$+2.11$
$(CH_3)_2S_2$	$n_1 \rightarrow \sigma_{SC}^*$	60	A	$+0.22$
		90	B	$+1.21$
	$n_1 \rightarrow \sigma_{SS}^*$	60	B	$+2.00$
		90	A	$+0.20$
$(CH_2)_3S_2$	$n_1 \rightarrow \sigma_{SC}^*$	35.5	A	$+2.79$
	$n_1 \rightarrow \sigma_{SS}^*$	35.5	B	$+0.98$

sulfur atoms, but the former departs from C_2 symmetry only because the calculations are done for an arbitrary conformation of the CH_3 rotamers at each end of the molecule; this conformation does not retain strict C_2 symmetry as the disulfide chromophore is skewed. This strong effect of rotamer conformation on rotational strengths has recently been discussed in relation to the diene chromophore (107).

In addition to the sensitivity to the dihedral skew angle, these calculations turn up other rather sensitive dependencies, possibly more sensitive than is warranted. For example the calculations show (Table 7.6), that for the left-handed 90-degree twisted H_2S_2 the first two transitions change type, symmetry, and sign of the rotational strength for a change of the angle between the S–S bond and the S–H bond.

Table 7.6   Dependence of the Rotational Strength of the Lowest Energy Transition of H_2S_2 on $\langle$SSH (104)

$\langle$SSH (degrees)	Transition Type	Symmetry	Rotational Strength (DBM)
92	$n_1 \rightarrow \sigma_{SS}^*$	A	-3.18
102	$n_1 \rightarrow \sigma_{SS}^*$	B	$+1.86$

There is no experimental evidence for this high sensitivity to small changes in the molecular parameters. It would appear to be best to hold in abeyance the desire to draw strong conclusions about the chiral activity of disulfide compounds until additional work is done. Note also that for the model 90-degree disulfide, H_2S_2, the rotational strengths are opposite in sign to those of the alkyl-substituted compounds. Since pure enantiomorphs of H_2S_2 do not exist, this has not been tested, but in the light of both experimental and theoretical observations on the effect of alkyl versus hydrogen substitution in other compounds, it is not unreasonable to expect that this may be a valid prediction. In the meantime, it is reasonably clear, both from experimental observations (101, 108, 109) and the theoretical calculations, that the longest wavelength band of chiral alkyl disulfides skewed to fairly small angles (less than 35 degrees) will have a large positive rotational strength if the skew sense is right-handed and a corresponding negative rotational strength for the left-handed enantiomorph.

7.3   CONJUGATED CHROMOPHORES

We do not wish to make any very special distinction between the conjugated and nonconjugated chromophores. Historically, the conjugated chromophores were the first to be treated in terms of their intrinsic dissymmetry, but this was only because it is easy to perceive that the nuclear framework of the four-or-more-atom systems over which the conjugation extends can be distorted to chiral symmetry even in zeroth order—no perturbation of the charge distribution by an asymmetric environment is necessary. We will examine in some detail the diene chromophore and in an abbreviated fashion the amide and benzenoid chromophores.

7.31 The Diene Chromophore

The first theoretical treatment (43b) of the chiral activity of the diene chromophore provides an informative example of the role detailed consideration of spectroscopic states can take in unraveling the complexities of chiral observations. In order to explain the puzzling distribution of + and − signs of the optical rotation of four compounds of the ergosterol series (Fig. 7.23), there had to be developed an explanation that is independent

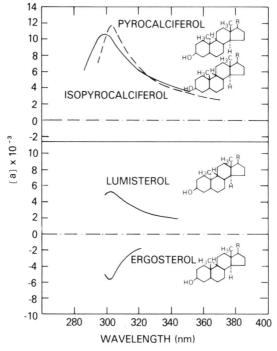

Fig. 7.23 Optical rotatory dispersion of ergosterol and its stereoisomers (43b). CD spectra of the entire diene absorption band in these compounds have been reported by H. J. C. Jacobs (Thesis, Rijksuniversiteit, Leiden, 1972).

of the concept of the additivity of contribution from chiral carbon atoms. Of these compounds, two–ergosterol and lumisterol—have oppositely signed optical activity, which accords with the additivity concept; that is, they are antipodal with respect to two optical centers. However, calciferol and isopyrocalciferol, which are likewise antipodal with respect to the same centers, have the *same* sign of the optical activity and also the same sign as lumisterol with which they are both antipodal with respect to one of these centers. The realization that in this spectral region the optical rotation exhibited a Cotton effect coincident with the diene absorption band led to an examination of the relationship between the stereochemical configuration of the diene and the properties of the optically excited states of the chromophore (110). As we shall see, the resulting "diene chirality rule" for the long wavelength transition of the diene chromophore has become inadequate to explain the experimental behavior of all chiral diene chromophores, but only because the original treatment was based on too limited a concept of what constituted that chromophore. In its original (Hückel) form the highest energy molecular orbital in the ground state configuration is obtained by a simple linear combination of the $2p$ atomic orbitals on the four carbon atoms of the "diene," a treatment that is discussed in most texts an organic quantum chemistry.

Dienes are the prototype conjugated chromophore because even in the simplest approximation, the charge of the bonding valence electrons is spread over more than one bond and includes the bond or bonds intervening between nominal double bonds. In the simplest diene, butadiene, there is ample evidence from infrared, Raman, and microwave investigations that the nuclear framework is planar and that the centrosymmetric trans form is the predominant one at room temperature, probably in excess of 99 percent. The molecule therefore belongs to the optically inactive C_{2h} point group. The cis planar form with C_{2v} symmetry does exist, but more important, it exists in a large variety of ring compounds in which the four-carbon nuclear framework is twisted to a greater or lesser extent out of planarity and thus becomes intrinsically dissymmetric. The ergosterol series noted above belongs to but one large class of such compounds, the 1,3-cyclohexadienes. A number of nonplanar acyclic dienes exist because of the steric hindrance of substituents, but these have not been optically resolved. However, it is instructive to look at their ordinary achiral spectra (Fig. 7.24). At room temperature *trans*-butadiene has a strong diffuse absorption band with a maximum near 47,800 cm^{-1} (201 nm), on which are superimposed numerous weaker and sharper transitions; some of these are vibrational or vibronic structure of the strong band, and others result from weakly allowed electronic (Rydberg) transitions. In 2,3-dimethylbutadiene, which is still a planar molecule, the strong band has moved

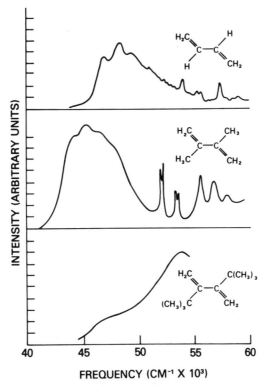

Fig. 7.24 The vapor phase spectra of *trans*-butadiene (upper), *trans*-2,3-dimethylbutadiene (middle), and 2,3,-di-*t*-butylbutadiene (bottom) (2).

down in energy to about 45,500 cm^{-1}, presumably because of the conjugative effects of the methyl substituents. At low temperature in a solid film where molecular interactions broaden and deintensify the sharp Rydberg transition, in addition to the strong band near 201 nm a relatively weak but distinct valence transition is observed in butadiene at 57,000 cm^{-1} (175 nm). Additional substitution of bulky groups at the 2,3 position, for example, di-*t*-butyl groups, results in severe steric hindrance that skews the diene system out of plane by quite a large angle. The long-wavelength band has now shifted back to higher energy at 47,900 cm^{-1} and appears as a shoulder of a stronger band at 53,800 cm^{-1}, despite the presumably even stronger conjugative effect of the tertiary butyl groups as compared to that of the methyl groups in the 2,3-dimethyl compound. These frequency shifts are entirely compatible with even the simplest theoretical treatments (111, 112, 43c), which predict that successive skewing of the diene reduces the conjugation. In the limit of a 90-degree twist, the nominal double bonds become the equivalent of two independent ethylene molecules insofar as the energy of the transition is concerned. Note, however, that the real molecule with only a twofold rotation axis and the identity for symmetry elements would be chiral, and therefore optically active.

The earliest theoretical Hückel-type treatments of the achiral diene chromophore predicted that the π system would give rise to four excited $\pi\pi^*$ states resulting from promotion of electrons from two filled π orbitals to two unfilled ones. If we ingnore the normalization factor, the π^* molecular orbitals are

$$\langle \pi_-^* | = c_1(2p_z) - c_2(2p_z) + c_2(2p_z) - c_1(2p_z), \qquad (7.300)$$

$$\langle \pi_+^* | = c_1(2p_z) - c_2(2p_z) - c_2(2p_z) + c_1(2p_z), \qquad (7.301)$$

where c_1 and c_2 in the Hückel treatment are the coefficients that describe the magnitude of the contribution of the $2p_z$ orbitals of the outer and inner carbons of the diene system, respectively. The highest energy filled MO in the ground state is

$$\langle \pi_- | = c_1(2p_z) + c_2(2p_z) - c_2(2p_z) - c_1(2p_z). \qquad (7.302)$$

In this approximation the first two transitions are represented by the orbital promotions $\pi_- \rightarrow \pi_+^*$ and $\pi_- \rightarrow \pi_-^*$. From the appropriate direct products of the orbital symmetries, the upper singlet states (core $+\pi_-^*$ and core $+\pi_+^*$ configurations) have the following symmetry for the trans, cis, and skewed (cisoid or transoid) compounds:

	Trans (C_{2h})	Cis (C_{2v})	Skewed (C_2)
$\pi_- \rightarrow \pi_-^*$	B_u	B_2	B
$\pi_- \rightarrow \pi_+^*$	A_g	A_1	A

Examination of the group character tables (Appendix I) shows that in the trans compounds the second transition (^1A_g) is electronically forbidden both in chiral and achiral absorption; in the cis compounds the second transition ($^1A_1$) is achirally allowed although, of course, chirally forbidden by the symmetry plane; in the skewed compounds the second transition (^1A) is allowed in both ordinary achiral absorption and in CD. The first transition to the ^1B_u, $^1B_2$, and ^1B states is allowed in achiral absorption in all three compound types. Now we return to the spectra of planar trans-butadiene and skewed-2,3-dibutylbutadiene in Fig. 7.24 and recall that in the spectrum of the unsubstituted butadiene at 25°K, a very weak band

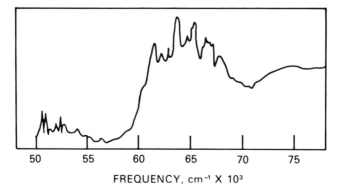

FREQUENCY, cm⁻¹ X 10³

Fig. 7.25 Vacuum ultraviolet spectrum of cyclopentadiene (2).

characteristic of a vibronically allowed, electronically forbidden valence transition is observed at 57,000 cm^{-1}, some 9000 or 10,000 cm^{-1} higher in energy than the first strong singlet transition. In the skewed molecule, the second transition at 53,800 cm^{-1} appears to be even stronger than the first. These two observations are thus in accord with the simple MO theory and suggest the assignment of ^1B and ^1A, respectively, to these absorption bands. Furthermore, if we examine the spectrum of a cis compound (Fig. 7.25), for example, cyclopentadiene, again a relatively strong band appears at about 65,000 cm^{-1}, some 12,000 or 13,000 cm^{-1} above the weaker band near 52,500 cm^{-1}. Small wonder that the simple Hückel MO theory or slight modifications were considered adequate to explain the lower excited state of dienes and were taken over directly into the theoretical treatment of chiral activity.

The application to the chiral activity of dienes proceeds by calculating the electric and magnetic transition moments and their product, using the molecular orbital wavefunctions in Eqs. 7.300–7.302 with the appropriate Hückel or modified Hückel values of the coefficients c_1 and c_2. The skew geometry of the system is incorporated by taking the origin of the coordinates at the center of the central diene bond, setting the four carbons in the xy plane for both the cis and trans configurations, and defining deviations from the plane in terms of the dihedral angle θ (refer to Fig. 7.12). Any other origin could be chosen, since the choice of the ∇ form of the operators, $|\nabla|$ and $|r \times \nabla|$, ensures origin independence of the calculated rotational strength. The projections of the bonds of the skewed diene $(0 < \theta < 180$ degrees) on the three coordinate planes, which are required to calculate the components of the transition moments, are obtained by the use of the midpoints of the bonds $\bar{x}_{lm}$, $\bar{y}_{lm}$, $\bar{z}_{lm}$ and the projection cosines, $\cos x_{lm}$, $\cos y_{lm}$, $\cos z_{lm}$, where lm indexes the four-carbon atoms. For the

$\pi_- \to \pi_-^*$ transition, the electric and magnetic moments are

$$\langle \pi_- | \nabla | \pi_-^* \rangle = -4c_1c_2(\mathbf{j}\cos y_{34} + \mathbf{k}\cos z_{34})\langle \nabla_{34} \rangle + 2c_1^2\mathbf{j}\langle \nabla_{23} \rangle,$$

(7.303)

$$\langle \pi_- | \mathbf{r}\times\nabla | \pi_-^* \rangle = 4c_1c_2\big[\mathbf{j}(\bar{z}_{34}\cos x_{34} - \bar{x}_{34}\cos z_{34})$$

$$+ \mathbf{k}(\bar{x}_{34}\cos y_{34} - \bar{y}_{34}\cos x_{34})\big]\langle \nabla_{34} \rangle$$

(7.304)

where $\langle \nabla_{34} \rangle$ is the absolute value of $\int (2p_z)\nabla(2p_z)\,d\tau$ evaluated at the $c_3 - c_4$ bond distance (113).

It is now possible to see how the absolute configuration determines the sign of the rotational strength of this transition. When the vector product of the moments, $\langle \pi_- | \nabla | \pi_- \rangle \cdot \langle \pi_-^* | \mathbf{r}\times\nabla | \pi_- \rangle$ is inserted in the Rosenfeld equation, the rotational strength for the $\pi_- \to \pi_-^*$ transition becomes

$$R_{\pi_- - \pi_-^*} = \frac{h^3 e^2}{2m^2 c\mathcal{E}_{\pi_- - \pi_-^*}}\big[16c_1^2 c_2^2(\bar{x}_{34}\cos z_{34}\cos y_{34} - \bar{y}_{34}\cos z_{34}\cos x_{34})\langle \nabla_{34} \rangle^2\big]$$

(7.305)

The signed quantities are the midpoints and the projection cosines, which specify the absolute configuration. Note that these enter as triple products and as a consequence cannot have the same sign for all values of the parameters; the signs must change somewhere in the 360-degree rotation angle of θ. Second, the presence of $\cos z_{34} = (z_4 - z_3)/R_{34}$ in each of the two terms ensures that the rotational strength will be zero for the planar compounds because the z coordinates are zero in the xy plane. (R_{34} is the distance between the origin of the coordinates and the midpoint of the 3,4 bond, and z_4 and z_3 are the coordinates of carbons 3 and 4.) In fact the geometric parameters are further reducible to

$$(\bar{x}_{34}\cos z_{34}\cos y_{34} - \bar{y}_{34}\cos z_{34}\cos x_{34}) = \frac{x_4 y_3 z_4}{2R_{34}^2}.$$

(7.306)

It is interesting that this simple expression reflects the fact that a perturbation along the z coordinate would be sufficient to induce optical activity in the trans C_{2h} compound, with the x and y coordinates unchanged, just as predicted by Schellman's symmetry rules (Table 6.1). Moreover, a rule for the sign of the optical activity associated with the $\pi_- \to \pi_-^*$ transition of skewed dienes falls immediately from this expression. The skew takes on the sense of a right- or left-handed helix, depending on whether θ is negative or positive as observed along the plane formed by the c_1-c_2 and c_2-c_3 bonds. The signs of the coordinates and their products in the two cases are

Coordinate	Right-Handed Skew	Left-Handed Skew
x_4	+	+
y_3	−	−
z_4	−	+
$x_4 y_3 z_4$	+	−

Since this product determines the sign of R in Eq. 7.305, the treatment yields the rule that *the longest wavelength valence transition of the diene chromophore will exhibit positive chiral activity if the diene is skewed in the sense of a right-handed helix and negative activity if skewed in the opposite sense.* A very large number of compounds containing dienes have been observed to obey this rule (43d). Moreover, some of the quantitative predictions of the rotational strength have been remarkably accurate (114). Two examples are illustrated in Fig. 7.26, where the calculated optical activity of the longest wavelength ultraviolet transition of (+)*trans*-9-methyl-1,4,9,10-tetrahydronapthalene and of the first two transitions of lumisterol are compared with the measured values (115).

Despite these successful applications, it has become clear that approximating the diene chromophore by the four-carbon π system of electrons is

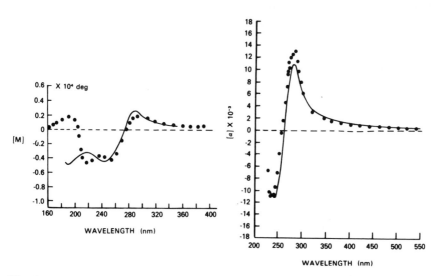

Fig. 7.26 The optical rotatory dispersion of (right) (+)-*trans*-9-methyl-1,4,9,10-tetra-hydronapthalene and (left) lumisterol observed (...) and calculated (——).

insufficient to explain the chiral activity of dienes in all the compounds in which they occur. The first reported exception to the diene rules was the transoid diene, $\Delta^{6,8,(14),22}$ ergostatriene-3β-ol acetate, which has a negative trough in the ORD spectrum at 275 nm rather than the positive one predicted on the basis of the diene skew sense (116). A number of other examples of ambiguous or incompatible results have since appeared. Burgstahler and his associates (117) attacked the problem empirically by examining the correlation between the absolute stereochemical relationships of substituents allylic to the nominal double bonds of the diene and the chiral activity. The resulting allylic chirality rule, which was formulated in terms of the sense of skew of the allylic substituents with respect to the double bonds of the diene, also proved not to be universally applicable. Their attribution of chiral effects to the allylic substituents was nevertheless an important clue to the difficulty and brought the problem into sharp focus. We have already seen that the algebraic formulation of the diene rule (Eqs. 7.305–7.306) carries in it the implication that the $\pi_- \rightarrow \pi_-^*$ transition of a planar compound perturbed by a substituent out-of-plane could be chirally active. The empirical correlations with allylic substituent effects seemed to be making the same implication. A significant clue recently appeared in the report of different signs of the CD of the long-wavelength band of three compounds in the steroidal $5\alpha - 1,3$,-dienes (Fig. 7.27), in all of which the skew sense of the four principal carbon atoms of the diene is the same (118, 119). Using the cyclobutadienes (I–IV of Fig. 7.27) as models for the three steroids and palustric acid, Rosenfield and Charney applied an all-electron semiempirical SCF method, CNDO/S, to the calculation of the oscillator and rotational strengths (107). The results, shown in Table 7.7, are clearly in accord with the

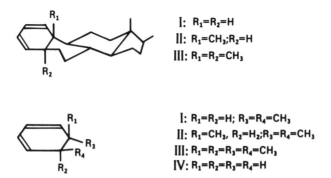

I: $R_1=R_2=H$
II: $R_1=CH_3; R_2=H$
III: $R_1=R_2=CH_3$

I: $R_1=R_2=H; R_3=R_4=CH_3$
II: $R_1=CH_3, R_2=H_2; R_3=R_4=CH_3$
III: $R_1=R_2=R_3=R_4=CH_3$
IV: $R_1=R_2=R_3=R_4=H$

Fig. 7.27 Steroidal 5α-1,3-dienes (118) and the model cyclohexadienes for which the calculations of oscillator and rotatory strength reported in Table 7.7 (107) are given.

Table 7.7   CNDO/S Calculation of Skewed Dienes

Compound	f (calculated)[a]	$[R]$ (calculated)[b]	$[R]$ (observed)[b]
I	0.19	+ 12.0	+ 14.9
II	0.18	− 6.7	− 8.4
III	0.17	− 27.0	− 37.0
IV	0.19	− 2.6	− 3.7

[a]The observed value of the oscillator strength for this transition in 1,3-cyclo-hexadiene (*I*) is $f = 0.14$.

[b]The results reported in reference 107 and reproduced in this table are in units of "reduced rotational strength" $[R]$. $[R] = 1.073 \times 10^2 R$ (in units of Debye-Bohr magnetons).

observations for all four compounds, but only for three of them (II, III, and IV) does the original Hückel-treatment-based "diene rule" predict the correct sign. Clearly the inclusion of valence electrons centered on atoms other than the four carbons of the formal diene has made a difference; the chromophore extends beyond the dienic carbon atoms!

There were several additional interesting conclusions drawn from the CNDO/S calculations. For one, the asymmetric disposition of a methyl substituent out-of-plane of planar acyclic *cis*-1,3-pentadiene, which may appear to be a minor perturbation of the planar system, is predicted to induce a very large rotational strength for this transition, of the order of 0.36 DBM in this otherwise achiral molecule. We have noted that even the Hückel treatment implied induced chiral activity by an out-of-plane (*z* direction) perturbation, but rigorously since only the π electrons are included, no rotational strength would actually be calculated for this system by that treatment. For another, a very strong effect on the chiral strength is predicted by the CNDO/S treatment for different conformational rotamers. We have seen this effect in (+)-S-2-butanol. In (-)-α-phellandrene, a cyclohexadiene with two ring conformations and with an allylic isopropyl substituent which can rotate about the bond connecting it to the ring, a quantitative prediction of the temperature dependence of the CD based on a Boltzman-weighted distribution of the two conformers and three minimum energy rotamers is in agreement with the experimental observations (107). The results of Table 7.8 show that the rotational strength of the rotamers of the quasi-equatorial conformer for which the diene moiety is skewed in a right-handed sense (positive by the "diene rule"), is a tenth or less than that of the rotamers of the quasi-axial conformers (negative by the "diene rule"). Nevertheless, the energy difference between these conformers is such that at room temperature and below, the measured CD is closer to that of the quasi-equatorial conformer

Table 7.8.   CNDO/S Calculation of the Rotatory Strength[a] of α-phellandrene.

Rotamer	Quasi-axial Conformer	Quasi-equatorial Conformer
A	−25.5	−1.0
B	−9.1	−1.6
C	−29.6	−1.5

Temperature Dependence

	At −20°C	At −150°C
Calculated	−7.1	−4.9
Observed[b]	−8.5	−2.0

[a]In units of reduced rotational strength [R].
[b]The value at −20°C is the average of observations by Horseman and Emeis (Tetrahedron 22, 167 (1966)], Snatzke, Kovats and Ohloff [Tetrahedron Lett. 38, 4551 (1966)] and Ziffer, Charney and Weiss [J. Amer. Chem. Soc. 83, 4660 (1961)] and the value at −150°C is the value obtained by Horseman and Emeis, ibid.

because of the equilibrium distribution of the species at these temperatures. Note also the large differences in rotational strength predicted between rotamers. These results not only demonstrate the magnitude of such effects but suggest that the measurement of the temperature dependence of CD may be used to determine rotamer and conformer equilibria and barrier heights.

These strong vicinal effects make it difficult to establish a general rule applicable to all dienes. In many cases, but obviously not in all, the original diene rule will continue to predict correctly the achiral sign of the low-energy $\pi{\rightarrow}\pi^*$ transition of dienic compounds. If either from the calculation of a sufficient variety of special cases or from empirical correlations, it becomes possible to circumscribe diene chirality, predictions of the signs of the rotational strength will not require individual calculation in almost all cases. At this time, it appears that in chiral cisoid cyclohexadienes, the sign of the rotational strength of the low-energy band will be the same as that predicted by the Hückel treatment, except when axial hydrogen substituents occupy the 1′, 4′ positions, or when substituent chiral rotamers are constrained to rather special conformations.

In this discussion of the chiral activity of diene chromophores, we have concentrated on the activity associated with the lowest energy valence transition. Both the theory (120–122) and experimental analysis (123–126) of the transitions of butadiene predict the existence of other transitions in this and slightly higher energy regions of the spectrum. The high-resolution polarized or chiral spectra necessary to clarify both the achiral and chiral activity, especially in complex dienes, are not yet available.

7.32   The Phenyl Chromophore

Despite the ubiquitous nature of the 260-nm ultraviolet absorption band of compounds with aromatic rings, its origin is in an electronically forbidden transition, both electric and magnetic dipole forbidden. This double forbiddenness for benzene does not prevent it from being associated with characteristic CD and ORD bands when the phenyl group occurs in an optically active molecule that has a rigid or conformationally restricted structure. Symmetry has been the major tool in unraveling the origin of this electronic transition and of the higher energy transitions near 200 and 180 nm, as well. Benzene itself, with a center of symmetry and three types of reflection planes, belongs to the optically inactive D_{6h} point group. The character table for this group shows that there are no collinear components of the magnetic and electric dipole moments associated with states of any symmetry in this group and that the allowed electronic transitions are out-of-plane (z direction) to states of A_{2u} symmetry and in-plane (xy direction) to degenerate states of E_{1u} symmetry. These degenerate states constitute circular absorbers in the sense that the transitions have the same intensity for light polarized in any direction in the planes. Transitions to allowed A_{2u} and E_{1u} states should, of course, be intense. However, the earliest spectra of benzene and its phenyl analogs were found to have only a very weak vibrationally structured band in the accessible region of the

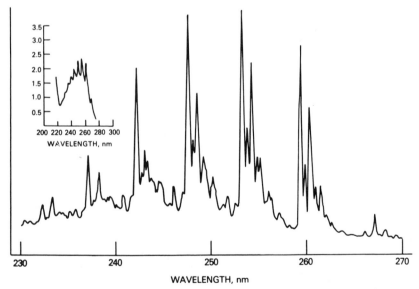

Fig. 7.28 Spectra of the weak 256-nm band of benzene in solution (insert) and as vapor. Intensity units for the vapor spectrum are arbitrary.

ultraviolet (Fig. 7.28). Since the maximum molar extinction coefficient of this band is less than 20 in benzene, it cannot be one of the electronically *allowed* transitions for which extinction coefficients are of the order of 10^3 to 10^5. The symmetry analysis of the vibronic structure eventually showed that the electronic symmetry is B_{2u}, forbidden in D_{6h} chromophores, but made partially allowed (E_{1u}) by vibronic (Herzberg-Teller) interaction with vibratons of e_{2g} symmetry (127); $e_{2g} \times B_{2u} = E_{1u}$. [The lowercase e here has been used to distinguish the vibrational from the electronic symmetry (128).] Other vibrations combine as well; the result is the highly structured benzene band, whose vapor spectrum is also shown in Fig. 7.28. While the assignment of this band is well accepted (129), there has been some question about the origin of bands at higher frequencies. The most generally accepted assignments of the next two features in the spectrum are the forbidden B_{2u} for the 200-nm band and the allowed E_{1u} as the upper state for the 180-nm band (2). To induce natural optical activity, the D_{6h} symmetry must be broken. Monosubstitution with a symmetric substituent is insufficient, since this only reduces the symmetry to C_{2v}. On the other hand, monosubstitution with an optically active group completely destroys the symmetry. In this case, the benzene B_{2u} transition will appear in the

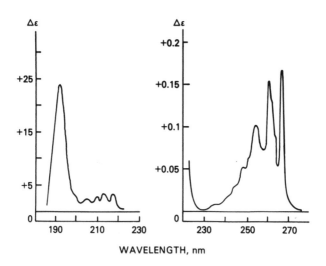

WAVELENGTH, nm

Fig. 7.29 The circular dichroism of the first three valence transitions of (+)-(S)-2-phenyl-3,3-dimethylbutane (134). Note the change in intensity between the CD of the 260-nm band and that of the 210- and 190-nm bands with the weak CD of the former reflecting the electronic forbiddeness of the B_{2u} band. The rotation of the dimethylbutane about the bond to the ring is severely restricted. An interesting and somewhat unusual aspect of this CD spectrum is that all three bands have the same sign. Since the sum of rotational strength over all transitions must be zero, this requires that there must exist equivalently strong negative bands in the nearby vacuum ultraviolet region.

CD spectrum if the optically active group is prevented from freely rotating about the single bond with which it is attached to the phenyl group. When a fixed conformation is taken up, opportunities arise for one or another of the various coupling mechanisms to produce chirally active bands (Fig. 7.29). Experimentally the chiral activity of many such compounds was studied by Crabbé and Klyne (130), Verbit (131), Salvadori and his colleagues (132), and Snatzke and Ho (133). The first systematic effort to treat these systems theoretically was made by Hooker and Schellman (134). The matrix method they used was developed concurrently with the attempts to deal in a detailed way with the μ-m mechanism (Sections 4.5 and 9.3). It is described in considerable detail in a paper on the chiral activity of amides (135); its application to the aromatic chromophore in tyrosine is given in the 1970 Hooker and Schellman paper (134). The method is no more complex than the detailed calculations required by the Tinoco treatment using Eq. 4.202, but the computations can get exceedingly complicated as the number of atoms and the number of interacting transitions get large. In the application to tyrosine, a large number of possible conformers complicate the problem; either conformational distributions with respect to the angles ψ, ϕ, χ_1, and χ_2 (Fig. 7.30) must be calculated, or the chiral activity must be computed at all conformational angles and experimentally determined conformational equilibria used to relate theory to experiment. In the work on tyrosine, a compromise was made in which the most probable (low-energy) conformations were calculated or estimated, and the chiral activity computed for this range of conformations. The problems introduced in this treatment are no different from others we have encountered in this chapter, and we call attention to it here primarily because the original papers (134, 135) present a rather detailed account of the way in which the parameters necessary to calculate the interaction potentials and the optical factors, both for the Tinoco formalism (Eqs. 4.400–4.406, for example) and for the matrix method, are obtained.

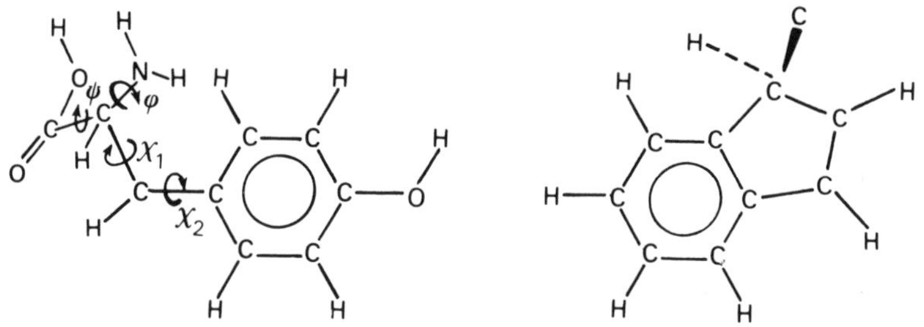

Fig. 7.30 Left: Tyrosine showing the conformational angles defining the possible rotational conformers. Right: (S)-(+)-1-methylindan.

Obviously, the problem is, or should be, much simpler in a rigid molecule. A case in point is 1-methylindan (Fig. 7.30). Although many conformations are possible with respect to the geometric relation between the plane of the phenyl ring and the number 2 carbon, the asymmetric substitution of the methyl group on the C-1 carbon atom forces the molecule into one predominant conformation with the methyl group, either axial or equatorial (136). The circular dichroism of (S)-(+)-1-methylindan has been measured from 280 to 170 nm by Allen and Schnepp and a perturbation theory calculation of the rotational strength compared to the experimental observation (137). Even a cursory comparison of spectra of this compound reveals that the CD and isotropic absorption bands in the 260-nm region (B_{2u} of benzene) and 185-nm region (E_{1u} of benzene) are quite similar to each other and probably dominated by the same mechanism in each case: by vibronic borrowing of the electric dipole intensity (with magnetic dipole for CD) in the case of the B_{2u} transition and by the *allowed* electric dipole in the case of the E_{1u} band (with magnetic dipole borrowing for the CD). In the case of the E_{1u} band, it is not at all surprising that the rotational strength is dominated by the dipole coupling mechanism (Table 7.9). If we look back eq. 4.202, we see that it is only the dipole coupling (Eq. 4.202-6) thatat contains the electric dipole transition moment μ_{i0a} of the band in question (the E_{1u} transition moment on the benzene chromophore) as a *required* factor in the rotational strength contribution. Thus the CD will be strong only if the absorption is strong, as it is for this allowed transition. In the case of the B_{2u} band, the situation is a little more ambiguous, since either the electric transition moment μ_{i0a} or the magnetic transition moment m_{i0a} is required in each of the four factors of Eq. 4.202-2 and 4.202-3 of the one-electron mechanism, but neither is allowed for B_{2u} in D_{6h} symmetry. However, in 1-methylindan the symmetry is no longer strictly D_{6h}; since the phenyl group is ortho

Table 7.9   Rotational Strengths for Electronic Transitions in (S)-(+)-1- methylindan[a]

Transition[b]	$\mu^c \times 10^{19}$	Rotational Strength (cgs) $\times 10^{41}$			
		One-Electron	Dipole Coupling	Total	Experimental
B_{2u}	5.1	4.1	2.4	2.4	1.6
B_{1u}	5.7	9.9	−2.6	7.3	9.5
E_{1u}	13.3	0.0	260	260	57

[a] Adopted from reference 137.
[b] Benzene parentage symmetry.
[c] See reference to Caldwell and Eyring in reference 137.

disubstituted, its behavior should be more like C_{2v} or even C_2 in which the corresponding B_2 transition is both electric and magnetic dipole allowed. Because the π system is only slightly perturbed by the disubstitution, the allowedness will be weak, but sufficient to permit rotatory strength to accumulate from one or more of the four terms of the one-electron mechanism, and probably also from the dipole coupling term. These expectations are in good accord with the calculations (Table 7.8).

The asymmetric perturbation in all cases comes primarily from the interactions of the benzene transitions with those of the ethane group consisting of the methyl carbon and C-2 of the cyclopentane ring of the methylindan.

The CD bands of the B_{1u} transition in the 210-nm region do not mimic the isotropic absorption at all. In the 195-nm region the band positions correlate fairly well in the two spectra, with the CD having a negative and positive peak near the peaks of the isotropic absorption. Allen and Schnepp suggest that the CD features in the 210-nm region are derived from an E_{1g} electronic state in addition to the B_{1u} transition, and in the 195-nm region from an equally elusive E_{2g} transition. This analysis requires that the contribution of the E_{1g} transition be too small to be detectable in the isotropic absorption. The arguments for these suggestions are not conclusive but do receive considerable support from other observations and calculations quoted in the paper.

The rotational strengths of the transitions of S(+)-1-methylindan have also been calculated by the direct quantum mechanical CNDO/S method (136). In these calculations three possible conformers were assumed: the planar conformation and the two conformations with the five-membered ring bent so as to make the methyl group either axial or equatorial. The results are qualitatively encouraging even if quantitatively disappointing. The reduced rotational strength for the three states listed in Table 7.8 vary slightly with conformation, but those of the more probable axial conformer are $+0.44$, $+0.58$, and $+26.09 \times 10^{-41}$ cgs units, respectively, for the transitions of B_{2u}, B_{1u}, and E_{1u} benzene parentage (the latter value being the sum of the two $\pi \rightarrow \pi$ components of the E_{1u}). These values are considerably lower than the experimental values, but of the correct sign and relative magnitudes. More significantly, the calculations predict that there are two $\sigma \rightarrow \pi^*$ transitions of opposite sign ($[R] = -4.49$ and $+4.30 \times 10^{-41}$ cgs units) juxtaposed between the B_{1u} and E_{1u} transitions, which do accord with spectral features in this region in the CD spectrum and which the perturbation treatment is incapable of explaining without the ad hoc hypotheses of the existence of additional transitions.

In this discussion of the phenyl chromophore, symmetry has been used to designate the benzene parentage of the transitions, but symmetry concepts can be applied more directly to the chirality of benzene derivatives. One of the more obvious ways involves the use of Schellman's rules (Section 6.4) to specify the number of sectors in the space surrounding the chromophore in which a perturbing atom or group can provide the asymmetric potential necessary to induce chiral activity by the one-electron mechanism. The perturbing potential for the D_{6h} symmetry of benzene produces 24 sectors (Fig. 6.01). Both the mono substituted and ortho disubstituted benzene derivative [(S-(+)-1-methylindan being an example of the latter] reduce the symmetry to C_{2v} for which the necessary asymmetric potential leads to a quadrant rule (4 sectors). Aside from the subtle factors that critically determine the shape of the nodal surfaces, these rules are grossly approximate at best because the lower energy transitions of the phenyl chromophore are weak and must gain chiral strength by borrowing intensity. Limited success, however, has been obtained by empirical derivation of sector rules for these systems by Snatzke et al. (133). The rules are discussed in detail in the papers by Snatzke, Kajtar, and Snatzke (138) and by Snatzke, Kajtar and Werner-Zamojska (133). Experimentally, but reminiscent of the more formal symmetry rules, the asymmetric influences on the phenyl group are proportioned in spheres that essentially represent distances from the center of the aromatic ring, the chiral first sphere belonging to dissymmetric systems such as the one present in diphenyl derivatives. The chiral "third sphere" (or fourth, which

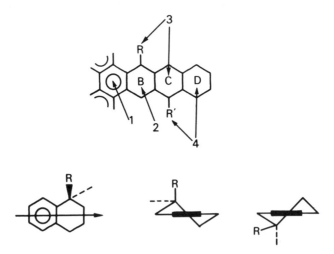

Fig. 7.31 Schematic representation of the division of an aromatic polycyclic compound into vicinal spheres (upper) and the example of a third-sphere substituent on a tetralin (lower). In the projections of the tetralin on the lower right (in the direction of the arrow), the heavy bar represents the benzene ring. (Adapted from Fig. 1 and 5 in reference 138.)

follows the same rules) involves substituents once removed from direct substitution on the aromatic rings. Thus in the tetralins (Fig. 7.31), the substitutents to the B ring or those further removed fall into this category. For the bands of B_{2u} and B_{1u} parentage, respectively [^2L_B and ^2L_A in the Platt notation (129) used by Snatzke et al.], the sectors are divided as illustrated in Fig. 7.32, with the signs designated for the upper sectors. (Note that this system of "chiral spheres" results in the formation of 16 sectors for the B_{2u} transition compared to the 24 designated by the one-electron symmetry perturbation for D_{6h} molecules and 4 for C_{2v} molecules.)

An illustration of one successful application is the different effect of an epimeric change on two different phenyl group transitions. Adjacent to the sector diagrams in Fig. 7.32 are two compounds containing epimeric centers which could be expected to affect the rotational strengths of the CD bands. The C-3 carbon and the methoxyl group and H atom substituents at this epimeric center lie almost astride a nodal plane for the B_{2u} transition, while they lie in the middle of a sector for the B_{1u} transition.

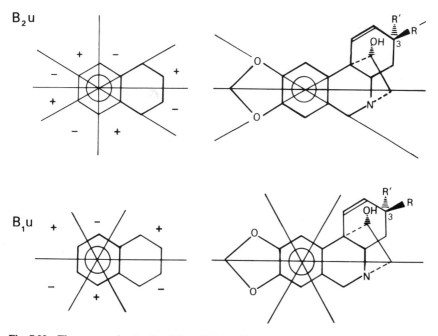

Fig. 7.32 The sector rules for the chiral third-(and fourth-) sphere substituents for the B_{2u} and B_{1u} parentage bands with examples of haemanthamine ($R=OCH_3$, $R'=H$) and crinamine ($R=H$, $R'=OCH_3$). (Adapted from Fig. 11 and 12 in reference 138.)

Table 7.10   Asymmetric Perturbation of the Phenyl Chromophore (139)

Compound	Rotational Strength (cgs × 10^{41})	
	B_{2u}	B_{1u}
I haemanthamine	+3.42	−3.96
II crinamine	+3.51	−2.84

The groups are somewhat remote from the aromatic ring; so their vicinal effect will not be large in either case. But if these sector rules are valid the effects, if any, should be considerably larger for the B_{1u} than for the B_{2u} transitions. The experimental rotational strengths of the epimers are, in fact, observed to be much closer in the B_{2u} than in the B_{1u} band (Table 7.10).

7.33   The Amide Chromophore

Although there is no classical formal conjugation between π systems, the amide group has long been considered a chromophore that extends, at least, over the oxygen, carbon, and nitrogen atoms. The amount and variety of the work that has been done on its spectral properties testifies to its importance both as a reactive chemical and as the unit building block in proteins. Nevertheless, our review of its achiral and chiral properties will be somewhat cursory because no concepts or problems not already treated earlier in this chapter arise in relation to the amide or peptide chromophore.

Of the various electronic transitions that have been attributed to amides and peptides, two have been especially important in the analysis of the chiral spectra of polypeptides and proteins. One of these is the lowest energy transition, a generally weak one attributed to the $n \rightarrow \pi^*$ transitions, largely involving the nonbonding electrons of the carbonyl oxygen, but as in the ketones, sufficiently distributed over adjacent atoms to be very sensitive to their configuration. Calculations designed to test this sensitivity have been done for a variety of substituted diketopiperazines (Fig. 4.05) and pyrrolidones (Fig. 7.33) (140–145). The other transition to which a significant role in the circular dichroism and optical rotatory dispersion of polypeptides and proteins has been attributed is a $\pi \rightarrow \pi^*$ type transition, designate ↓, as in ethylene, $N \rightarrow V_1$. In the ordinary absorption spectrum in solution or vapor at room temperature, the $n \rightarrow \pi^*$ transition is not well or not at all resolved in the simple amides and peptides; it lies close to and is buried under the strong absorption band in the 1950–1700 Å region attributed to the $\pi \rightarrow \pi^*$ transition. Except for its presence in the polarized

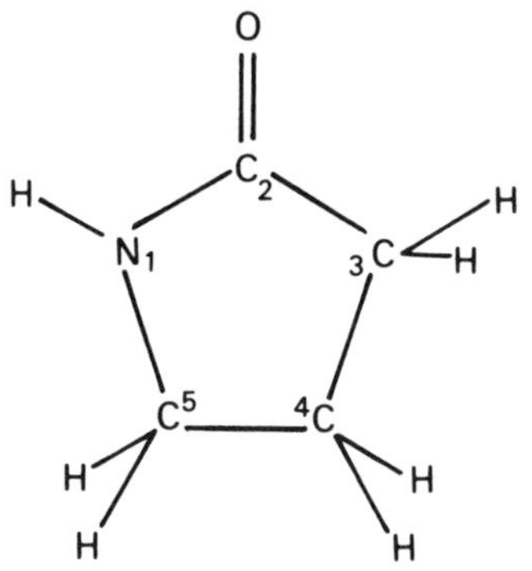

Fig. 7.33 The pyrrolidone ring system. Substitution of one H atom at C-3, C-4, or C-5 produces a chiral molecule with amide (cyclic) transitions.

single-crystal spectrum of myristamide (146), it appears to have been first observed as a weak Cotton effect in the rotatory dispersion curve of L-3-aminopyrrolid-2-one in acetonitrile and dioxane solutions by Littman and Schellman (143) in 1965, although its presence in polypeptides had long been inferred. In a subsequent experimental and theoretical study of the electronic states of the amide group by Basch, Robin, and Kuebler, the $n \rightarrow \pi^*$ transition was calculated to be the lowest energy singlet transition in formamide, difficult to see in the vapor spectrum (147) because of the broad structured $\pi \rightarrow \pi^*$ band (148). This study also came up with another interesting observation: A moderately weak band found in the vapor spectrum of 1-methyl-pyrrolid-2-one and trifluoroacetamide between the very weak $n \rightarrow \pi^*$ and strong $\pi \rightarrow \pi^*$ bands coincided with the position calculated for a $\pi \rightarrow \sigma^*$ transition, the σ^* molecular orbital consisting largely of the $3p$ atomic orbitals on the carbon and oxygen atoms with smaller contributions from the $3p$ nitrogen and carbon valence shell orbitals. The locations of these bands vary considerably with substituents, as might be expected, but in general the $n \rightarrow \pi^*$ is in the vicinity of 220 to 225 nm, the intense $\pi \rightarrow \pi^*$ at 170–195 nm, with the weak ($\pi \rightarrow \sigma^*$?) transition between them, at least in trifluoroacetamide and 1-methyl-2-pyrrolidone (148), and a $\sigma \rightarrow \pi^*$ or other Rydberg-like transition at higher energy, near 160 nm. Since the circular dichroism spectra of the helical polypeptides (see Fig. 8.06) show three bands in this region, the question arises as to whether they derive directly from these transitions, or whether in accord with the predictions of exciton theory, the second, third, and a predicted fourth arise from the exciton components of the $\pi \rightarrow \pi^*$ transitions. We examine this question in Section 8.44.

In all analyses of optical activity, the directions of the transition moments, as well as their magnitude and spectral location, are determinants of the rotational strength. In the case of exciton bands in crystals and repeating polymers, these enter in still another way. We shall see in Chapter 8 that the rotational strength of the components of the exciton bands (Eqs. 8.215 and 8.216) arise from transition moment components, given there in the equivalent form of components of linear and angular momenta (rather than as components of ∇ and $r \times \nabla$, polarized along orthogonal axes in the helix (Eqs. 8.202–8.204 and 8.211–8.213). The polarization directions of the amide transitions and the orientation of the amide groups are therefore determinants of the circular dichroism of the polypeptides and proteins. We will discuss the larger molecules in Section 8.44. For the present, we consider what is known about the polarization directions in the isolated chromophore.

1. The $n \rightarrow \pi^*$ transition: Calculations, all of which place this as the lowest energy transition, require that the electric dipole transition moment be polarized out-of-plane (142, 145, 148, 149). Experimentally, two results are available; those of Petersen and Simpson (146) differ slightly from those of Kaya and Nagakura (148), but the experiments were done on different molecules, myristamide and diketopiperazine, respectively. Both found out-of-(amide) plane polarization, but in myristamide, some of the intensity was found to be in-plane and attributed to vibronic borrowing.

2. The $\pi \rightarrow \pi^*$ transition: This is universally calculated as being in the plane of the amide; the calculations and the experimental data (146, 148) both support this assignment. The angle that the transition moment makes with the nitrogen-carbon bond undoubtedly varies slightly with the substituents. The experimentally determined angle for myristamide measured from the N–C bond with the origin at the center of the bond is 17.9 degrees in the direction of the oxygen atom. Calculations on formamide (148) have given angles between 31 and 45 degrees. Since it is the coupling of this transition moment that is predicted to give rise to exciton splitting in polypeptides, the value is a significant parameter.

3. The $n \rightarrow \sigma^*$ (?) transition: The question mark refers to the fact that as yet there is no universal agreement on the assignment of this transition. The experimental evidence (148) is that a transition of intermediate energy exists between the $n \rightarrow \pi^*$ and $\pi \rightarrow \pi^*$ transitions and that it is polarized in-plane. Except for the calculations of Basch, Robin, and Kuebler, who make the in-plane $n \rightarrow \sigma^*$ assignment, most calculations predict a $\pi \rightarrow \sigma^*$ or $\sigma \rightarrow \pi^*$ out-of-plane transition in this region and predict the $n \rightarrow \sigma^*$ transition to come at higher energy. A conclusive test of the assignment and its effect on the chiral activity has yet to be made.

7.4   OTHER CHROMOPHORES

There are, of course, many more optically active chromophores than we have treated in this chapter. One very large class that deserves a separate treatment is that of the transition metal complexes, where the electrons of a still higher orbital quantum number, the d electrons, take part in the optical transitions. The analysis of the optical states of these compounds may, in fact, be greatly aided by measurements of the chiroptical activity because of the considerable variation in possible magnitudes of the changes in orbital magnetic moments in these transitions. Once again, the pioneering theoretical approach to their chiral activity is due to Moffitt (150). More recent studies are those of Richardson (151) and Mason (152) and their colleagues, Piper and Karipides (153), and most recently, that of Schipper (154). Two of the more recent papers that examine very specific complexes of cobalt and manganese are, respectively, those of House and Visser (155) and LeMoigne, Dabard, and Le Plouzennec (156), but a large literature exists.

Still other chromophores have been studied but are generally less adequately understood from the point of view of the optical states involved. A very extensive bibliography, extending to 1971, of chromophores on which chiral-optical measurements have been made was compiled by Crabbé (157). A few of the more recent papers from which prior references can be obtained are those on the following chromophoric systems or compound types: N-nitroso (158), sugars (159), arylamides (160), lactones and lactams, as well as peptides (161), aliphatic amines (162), pyrazines (163), glycopyranosides (ring oxygen compounds) (164), cyanates and related groups (165), and papers on disulfides (166), and on dipeptides (167, 168), which were not included in the discussion of these chromophores in Sections 7.24 and 7.33.

7.5   SUMMARY

In this chapter, the relationship between theory and experiment has been applied to specific chromophores. By now it should be clear that neither the perturbation methods nor the direct but approximate quantum methods can be depended on to give highly accurate predictions of the magnitude of the chiral activity. In Table 7.8, for example, we saw that in a relatively simple molecule, the perturbation method predicts rotational strengths as high as four times those observed. Even the more recent quantum calculations in some cases give magnitudes almost an order of magnitude too low. In the case of the low-energy diene transition, the results (Table 7.7) are better; the worst disagreement based on a CNDO/S calculation is of the order of 30 percent. In both methods relatively severe approximations have to be made in describing the chromophore. Generally these take the form of truncating the molecule: In the quantum method this usually means limiting the calculation to a model compound containing the nearest or most asymmetrically disposed atoms; in the perturbation method it takes the form of limiting the number of interacting groups to the most vicinal or asymmetrically disposed. In both methods the number of transitions or states considered is usually severely restricted, more so in the perturbation method. Nevertheless, these procedures have made and will continue to make very significant contributions to our ability to analyze the chiral activity of specific molecular systems both qualitatively and semiquantitatively. Without them our ability to predict the activity in new compounds or to understand the activity of repeating polymers and large molecules would still be severely limited.

REFERENCES AND NOTES

1. R. B. Woodward, *J. Am. Chem. Soc.*, **63**, 1123 (1941); *ibid.*, **64**, 72, 76 (1942).

2. M. B. Robin, *Higher Excited States of Polyatomic Molecules*, Vols. 1 and 2, Academic Press, New York, 1975.

3. The concept of charge population densities is described in most books on quantum chemistry calculations. In LCAO-MO schemes it is usually taken as the square of the atomic orbital coefficients in the molecular orbital (MO) and, of course, refers only to the charge associated with a particular MO, the sum of whose charges is normalized to unity. Consequently, it represents in the simplest approximation the fraction of the charge residing in the vicinity of a particular atom of that MO.

4. From the point of view of molecular dynamics, this is not quite true. The rotation of the methyl groups must then be taken into account, but this is of importance only in high-resolution infrared spectra. See, for example, J. Hougen, *J. Chem. Phys.*, **37**, 1433 (1962); *ibid.*, **38**, 358 (1963). The rotatory strengths of *fixed* chiral rotamers of acetone have, in fact, recently been calculated by E. E. Ernstbrunner, M. R. Giddings, and J. Hudec, *J. Chem. Soc. Chem. Commun.*, 953 (1976).

5. Transitions of the ketones at higher energies have been studied experimentally. For example, see S. Feinlieb and F. A. Bovey, *Chem. Commun.*, 978 (1968); O. Schnepp, D. R. Pearson, and E. Sharman, *Chem. Commun.*, 545 (1970); and D. N. Kirk, W. Klyne, W. P. Mose, and E. Otto, *Chem. Commun.*, 35 (1972).

6. Y. H. Pao and D. P. Santry, *J. Am. Chem. Soc.*, **88**, 4157 (1966); Note that in this calculation for cyclohexanone, only about 37 percent of n orbital charge density arises from the $2p$ orbital of oxygen, the rest of it coming from the three adjacent carbon atoms and nearby $1s$ orbitals of hydrogens. Nevertheless, the symmetry of the situation is preserved, and the analysis that follows remains valid from a group-theoretical point of view.

7. T. D. Bouman and D. Lightner, *J. Am. Chem. Soc.*, **98**, 3145 (1976).

8. For early examples of the work of Djerassi's group on saturated ketones, see (a) C. Djerassi and W. Closson, *J. Am. Chem. Soc.*, **78**, 3761 (1956); (b) C. Djerassi, W.

Closson, and A. E. Lippman, *J. Am. Chem. Soc.*, **78**, 3163 (1956); (c) C. Djerassi, R. Riniker, and B. Riniker, *J. Am. Chem. Soc.*, **78**, 6362 (1956).

 9. Klyne's important work relating to this problem started somewhat earlier, but the measurements were made at longer wavelengths where no Cotton effects are observed. See W. Klyne, *J. Chem. Soc.*, 2916 (1952); *ibid.*, 3072 (1953).

10. The work of the group under L. Velluz and M. Legrand also contributed significantly but came somewhat later. See, for example, L. Velluz and M. Legrand, *Angew. Chem.*, **73**, 603 (1961); L. Velluz and J. Mathieu, *Bull. Soc. Chim. (Fr.)*, 1679 (1961).

11. T. M. Lowry, *Optical Rotatory Power*, Longman, Green and Co., London, 1935, p. 411; reprinted by Dover Publications, New York, 1964.

12. W. Moffitt, R. B. Woodward, A. Moscowitz, W. Klyne, and C. Djerassi, *J. Am. Chem. Soc.*, **83**, 4013 (1961).

13. Some of the theoretical work, especially that on the theory of the rotatory power of helical polymers, is quite outside this realm.

14. W. J. Kauzmann, J. E. Walter, and H. Eyring, *Chem. Rev.*, **26**, 339 (1940). The fact that the geometry of the cyclopentanone ring chosen for the calculation in this paper was incorrect does not vitiate the intrinsic contribution, which was to demonstrate quantum mechanically that a simple isotropically polarizable perturber in an asymmetric geometry could induce optical activity in a symmetric chromophore.

15. Quoted in reference 16.

16. A. Moscowitz, in *Advances in Chemical Physics*, I. Prigogine (Ed.), Interscience, New York - London, 1962, Vol. 4, pp. 67–112. This tersely written paper is so well worth reading that we will not reproduce the entire argument here.

17. C. Djerassi and G. W. Krakower, *J. Am. Chem. Soc.*, **81**, 237 (1959).

18. (a) W. Klyne, in *Optical Rotatory Dispersion and Circular Dichroism in Organic Chemistry*, G. Snatzke (Ed.), Heyden & Sons, London, 1967; (b) P. Crabbé, *Optical Rotatory Dispersion and Circular Dichroism in Organic Chemistry*, Holden-Day, San Franciso, London, Amsterdam, 1965; (c) C. Djerassi, *Optical Rotatory Dispersion, Applications to Organic Chemistry*, McGraw-Hill Book Co., New York, 1960; (d) W. Klyne, in *Advances in Organic Chemistry*, Vol. 6, R. A. Raphael, E. C. Taylor, and H. Wynberg (Eds.), Interscience, New York, 1960; (e) L. Velluz, M. Legrand, and M. Grosjean, *Optical Circular Dichroism*, Verlag Chemie, West Germany, and Academic Press, New York and London, 1965.

19. C. Djerassi, J. Osiecki, R. Riniker, and B. Riniker, *J. Am. Chem. Soc.*, **80**, 1216 (1958).

20. A convenient, but not especially accurate, measure of the chiral strength associated with the Cotton effect on an ORD curve is provided by the molar amplitude or "amplitude" as it is frequently termed. In units of $[\theta]/100$, it is the value of the absolute difference between the molar rotations of the peak and trough of the Cotton effect.

21. C. Djerassi and J. E. Gurst, *J. Am. Chem. Soc.*, **86**, 1755 (1964).

22. For references on the nonplanarity of cyclopentane, refer to the paper by W. Klyne, *Tetrahedron*, **13**, 27 (1961). This issue of *Tetrahedron*, devoted to the memory of John G. Kirkwood and William Moffitt, is entirely concerned with the subject of rotatory dispersion. Conformational equilibria in six-membered rings have been treated by M.J. T. Robinson, *Tetrahedron*, **30**, 1971 (1974); This issue of *Tetrahedron* is a commemorative issue in honor of the Van't Hoff-Le Bel Contributions to Stereochemistry.

23. P. M. Bourne and W. Klyne, *J. Chem. Soc.*, 2044 (1960).

24. C. Ouannes and J. Jacques, *Bull. Soc. Chim. (Fr.)*, 3601 (1965).

25. From reference 18b, p. 166, based on information from Dr. K. Kuriyama, Shiongi Research Laboratory, Osaka, Japan.

26. C. Djerassi, W. Klyne, T. Norin, G. Ohloff, and E. Klein, *Tetrahedron*, **21**, 132 (1965).

27. K. Schaffner and G. Snatzke, *Helv. Chem. Acta*, **48**, 347 (1965).

28. D. A. Lightner and W. A. Beavers, *J. Am. Chem. Soc.*, **93**, 2677 (1971).

29. J. F. Tocanne and R. G. Bugman, *Tetrahedron*, **28**, 373 (1972); J. F. Tocanne, *Tetrahedron*, **28**, 389 (1972).

30. S. F. Mason, *Mol. Phys.*, **5**, 343 (1962).

31. C. A. Emeis and L. J. Oosterhoff, *J. Chem. Phys.*, **54**, 4809 (1971).

32. M. Jungen, H. Labhart, and G. Wagniere, *Theor. Chim. Acta*, **4**, 305 (1966).

33. O. E. Weigang, Jr., and E. C. Ong, *Tetrahedron*, **30**, 1783 (1974).

34. R. M. Lynden-Bell and V. R. Saunders, *J. Chem. Soc.*, A, 2061 (1967).

35. (a) G. H. Dieke and G. B. Kistiakowsky, *Phys. Rev.*, **45**, 4 (1934); (b) G. W. Robinson, *Can. J. Phys.*, **34**, 699 (1956); (c) J. C. D. Brand, *J. Chem. Soc.*, 858 (1956); J. H. Calloman and K. K. Innes, *J. Mol. Spectrosc.*, **10**, 166 (1963).

36. (a) R. J. Buenker and S. D. Preyerimhoff, *J. Chem. Phys.*, **53**, 1368 (1970); (b) W. C. Johnson, Jr., *J. Chem. Phys.*, **63**, 2144 (1975).

37. H. E. Howard-Luck and G. W. King, *J. Mol. Spectrosc.*, **36**, 53 (1970); but see also W. D. Chandler and L. Goodman, *J. Mol. Spectrosc.*, **36**, 141 (1970), where it is claimed that there is appreciable z-axis intensity in this compound.

38. A. D. Walsh, *J. Chem. Soc.*, 2260 (1953).

39. V. T. Jones and J. B. Coon, *J. Mol. Spectrosc.*, **31**, 137 (1969).

40. H. P. J. M. Dekkers and L. E. Closs, *J. Am. Chem. Soc.*, **98**, 2210 (1976).

41. See, for example, the discussion of phenylalanine by J. H. Horowitz, E. H. Strickland, and C. Billup, *J. Am. Chem. Soc.*, **91**, 184 (1969), and conjugated diketones by E. Charney and L. Tsai (105).

42. A. Moscowitz, Ph.D. Thesis, Harvard University, 1957.

43. (a) A. R. Deen and H. C. Jacobs, *K. Ned. Akad. Wet. (Amsterdam)*, **64**, 313 (1961); (b) A. Moscowitz, E. Charney, U. Weiss, and H. Ziffer, *J. Am. Chem. Soc.*, **83**, 4661 (1961); (c) E. Charney, *Tetrahedron*, **21**, 3127 (1965); (d) U. Weiss, H. Ziffer, and E. Charney, *Tetrahedron*, **21**, 3105 (1965).

44. F. S. Richardson, D. D. Shillady, and J. E. Bloor, *J. Phys. Chem.*, **75**, 2466 (1971).

45. J. A. Pople, D. L. Beveridge, and P. Dobosh, *J. Chem. Phys.*, **47**, 2026 (1967).

46. J. Del Bene and H. H. Jaffe, *J. Chem. Phys.*, **48**, 1807, 4050 (1968); *ibid.*, **49**, 1221 (1968); *ibid.*, **50**, 1126 (1969).

47. J. Hudec, *Chem. Commun.*, 829 (1970); M. T. Hughes and J. Hudec, *Chem Commun.*, 805 (1970); G. P. Powell and J. Hudec, *Chem. Commun.*, 806 (1970); E. E. Ernstbrunner and J. Hudec, *J. Am. Chem. Soc.*, **96**, 7106 (1974).

48. For extensive discussions see C. C. J. Roothan, *Rev. Mod. Phys.*, **23**, 69 (1951), or modern quantum chemistry texts for general discussions of SCF methods; for example, a critical discussion of CNDO and related methods may be found by J. N. Murrell and A. J. Harget, in *Semi-Empirical Self-Consistent-Field Molecular Orbital Theory of Molecules*, Wiley-Interscience, New York, 1972.

49. G. Snatzke and G. Eckhardt, *Tetrahedron*, **24**, 4543 (1968); G. Snatzke, B. Ehrig, and H. Klein, *Tetrahedron*, **25**, 5601 (1969).

50. E. E. Ernstbrunner, M. R. Giddings, and J. Hudec, *J. Chem. Soc. Chem. Comm.*, 953 (1976); M. R. Giddings, E. E. Ernstbrunner, and J. Hudec, *J. Chem. Soc. Chem. Comm.*, 954, 956 (1976).

51. J. A. Pople and G. A. Segal, *J. Chem. Phys.*, **44**, 3289 (1966).

52. D. N. Kirk, *J. Chem. Soc.*, Perkin I., 2171 (1976).

53. See the references to Biot's memoirs in reference 11, p. 270.

54. S. F. Boys, *Proc. R. Soc.*, (*London*) **A144**, 655, 673 (1934).

55. R. S. Mulliken, *J. Chem. Phys.*, **8**, 383 (1940).

56. P. A. Guye and A. Gautier, *C.—R.*, **119**, 740, 741, 953 (1894).

57. Reference 11, p. 274.

58. W. Kauzmann, F. B. Clough, and I. Tobias, *Tetrahedron*, **13**, 57 (1961).

59. D. H. Whiffen, *Chem. Ind.*, 964 (1956).

60. J. H. Brewster, *J. Am. Chem. Soc.*, **81**, 5475, 5483, 5493 (1969); J. H. Brewster, *Tetrahedron Lett.*, 23 (1959); J. H. Brewster, *Tetrahedron*, **13**, 106 (1961). See especially the last of these for an excellent summary.

61. E. U. Condon, J. Walter, and H. Eyring, *J. Chem. Phys.*, **5**, 753 (1937); E. Gorin, W. Altar, and H. Eyring, *J. Chem. Phys.*, **6**, 824 (1938).

62. J. L. Mateos and D. J. Cram, *J. Am. Chem. Soc.*, **81**, 2756 (1959).

63. D. R. Salahub and C. Sandorfy, *Chem. Phys. Lett.*, **8**, 71 (1971).

64. The Rydberg transition can gain rotational strength by mixing with non-Rydberg parts or in the language of perturbation theory, by the interaction with perturbing groups through the static and dynamic perturbations described by the appropriate terms of Eq. 4.202. For a full discussion of the application of perturbation theory to the chiral activity of the hydroxyl chromophore, see the paper by Snyder and Johnson (66).

65. D. N. Kirk, W. P. Mose, and P. M. Scopes, *J. Chem. Soc. Chem. Comm.*, 81 (1972).

66. P. A. Snyder and W. C. Johnson, Jr., *J. Chem. Phys.*, **59**, 2618 (1973).

67. J. G. Kirkwood, *J. Chem. Phys.*, **5**, 753 (1937).

68. I. Tinoco, Jr., *Adv. Chem. Phys.*, **4**, 119 (1962).

69. Snyder and Johnson (66) phrase their conclusions in reverse form. They assume these conformations and spectroscopic assignments and draw conclusions about the nature of the interactions that give rise to the chiroptical activity.

70. J. Applequist, *J. Am. Chem. Soc.*, **95**, 8258 (1973); J. Applequist, *J. Chem. Phys.*, **58**, 4251 (1973); see also H. DeVoe, *J. Chem. Phys.*, **43**, 3199 (1965).

71. C. P. Snow and C. B. Allsopp, *Trans. Faraday Soc.*, **30**, 93 (1934).

72. W. C. Price and W. T. Tutte, *Proc. R. Soc. (London)* **A174**, 207 (1940).

73. R. S. Mulliken, *Phys. Rev.*, **41**, 751 (1932).

74. P. G. Wilkinson and R. S. Mulliken, *J. Chem. Phys.*, **23**, 1895 (1955).

75. R. McDiarmid and E. Charney, *J. Chem. Phys.*, **47**, 1517 (1967).

76. A. D. Walsh, *J. Chem. Soc.*, 2325 (1953).

77. R. S. Mulliken, *Rev. Mod. Phys.*, **14**, 265 (1942).

78. A. Moscowitz and K. Mislow, *J. Am. Chem. Soc.*, **84**, 4605 (1962).

79. R. McDiarmid, *J. Chem. Phys.*, **50**, 2328 (1969).

80. M. Yaris, M. Moscowitz, and R. S. Berry, *J. Chem. Phys.*, **49**, 3150 (1968); for the absolute configuration of *trans*-cyclooctene, see A. C. Cope and A. S. Mehta, *J. Am. Chem. Soc.*, **86**, 5626 (1964).

81. The most recent calculation appears to be that of L. E. McMurchie and E. R. Davidson, *J. Chem. Phys.*, **66**, 2959 (1977); see also references 82 and 83 and the references contained therein.

82. R. J. Buenker and S. D. Peyerimhoff, *Chem. Phys.*, **9**, 75 (1975); see also the most recent discussion of this and of the general problem of the interaction of Rydberg and valence states by R. S. Mulliken (84).

83. T. D. Bouman and A. E. Hansen, *J. Chem. Phys.*, **66**, 3460 (1977).

84. R. S. Mulliken, *Chem. Phys. Lett.*, **46**, 197 (1977).

85. M. B. Robin, H. Basch, N. A. Keubler, B. E. Kaplan, and J. Meinwald, *J. Chem. Phys.*, **48**, 5037 (1968).

86. C. E. Levin and R. Hoffmann, *J. Am. Chem. Soc.*, **94**, 3446 (1972).

87. M. Yaris, A. Moscowitz, and R. S. Berry, *J. Chem. Phys.*, **49**, 3150 (1968).

88. M. G. Mason and O. Schnepp, *J. Chem. Phys.*, **59**, 1092 (1973); also see R. D. Bach, *J. Chem. Phys.*, **52**, 6423 (1970); and O. Schnepp, E. F. Pearson, and E. Sharman, *Chem. Commun.*, 545 (1970).

89. A. F. Drake, *J. Chem. Soc. Chem. Comm.*, 515 (1976).

90. See footnote, reference 83.

91. J. Hudec and D. N. Kirk, *Tetrahedron*, **32**, 2475 (1976).

92. A. Yogev, D. Amar, and Y. Mazur, *Chem. Commun.*, 339 (1967); these authors were the first to attribute specifically the signs of chiral effects in the transitions of "planar" π systems to their stereochemical relation with asymmetrically placed allylic substituents.

93. A. I. Scott and A. D. Wrixon, *Chem. Commun.*, 1182 (1969); A. I. Scott and A. D. Wrixon, *Tetrahedron*, **26**, 3695 (1970).

94. A. W. Burgstahler and R. C. Barkhurst, *J. Am. Chem. Soc.*, **92**, 7601 (1970).

95. N. H. Anderson, C. R. Costin, D. D. Syrdal, and D. P. Svedberg, *J. Am. Chem. Soc.*, **95**, 2049 (1973).

96. In interpreting the literature great care should be exercised with respect to the orientation of the Cartesian coordinates relative to the planes of the olefinic bond. The choice of the coordinates cannot, of course, affect the results, but in D_{2h} symmetry much confusion can arise (and has) because the subclassification of symmetry elements B_{1g}, B_{2g}; and B_{3g} to polarizations along particular Cartesian axes depends on the choice of the coordinate system.

97. D. W. Sears and S. Beychock, in *Physical Principles and Techniques of Protein Chemistry*, Pt. C, S. J. Leach (Ed.), Academic Press, New York, 1973, Chapter 3 and references cited therein.

98. C. Bergson, *Arch. Kem.*, **12**, 233 (1958); *ibid.*, **18**, 409 (1962).

99. J. Linderberg and J. Michl, *J. Am. Chem. Soc.*, **92**, 2619 (1970).

100. R. W. Woody, *Tetrahedron*, **29**, 1273 (1973).

101. M. Carmack and L. A. Neubert, *J. Am. Chem. Soc.*, **89**, 7134 (1967).

102. H. Ziffer, U. Weiss, and E. Charney, *Tetrahedron*, **23**, 3881 (1967); A. F. Beecham and A. M. Mathieson, *Tetrahedron Lett.*, 3139 (1966); R. Nagarajan and R. W. Woody, *J. Am. Chem. Soc.*, **95**, 7212 (1973).

103. The dissymmetric chromophore treatment has its explicit basis in the paper by W. Moffitt and A. Moscowitz, *J. Chem. Phys.*, **30**, 648 (1959), and the Ph.D. thesis of A. Moscowitz, Harvard University, 1957, where the earliest treatment of hexahelicene is given. A brief description and a slightly modified Hückel MO treatment of butadiene are given in reference 43c.

104. J. Webb, R. W. Strickland, and F. S. Richardson, *J. Am. Chem. Soc.*, **95**, 4775 (1973).

105. See reference 43c; see also E. Charney and L. Tsai, *J. Am. Chem. Soc.*, **93**, 7123 (1971).

106. G. Wagniere and W. Hug, *Tetrahedron*, **27**, 4765 (1971).

107. J. S. Rosenfield and E. Charney, *J. Chem. Phys.*, **99**, 3209 (1977); E. Charney, C-H Lee, and J. S. Rosenfield, new results communicated by the authors.

108. G. Claesson, *Acta Chem. Scand.*, **22**, 2429 (1968).

109. R. M. Dodson and V. C. Nelson, *J. Org. Chem.*, **33**, 3966 (1968).

110. U. Weiss, H. Ziffer, and E. Charney, at the National Institutes of Health, measured the rotation of lumisterol, calciferol, pyrocalciferol, and ergosterol in the spectral region above 250 nm and noted that the anomaly discussed in the text occurred in the vicinity of an extremum caused by the strong diene transition near 270 nm. A. Moscowitz, who with W. Moffitt had recently treated (103) the dissymmetric hexahelicene chromophore, recognized that the diene transitions could develop oppositely signed optical activity if the four-atom π systems were skewed in opposite senses. A Hückel calculation (43b) of the rotational strength of the lowest energy $\pi \rightarrow \pi^*$ transition of butadiene, with the geometry chosen to mimic that of $(+)$-*trans*-9-methyl-1,4,9,10 tetrahydronapthalene, gave astonishing agreement with the magnitude of the observed optical rotation of that compound (see Fig. 7.26). To some extent methyl tetrahydronapthalene was a fortuitous choice because, as we will see, the situation is rather more complex than is apparent from the relatively simple "diene rule" postulated as a result of this investigation. [Note that A. R. Deen and H. C. Jacobs (43a) developed a similar treatment at about the same time. A more complete discussion of the treatment than that in the text, together with an analysis of the dependence of the achiral absorption intensity and the rotational strength of several of the low-lying π transitions and of the energy of the lowest transition in this approximation, is given in reference 43c.] Nevertheless, the basic experimental fact had been established and theoretically justified: A simple chromophore twisted so that its nuclear framework would have the symmetry characteristics of an optically active point group would give rise to Cotton effects associated with those of its electronic transitions for which there are allowed collinear components of electric and magnetic dipole transition moments. This had in fact already been demonstrated for hexahelicene but had not yet been published. The publication of the diene work gave additional impetus to the already growing conviction that chiral observations made in spectral absorption bands would be important in structural and spectroscopic investigations.

111. H. Wynberg, A. de Groot, and D. W. Davies, *Tetrahedron Lett.*, 1083 (1963).

112. H. Suzuki, *Electronic Absorption Spectra and Geometry of Organic Molecules*, Academic Press, New York, 1967, Chapter 16.

113. M. Kotani, *Table of Molecular Integrals*, Maruzen Co., Tokyo, 1955.

114. Calculations of CD and ORD Cotton effects require the specification of four parameters; magnitude, sign, wavelength, and bandwidth. Frequently the latter two are taken from empirical data in achiral absorption which differ significantly from the same parameters in CD and ORD spectra only in special situations. Consequently, it is not surprising that agreement between calculated and observed curves are good in this respect.

115. The agreement in Fig. 7.26 is better than in the original calculation because of an error in the dihedral angle of lumisterol used in that calculation. See also the calculations using perturbation and one-dimensional gas models by M. Maestro, R. Moccia, and G. Taddei, *Theor. Chim. Acta*, **8**, 80 (1967), and H. J. Nolte and V. Buss, *Tetrahedron*, **31**, 719 (1975).

116. E. Charney, H. Ziffer, and U. Weiss, *Tetrahedron*, **21**, 3121 (1965).

117. A. W. Burgstahler, D. L. Boger, and N. C. Naik, *Tetrahedron*, **32**, 309 (1976), and reference therein.

118. A. W. Burgstahler, L. O. Weigel, and J. K. Gawronski, *J. Am. Chem. Soc.*, **98**, 3015 (1976).

119. H. Paaren, R. M. Moriarity, and J. Flippen, *J. Chem. Soc. Chem. Commun.*, 114 (1976).

120. S. Shih, R. J. Buenker, and S. D. Peyerimhoff, *Chem. Phys. Lett.*, **14**, 301 (1972).

121. T. H. Dunning, R. P. Hosteny, and I. Shavitt, *J. Am. Chem. Soc.*, **95**, 5067 (1973).

122. T. D. Bouman and A. E. Hansen, *Chem. Phys. Lett.*, **53**, 160 (1978).

123. B. S. Hudson and B. E. Kohler, *Annual Reviews of Physical Chemistry*, Vol. 25, Annual Reviews, Inc. Palo Alto, 1974, pp. 437 ff., and references therein.

124. R. McDiarmid, *Chem. Phys. Lett.*, **34**, 130 (1974).

125. B. Mallik, K. Meljain, K. Mandel, and T. N. Misra, *Indian J. Pure Appl. Phys.*, **13**, 699 (1975).

126. K. P. Gross and O. Schnepp, personal communication.

127. M. Goppert-Mayer and A. L. Sklar, *J. Chem. Phys.*, **6**, 645 (1938).

128. An entirely alternate notation for designating the aromatic transitions was proposed by J. R. Platt, *J. Chem. Phys.*, **17**, 484 (1949), and has been frequently used. In this notation, the first singlet valence transitions of benzene are designated, $^1A \rightarrow ^1L_B$, $^1A \rightarrow ^1L_A$, and $^1A \rightarrow ^1B$, corresponding, respectively, to the $^1A_g \rightarrow ^1B_{2u}$, $^1A_g \rightarrow ^1B_{1u}$, and $^1A_g \rightarrow ^1E_{1u}$ in the Schonflies notation used here.

129. T. M. Dunn, in *Chemical Studies on Chemical Structure and Reactivity*, J. H. Kidd (Ed.), Methuen, London, 1966.

130. P. Crabbé and W. Klyne, *Tetrahedron*, **23**, 3449 (1967).

131. L. Verbit, *J. Am. Chem. Soc.*, **88**, 5340 (1966), and subsequent papers.

132. P. Salvadori, L. Lardicci, R. Memicagli, and C. Bertucci, *J. Am. Chem. Soc.*, **94**, 8598 (1972).

133. G. Snatzke and P. C. Ho, *Tetrahedron*, **27**, 3645 (1971), G. Snatzke, M. Kajtar, and F. Werner-Zamojska, *Tetrahedron*, **28**, 281 (1972); and references to earlier papers therein.

134. T. M. Hooker, Jr., and J. A. Schellman, *Biopolym.*, **9**, 1319 (1970).

135. P. M. Bailey, E. R. Nielsen, and J. R. Schellman, *J. Phys. Chem.*, **73**, 228 (1969).

136. H. Dickerson and F. Richardson in a recent paper [*J. Phys. Chem.*, **80**, 2686 (1976)] calculated the energy difference between the two bent conformers as only 0.01 eV and interpreted rotational strengths on the basis that an equilibrium of approximately equal populations exists between the three conformers (two bent and one planar). However, the CNDO/S method is not adequate for total energy calculation, having been parameterized for transition densities. Even if the energy difference were this low, the existence of a high-energy conformation in the planar forms would provide a considerable barrier to the interconversion once formed. Small energy differences in calculations of this type must be viewed sceptically. For example, the number of electronic configurations included in the calculation can have stronger effects than the structural modifications. See references 107 and 122 and the paper by A. P. Volasov and V. A. Zubkov, *Theor. Chim. Acta*, **44**, 375 (1977).

137. S. D. Allen and O. Schnepp, *J. Chem. Phys.*, **59**, 4547 (1973). In an earlier unpublished work by D. J. Caldwell and H. Eyring, *Ann. Rev. Phys. Chem.*, **15**, 281 (1964), the one-electron contributions (Eqs. 4.202 and 4.203) to the chiral activity of 1-methylindan had been calculated. Allen and Schnepp calculated the dipole coupling contribution (Eqs. 4.202-4.206) and summed the contributions as shown in Table 7.8 in the text.

138. G. Snatzke, M. Kajtar, and F. Snatzke in *Fundamental Aspects and Recent Developments*

in Optical Rotatory Dispersion and Circular Dichroism; Proceedings of NATO Advanced Study Institute (1971), F. Ciardelli and P. Salvadori (Eds.), Heyden & Son, London, 1973.

139. Quoted in reference 138 from the paper by K. Kuriyama, T. Iwata, K. Moriyama, K. Kotera, Y. Hamada, R. Mitsui, and K. Takeda, *J. Chem. Soc.*, **B**, 46 (1967).

140. T. H. Hooker, Jr., P. M. Bailey, W. Radding, and J. A. Schellman, *Biopolym.*, **13**, 549 (1974).

141. P. M. Bailey, E. B. Nielsen, and J. A. Schellman, *J. Phys. Chem.*, **73**, 228 (1969).

142. F. S. Richardson and W. Pitts, *Biopolym.*, **13**, 703 (1974).

143. B. J. Litman and J. A. Schellman, *J. Phys. Chem.*, **69**, 978 (1965).

144. J. A. Schellman and S. Lifson, *Biopolym.*, **12**, 315 (1973).

145. R. E. Geiger and G. H. Wagniere, *Helv. Chim. Acta*, **58**, 738 (1975).

146. D. L. Petersen and W. T. Simpson, *J. Am. Chem. Soc.*, **77**, 3929 (1955); *ibid.*, **79**, 2375 (1957).

147. H. D. Hunt and W. T. Simpson, *J. Am. Chem. Soc.*, **75**, 4540 (1953); E. E. Barnes and W. T. Simpson, *J. Chem. Phys.*, **39**, 670 (1963).

148. H. Basch, M. B. Robin, and N. A. Kuebler, *J. Chem. Phys.*, **47**, 1201 (1967); *ibid.*, **49**, 5007 (1968). See also K. Kaya and S. Nagakura, *J. Mol. Spectrosc.*, **44**, 279 (1972), for a further discussion of the $n \rightarrow \sigma^*$ transition.

149. M. A. Robb and I. G. Csizmadia, *J. Chem. Phys.*, **50**, 1819 (1969); M. A. Robb and I. G. Csizmadia, *Theor. Chim. Acta*, **10**, 269 (1968).

150. W. Moffitt, *J. Chem. Phys.*, **25**, 1189 (1956).

151. R. W. Strickland and F. S. Richardson, *J. Phys. Chem.*, **80**, 164 (1976).

152. S. F. Mason and R. H. Seal, *Mol. Phys.*, **31**, 755 (1976); a good summary of work on coordination complexes to 1971 is given by Mason in the 1971 NATO meeting monograph (see Ref. 138).

153. T. S. Piper and A. Karipides, *Mol. Phys*, **5**, 475 (1962).

154. P. E. Schipper, *Chem. Phys.*, **23**, 159 (1977); P. E. Schipper, *J. Am. Chem. Soc.*, **100**, 1433 (1978).

155. D. A. House and R. S. Visser, *J. Inorg. Nucl. Chem.*, **38**, 1157 (1976).

156. F. Le Moigne, R. Dabard, and M. Le Plouzennec, *J. Organomet. Chem.*, **122**, 365 (1976).

157. P. Crabbé, in *Determination of Organic Structures by Physical Methods*, Vol. 3, F. C. Nachod and J. J. Zuckerman (Eds.), Academic Press, New York and London, 1971.

158. T. Polonski and K. Prajer, *Tetrahedron*, **32**, 847 (1976).

159. W. C. Johnson, Jr., *Carbohydr. Res.*, **58**, 9 (1977).

160. H. C. Price, M. Ferguson, and P. W. Alexander, *J. Org. Chem.*, **43**, 355 (1978).

161. E. C. Ong, L. C. Cusachs, and O. E. Weigang, Jr., *J. Chem. Phys.*, **67**, 3289 (1977).

162. C. Bertucci, C. Rosini, and R. Lazyaron, *Chim. Ind. (Milan)*, **59**, 453 (1977).

163. G. Snatzke and G. Hajos, *Jerusalem Symp. Quantum Chem. Biochem.*, **10**, 295 (1977).

164. E. Ohtaki, H. Meguro, and K. Tuzimura, *Tetrahedron Lett.*, **49**, 4339 (1977).

165. C. Toniolo in *Chemical Cyanates Their Thio Derivatioves*, Vol. 1, S. Patai (Ed.), John Wiley & Sons, Chichester, England, 1977.

166. W. L. Mattice, *J. Am. Chem. Soc.*, **99**, 2324 (1977).

167. H. Edelhoch, R. E. Lippoldt, and M. Wilchek, *J. Biol. Chem.*, **243**, 4799 (1968).

168. E. H. Strickland, M. Wilchek, J. Horwitz, and C. Billups, *J. Biol. Chem.*, **245**, 4168 (1970).

CHAPTER EIGHT

THE OPTICAL ACTIVITY OF POLYMERS

8.1 POLYMER: A ONE-DIMENSIONAL CRYSTAL OR A CONTINUOUS DIMER

It is possible to consider the optical properties of linear polymers as special cases of the optical properties of a crystal. A crystal is, after all, a three-dimensional polymer in which the forces that hold the monomers together are van der Waal's and electrostatic forces other than those that result in the formation of valence bonds. Van der Waal's forces do contribute importantly to polymer conformation, but only to the extent that aggregates may be classified as polymers do they contribute to the existence of polymers. This difference in the attractive forces is, however, more important in the physicomechanical properties of the polymer than in the optical properties; with respect to the ground state properties, the difference may be thought of as a semantic one, except for the special cases of the highly conjugated polymers where electron delocalization becomes important. Why then do we not rely solely on the optical properties of crystals and treat polymers as the special one-dimensional case? The answer is that the class of polymers, especially that of biological polymers but certainly not confined to these, represents an important class of substances, a knowledge of whose conformation in solution under a variety of environmental conditions is desired, and whose optical properties, especially chiroptical properties, are usually sensitively related to conformation. In addition, a linear polymer can be treated as an extension of a dimer. As a result, the treatments developed especially for the various coupled oscillator mechanisms and, in general, for optical activity induced by an asymmetric environment can be applied to the polymer problem. Nevertheless, as we will see, certain aspects of the crystal properties will appear importantly in the theoretical treatments (1–3).

The dependence of the rotational strength on the degree of polymerization of ribose adenine phosphate, discovered by Brahms, Michelson, and van Holde, is a most dramatic example of the chiral influence of conformation built into some polymeric systems (4). They observed that the rotational strength per monomer unit increases slightly as the degree of polymerization is increased (Fig. 8.00) and that at high pH this increase is monotonic, leveling off with a rotational strength 75 percent greater than that of the monomer after reaching a polymer size of about 20 or so monomeric units. At low pH there is an abrupt increase when the polymer reaches about 7 monomeric units, and the monotonic curve that follows also levels above 20 monomeric units but with a rotational strength over *three times* that of the monomer. The more detailed discussion of this result in Section 8.4 will examine its consistency with theoretical considerations. Here it is already evident that the mere fact of polymerization is sufficient to induce enhanced chirality in the system. The magnitude of the effect also suffices to tell us that at least at low pH, a very profound structural event must be occurring as a direct result of the polymer formation. Knowing what we do from x-ray diffraction results (5), that at this value of the acidity, highly polymeric polyribosephosphate adenylate polymers [(poly(A)] form double helices in which two polyadenylate strands wind around each other in a DNA-like structure, it is not difficult to conclude that the sharp increase in the rotational strength at 7 monomeric units per polymer molecule must be associated with the formation of the doubly

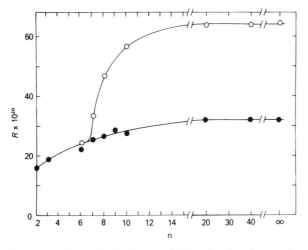

Fig. 8.00 The rotational strength of polymers of adenosine phosphate: n, degree of polymerization; $\bigcirc$, in acid solution; $\bullet$, in neutral or alkaline solution. (From reference 4, reproduced with permission.)

helical structure. It also makes plausible the suggestion that the smaller increase at high pH, where no such double helix has been found, may be due to the formation of a single stranded helical structure or some other rigid asymmetric unit.

The earliest and most elegant attempt to treat the polymer problem for the special case of helical symmetry, which represented at the time, as it still does, a very important class of biological polymers, was that of William Moffitt (1). Moffitt did, in fact, build on the exciton treatments of molecular crystals by Frenkel (6) and Davydov (7). But most important, he recognized two things: First, that helical symmetry would impose a very special set of selection rules on the allowedness and polarization of optical transitions, and second, that the valence bonding that created the polymer would not impose especially stringent conditions on the nature of the optical transitions, at least not insofar as strong optical transitions were concerned. Thus, he was able to foresee that most of the optical properties of the polymer could be simply and directly related to the optical transitions of the monomer molecules from which it was formed. True, the optical properties of the polymer and especially the chiral properties might be highly modified from those of the monomer, but they would bear a direct, traceable, formulable relation to them. Of course, the situation can arise in which the chemical interaction of the original monomers to form the polymer would result in a strong new chromophore not inherent in the monomer. But this situation may easily be incorporated into the theoretical framework simply by considering the polymer to be made of *model* monomers having the new chromophore rather than the actual chemical monomers from which it was synthesized. We will discuss how the recognition of the special effect of helical symmetry gives rise to the chiral-optical properties of helical polymers, and how induced optical activity is related to the transitions of the *symmetric* chromophores of a monomer; and we will examine the use of the theory to extract conformational information from the experimental data. Without exploring the many subtleties of the original treatments, we will try to develop the important theoretical ideas sufficiently to see what light they throw on the ability to understand the experimental observations.

The discussion is divided into the three sections that follow. In the first of these, the results of Moffitt's treatment of the optical properties of polymers are described. We will see that these results have been reexamined and reworked by others, as well as by Moffitt, who with Fitts and Kirkwood (8) found that Moffitt's original use of "reentrant boundary conditions" (see Section 8.2) in the Rosenfeld expression, made in order to account properly for the relation between the length of the polymer and the wavelength of the incident radiation was invalid. In more recent investigations, Loxsom (2, 9) and others have shown that the use of helical symmetry and reentrant boundary conditions are correct if the complete quantum transition probability integral is used and that the use of the Rosenfeld approximation is valid when the reentrant boundary conditions are not used (10, 11). In the second section we will discuss the application to polymers of Tinoco's formulation (12, 13) of the induction of chiral activity in asymmetric chromophores. In this section we will also examine some of the other major contributions to the theory of the optical activity of polymers, in particular, those of Rhodes (3, 14, 15), DeVoe (16), Schellman (17), and Philpott (18) and their respective colleagues. Finally, in the third section the theoretical framework is applied to the interpretation of the observed chiral activity of polymers and biological macromolecules. A very large amount of experimental research has been reported. No attempt is made to be inclusive; the small portion selected is based neither on its definitiveness nor on its quantity, but rather on the extent to which we can use it here to help gain insight into the problems.

8.2   HELICAL POLYMERS AS COUPLED OSCILLATORS: THE EXCITON METHOD

We are entitled to ask, "What is there about a polymer, and more specifically a helical polymer, that endows it with chiral-optical properties that are not the simple sum of the chiral properties of its monomers?" We have to recognize first, of course, that there are polymers for which indeed the chiral properties are insignificantly different from those of its monomers, null if the monomers are themselves optically inactive, or that have a rotational strength per monomer unit that is the observed rotational strength divided by the number of monomer units per polymer molecule. This expected and uninteresting result obviously (from the considerations of Chapters 4 and 7) comes about when the individual monomers in the polymer have little or no inclination to take up a particular conformation from the nearly infinite number of conformations possible by virtue of almost free rotation about the various inter- and/or intramonomeric bonds. However, as we saw in Section 8.1, this is not always the case; very substantial chirality can be induced in the polymer itself. Thus, for example, the planar optically inactive adenine molecule shows strong optical activity even as a dimer separated by the ribose phosphate group (Fig. 8.01), and as we saw in Fig. 8.00, this activity is still further enhanced by extension of the dimer into a trimer, tetramer, etc. So we must conclude that a polymer in which the monomers take up some fixed, or at least

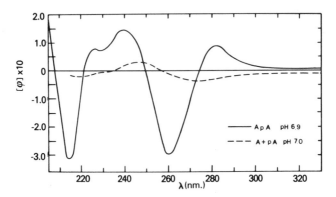

Fig. 8.01 Comparison of the optical activity of the adenine dimer, adenyl-(3′−5′)-adenosine with the sum of the optical activity of its constituent monomers, adenosine and 5′-adenylic acid. (70)

restricted, location in space relative to each other will exhibit optical activity independent of the existence of chiral activity in the individual monomers. The precise space organization or conformation of the atoms and molecules that make up the polymer is, of course, one of the types of information sought from the interpretation of measurements; the interpretations other than correlative are highly dependent on the theory of polymer optical activity. The conformations may be arbitrary, that is, without repeating elements of symmetry or with some repeating elements embedded in an overall matrix of some asymmetric organization—for example, the sections of a helical chain in many of the natural proteins. But frequently and with enormous significance for both their chemical, biological, or mechanical properties and their optical properties, the conformations may be of a very high degree of symmetry. It was this problem, the chiral properties of repeating polymers and specifically of helical polymers, that Moffitt treated in 1956 (1). Let us consider very briefly how the symmetry of repeating polymers can give clues to the optical properties; then we can return to examine Moffitt's treatment.

There are two general classes of polymers with repeating symmetry; those whose principal atoms or chromophores lie in a plane and those that take up positions in three-dimensional space and thus have a nonplanar surface or surface consisting of at least three planes. Of the latter, the most intriguing and beautiful is the helix. For the former, a reflection element of symmetry must be present, and we already know that systems having reflection symmetry cannot be chirally active; polymers are no exception to this rule. A glance at the character tables of Appendix I will show that for any group having an element of reflection symmetry, namely, C_i, C_{nh}, C_{nv}, D_{nh}, D_{nd}, S_s, T_d, and O_h, there are no representations that contain an electric- and magnetic-dipole-allowed transition along the same Cartesian coordinate. From the point of view of chiral-optical activity, therefore, such polymers are of little interest. Nominally planar polymers may have some conformational asymmetry or contain asymmetric centers (asymmetrically substituted tetrahedral carbon or silicon atoms, for example) that in effect eliminate the nominal symmetry and so induce some optical activity. No theoretical treatments have been specifically developed to treat these perturbations for polymer systems differently from the treatment for monomeric molecules that we examined in Chapter 4. The major difference is that the perturbed wavefunctions are polymer rather than monomer wavefunctions, but this part of the problem is identical with the case for helical polymers. Of the various theoretical methods developed for treating dimers and higher polymers, it is possible that methods developed by Schellman and his colleagues (17, 19) are more appropriate to these almost planar systems, but only a little work has been done in this area.

Let us consider the nonplanar polymer and more specifically one whose monomer chromophoric groups are arrayed in some helically symmetric array around a uniform axis. If we project these groups on a plane perpendicular to the helix axis, the resultant figure (Fig. 8.02) has axes of rotation and reflection—half as many reflection and rotation axes as there are points corresponding to the positions of the chromophoric groups and a rotation axis perpendicular to the plane with an axial index equal to the number of equivalent points arrayed around the center; in the case illustrated, there are eight such points. However, since in the helix, each of the points is not opposite an identical point (on the plane perpendicular to the helix axis), the rotation-reflection axes must be absent. Only the axis along the helix axis remains as an element of symmetry. Examining the character tables for *point* groups (Appendix I), we see that only the D_n and C_n groups have only rotation axes as symmetry elements other than the identity. Still closer examination shows that except for the groups C_2 and D_2, which bear no relation to the helical situation, all other point groups of C_n and D_n have one representation for which there is an axis along which optical activity can be generated and also a degenerate representation in which the electric and magnetic dipole transitions along the other two axes are mixed. Optical activity corresponding to these degenerate representations, as we determined in Chapter 6, can only be observed if something lifts the degeneracy, in this case, for moments perpendicular to the major axis. The C_n and D_n groups may thus be correlated to the *space* groups to which the symmetry of the helix will correspond. If we wish to make such a correlation, we can let z point in the direction of a translation along the helix axis. Then the new space group will have symmetry elements that

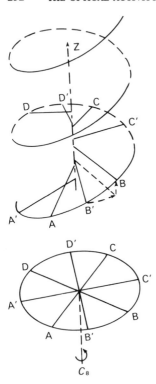

Fig. 8.02 The symmetry relationship between the repeating elements of a helix.

involve a so-called improper rotation, which is a rotation followed by a translation perpendicular to the rotation plane. In Fig. 8.02, this carries A' into A, B' into B, etc. It is easy to see that the step-by-step operation of the improper rotation will generate a helix. Suppose the points represent the center of gravity of transition moments of the polymeric chromophore groups; then each moment will have components in the helix with representation along the several axes, if it had components along these axes in the point group representation. But the *difference* is that by virtue of transition moments located asymmetrically removed from the plane perpendicular to the helix axis containing a particular moment, asymmetric rather than symmetric interactions may occur between the moments. The simplest type of interaction will result in addition of either electric or magnetic moments along the helix axis and in cancellation or addition perpendicular to the helix axis. Furthermore, the magnetic dipole transition mo nents in one group may interact with the electric dipole transitions momen.ʒ in an adjacent or nearby group. These interactions then can "lift" the degeneracy of the transitions inherent in the original transition moment of, say, the D_n site and cause the components perpendicular to the helix axis to split in energy (which, of course, is what we mean by "lifting" the degeneracy) and give rise to separate rotational (CD or ORD) bands. Finally, since the helix can have one of two directional senses, which of the nominally "degenerate" transitions is positive and which is negative will depend on the absolute sense of the helix rotation.

This, then, was the effect of helical symmetry that Moffitt clearly saw, that a transition strong enough to interact with an identical or similar transition on another group in a helix could give rise to three components, one of which in achiral absorption would be polarized along the helix axis and the other two perpendicular to it. The latter are at different energies if the intergroup interactions in the helix is strong enough to lift their degeneracy. The consequence that this set of selection rules had for chiral rotational strength, then, became a matter of defining appropriate wavefunctions and electric and magnetic dipole moment operators for a polymer, inserting these into the Rosenfeld equation, and calculating the resulting rotational strength. The selection rules demanded by the symmetry of the situation manifest themselves by the requirement that optical activity will occur only for those bands whose electric dipole transition moments $\mu_{0,\sigma K}$ do not vanish, that is, for bands with finite absorption coefficients (σK indexes the excited state of the polymer and is described more explicitly below). Now this is a rather strong requirement, not only because it defines the nature of the optical activity that arises from the transitions defined by symmetry, but also because it restricts the activity to polymer transitions that arise from strong monomer transitions. This means that this mechanism does not characterize the chiral activity arising from weak or nominally forbidden electric dipole transitions in the monomers even if they are enhanced in the polymers by the borrowing of electric or magnetic dipole intensity from other transitions through interactions in the helix. It will be left to Tinoco in his generalization (12) of the Moffitt-Fitts-Kirkwood approach to show how this latter effect occurs. The latter approach is a combined generalization of the Moffitt theory of exciton coupling and Kirkwood's coupled oscillator polarizability theory that Moffitt, Fitts, and Kirkwood published (8) in order to correct what was believed to be the error in Moffitt's original treatment referred to earlier in this chapter.

The rotational strength for an optical transition from the polymer ground state Φ_0, to the polymer excited state $\Phi_{\sigma K}$ (vide infra) is given in the usual way by

$$R_{0,\sigma K} = -\mu_{0,\sigma K} \cdot \mathbf{m}_{0,\sigma_K}. \tag{8.200}$$

In order to facilitate comparison with Moffitt's original treatment, the electric and magnetic transition moments will be rewritten in terms of the linear and angular momenta $\mathbf{P}_{\sigma K}$ and $\mathbf{L}_{\sigma K}$. It will be recalled that in Chapter 3, Section 3.4, we pointed out that frequently in treatments of optical activity the vector potential $\mathbf{A}$, of the light beam rather than the $\mathbf{E}$ and $\mathbf{H}$ fields are used to describe the incident radiation. Moffitt followed convention in this respect because the interaction potential he used to obtain perturbed polymer wavefunctions is proportional to $\mathbf{P} \cdot \mathbf{A}$ from which it became convenient to cast the transition moments μ and $\mathbf{m}$ in terms of the associated moments $\mathbf{P}$ and $\mathbf{L}$. It is important then in following Moffitt's treatment to remember that we are not discussing a different phenomenon, only an alternate description. For convenience and without loss of generality, we drop the subscript 0 designating the starting state.

$$R_{\sigma K} = \frac{e^2}{4\pi m^2 c^2} \nu_{\sigma K} \cdot \left[\mathbf{P}_{\sigma K} \cdot (\mathbf{L}_{\sigma K})^* \right]. \qquad (8.201)$$

The asterisk refers to the complex conjugate and $\mathbf{P}_{\sigma K} = \langle 0|\mathbf{P}|\sigma K \rangle$ and $(\mathbf{L}_{\sigma K})^* = \langle 0|\mathbf{L}|\sigma K \rangle^*$ in the units appropriate to Eq. 8.201. Now then, what Moffitt was able to demonstrate was that helical symmetry required that of all the possible values of the linear momenta attributable to the chromophoric electrons, only three fail to vanish identically. These three are written in terms of the radial, tangential, and vertical (to the helix axis) components of the momentum for the $0 \rightarrow \sigma K$ transition:

$$\mathbf{P}_{\sigma, M} = \frac{\sqrt{N}}{2} (P_{\sigma r} - iP_{\sigma t})(\mathbf{i} + i\mathbf{j}), \qquad (8.202)$$

$$\mathbf{P}_{\sigma, N-M} = \frac{\sqrt{N}}{2} (P_{\sigma r} - iP_{\sigma t})(\mathbf{i} - i\mathbf{j}), \qquad (8.203)$$

$$\mathbf{P}_{\sigma, N} = \sqrt{N} P_{\sigma v}\mathbf{k}. \qquad (8.204)$$

There are, of course, corresponding values for the angular momenta $\mathbf{L}_{\sigma K}$, but before examining them, let us see what the expressions for the allowed linear momenta alone tell us. The vectors $\mathbf{i}$ and $\mathbf{j}$ are unit vectors directed along the x and y polymer axes, respectively, and so perpendicular to the helix axis; $\mathbf{k}$ is a unit vector along the helix axis. N is the total number of monomers in the polymer. Since, as we will see, the angular momentum components are also proportional to $\sqrt{N}$, the rotational strength as defined in Eq. 8.201 is proportional to N and thus must be divided by N to obtain the rotational strength per monomer residue. Next, we see that there are two equal components, $\mathbf{P}_{\sigma, M}$ and $\mathbf{P}_{\sigma, N-M}$, which are the degenerate components directed perpendicular to the helix axis. Note here that they are of the same sign, although out of phase with each other. This is of no significance for absorption intensity, the polarized components of which are proportional to $|\mathbf{P}_{\sigma K}|^2$, $K = M, N - M$; but it is obvious that if there is any difference in the *sign* of the rotational strength associated with these two components, it will have to come from the angular and not the linear momentum. In these equations the index M stands for the number of turns of the helix. This parameter does not enter into the magnitude of the effect but is critically important in determining which of the many possible values of $\mathbf{P}_{\sigma K}$ are nonzero. This, in turn, is an important fact in the discussion of the validity of Moffitt's treatment, as we shall see shortly.

Perhaps this is a suitable point at which to discuss the nature of the wavefunctions used by Moffitt in this treatment. They are exciton wavefunctions appropriate to a linear crystal, and they enter, of course, because the linear and angular momenta (electric and magnetic transition moments) are, respectively,

$$\mathbf{P}_{\sigma K} = \int \Phi_0 \mathbf{P} \Phi_{\sigma K} \, d\tau, \qquad (8.205)$$

$$\mathbf{L}_{\sigma K} = \int \Phi_0 \mathbf{L} \Phi_{\sigma K} \, d\tau, \qquad (8.206)$$

where in the excited state $\Phi_{\sigma K}$, one of the monomer residues has been excited to the monomer excited state σ. In exciton theory, these excitations lead to a redistribution of charge in the polymer described by the product wavefunctions similar to those we saw in Section 4.2 (Eqs. 4.200 and 4.201). The ground state function is

$$\Phi_0 = \psi_{10} \psi_{20} \cdots \psi_{M0} \cdots \psi_{N0} \qquad (8.207)$$

and the normalized excited state function is

$$\Phi_{\sigma K} = \frac{1}{N} \sum_m \omega^{MK} \Phi_{m\sigma}, \qquad (8.208)$$

where $\omega = e^{2\pi/N}$, M has the values $1, 2, 3, \ldots, N$, and

$$\Phi_{m\sigma} = \psi_{10} \psi_{20} \cdots \psi_{M\sigma} \cdots \psi_{N0}. \qquad (8.209)$$

$\Phi_{m\sigma}$ differs from the ground state function by the fact that the chromophoric electron on the Mth residue has been excited to the monomeric state ψ_o. The significant point here is that writing the wavefunction in this way guarantees that we are describing ground and excited states in which the chromophoric electrons on each monomer interact to produce characteristic polymer states (excitons). This particular description, in which the excited state product wavefunction is modified by $\omega^{MK} = e^{2\pi iK}$ for $M = N$, results from Moffitt's adoption of a cyclic boundary condition that requires that a unit of the polymer repeat itself an integral number of times. As we shall see, this caused some problems. For the present it is sufficient to note that since any one of the N chromophoric electrons may be the one that is excited, there is an entire band of possible energy levels, each corresponding to the excitation of different electrons with different resulting values of the linear and angular momenta of the polymer as a whole. More specifically, each of the original monomer transitions $\psi_{m0} \rightarrow \psi_{m\sigma}$, which we have designated as $0 \rightarrow \sigma$, leads, as in a crystal, to a band of polymer states (exciton states) $\Phi_{\sigma K}$, given by Eq. 8.208. Cyclic boundary conditions applied to the symmetry of a helix limits the allowed states of this band to those with the values of K required by Eqs. 8.202–8.204, namely, to $K = M$, $N - M$, N. We shall not go through the details of the analysis of the energies of these states and therefore of their spectral frequencies, but intuitively it is possible to reach the same qualitative conclusions as are achieved by the application of perturbation theory to this problem. The state corresponding to $K = N$ is polarized (in absorption) parallel to the helix axis, while the states corresponding to $K = M$, $N - M$ are polarized perpendicular to it. By Eqs. 8.202–8.204 the moments developed for the latter two are equal but different from the first. The interaction energy with the fields that surround them is therefore different for the components polarized in different directions, both because of the different magnitudes of the moments and because the parallel and perpendicular moments "see" different fields; the molecular architecture is different in directions parallel and perpendicular to the helix axis. As a consequence, the polymer bands are at frequencies different from the isolated monomer transition frequency ν_σ. Moffitt demonstrates by a perturbation theory approach that these frequencies are given by

$$\nu_{\sigma K} = \nu_\sigma \pm \zeta_{\sigma K}, \tag{8.210}$$

where $\zeta_{\sigma K}$ is negative for the parallel band and positive for the degenerate perpendicular band. Before terminating this discussion, it may be wise to point out that this highly oversimplified description of exciton wavefunctions is designed only to support the fact that they are different from those of a noninteracting system. An excellent discussion is given in Davydov's book, *Theory of Molecular Excitons* (7).

Corresponding to the linear momentum terms of Eqs. 8.202–8.204, the significant terms in the angular momenta are given by

$$L_{\sigma,M} = \frac{\sqrt{N}}{2}(l_{\sigma r} - il_{\sigma t} + i\rho P_{\sigma t})(\mathbf{i} + i\mathbf{j}), \tag{8.211}$$

$$L_{\sigma,N-M} = \frac{\sqrt{N}}{2}(l_{\sigma r} + il_{\sigma t} - i\rho P_{\sigma t})(\mathbf{i} - k\mathbf{j}), \tag{8.212}$$

$$L_{\sigma,N} = \sqrt{N}(l_{\sigma v} + \rho P_{\sigma t})k, \tag{8.213}$$

where the $l_{\sigma r}$, $l_{\sigma v}$, and $l_{\sigma t}$ are the radial, vertical, and tangential angular momentum components and ρ is a geometric parameter of the helix. The first two of these expressions again correspond to the degenerate components perpendicular to the helix axis ($\mathbf{i}$ and $\mathbf{j}$ components). For substitution in Eq. 8.201, the complex conjugate of the angular momenta are formed by replacing $\mathbf{i} + i\mathbf{j}$ in Eq. 8.211 by $\mathbf{i} - i\mathbf{j}$, and vice versa for Eq. 8.212. The complex conjugate of Eq. 8.213 is equal to itself. $P_{\sigma,M} \cdot (L_{\sigma,M})^*$ and $P_{\sigma,N-M} \cdot (L_{\sigma,N-M})^*$ then *both* have the same factor $(\mathbf{i} + i\mathbf{j})(\mathbf{i} - i\mathbf{j}) = 2$ while $P_{\sigma,N} \cdot (L_{\sigma,N})^*$ has $k \cdot k = 1$ as a factor. It follows that the relative magnitudes and signs of the linear and angular momentum components in Eqs. 8.211 and 8.212 will determine the sign of the rotational strength associated with each of the degenerate components.

There are several ways of dividing the rotational strength resulting from these components. One is to sum the contributions of $P_{\sigma K} \cdot (L_{\sigma K})^*$ of the M and $N - M$ exciton states to give the rotational strength perpendicular to the helix axis:

$$R_{\sigma,M} + R_{\sigma,N-M} = \frac{e^2}{4\pi m_e^2 c^2 \nu_{\sigma M}}(P_{\sigma r}l_{\sigma r} + P_{\sigma t}l_{\sigma t} - \rho P_{\sigma t}P_{\sigma v}) \tag{8.214}$$

as distinct from the N component parallel to the helix axis:

$$R_{\sigma,N} = \frac{e^2}{4\pi m_e^2 c^2 \nu_{\sigma N}}(P_{\sigma v}l_{\sigma v} + \rho P_{\sigma t}P_{\sigma v}). \tag{8.215}$$

The measurement of these components on a system of oriented polymer molecules would be very helpful in analyzing the structural contributions. The isotropic optical rotation per residue due to the exciton band components M, $N - M$, and N at the frequency ν is given by (20)

$$[M_\sigma] = \frac{96\pi N_0}{hc}\sum \frac{\nu^2 R_{\sigma K}}{(\nu_{\sigma K}^2 - \nu^2)}. \tag{8.216}$$

If the interactions that lead to these expressions are insufficiently large because of the relative orientation of the monomer chromophores or because the distance between them is large, then the rotational strength $R_{o,K}$ reduces to just the sum of the residue rotational strength for the $0 \rightarrow \sigma$ transitions of the monomers. Thus the expression for exciton rotational strengths does include the intrinsic rotational strength of strong monomeric chromophores. It is important to remember, however, that the polymer rotation may consist of still additional contributions. For example, the intrinsic monomer transitions that are forbidden or only weakly allowed in electric dipole absorption may contribute to the rotation intrinsically or by induction through mechanisms other than the helix exciton.

In terms of the implications of the theory for understanding the origin of the chirally active transitions, that is, of the observed circular dichroism or molecular rotation near absorption bands, a different distribution of the contributions to the residue rotation can be made:

$$[M_\sigma] = [M_{\sigma i}] + [M_{\sigma h}] + [M_{\sigma j}]. \tag{8.217}$$

$[M_{\sigma i}]$ is the contribution of the intrinsic rotation of the monomers and is proportional to

$$\frac{\nu^2(\mathbf{P}_\sigma \cdot \mathbf{L}_\sigma)}{\nu_\sigma(\nu_\sigma^2 - \nu^2)}, \tag{8.218}$$

where $\mathbf{P}_\sigma$ and $\mathbf{L}_\sigma$ are the intrinsic momenta associated with the $0 \rightarrow \sigma$ monomer transition. The helix contribution from correlation between the electrons in different residues $[M_{\sigma h}]$ is proportional to

$$\frac{\nu^2 K_\sigma}{(\nu_\sigma^2 - \nu^2)^2}, \tag{8.219}$$

where K_σ is a measure of the rotational strength in terms of the radial, tangential, and vertical components of the polymer momenta and of the polymer geometry [see Eq. 71 of Moffitt's paper (1)]. $[M_{\sigma j}]$ is a first-order correction to the exciton interaction and has essentially the same frequency dependence as $[M_{\sigma i}]$, but not of $[M_{\sigma h}]$. The frequency dependence of the helical term results in a particularly important diagnostic tool for the helical interaction, which will be discussed in more detail in Section 8.4; note that $\nu^2/(\nu_\sigma^2 - \nu^2)^2$ is always positive, while the frequency term in Eq. 8.218 changes sign about the points where the measuring frequency ν equals ν_σ. The value of Eq. 8.219 may, of course, be either positive or negative, depending on the sign of K_σ. The resonance "catastrophe" form of the frequency dependence of Eqs. 8.217–8.219 requires that these expressions be applicable only for measurements made at frequencies further from the unperturbed frequency ν_σ of the monomer than the frequency difference between the parallel and perpendicular bands produced by the helical interaction.

This then is the essence of the Moffitt treatment, abstracted from the exquisitely detailed original analysis. It leads to the conclusion that strong monomer transitions in a repeating helical polymer give rise to separated absorption bands polarized parallel and perpendicular to the helix axis, that these bands are chirally active even if the monomer transitions are not, that the rotation results from exciton bands of the polymer which have their origin in the strong (electric-dipole-allowed) monomer bands, that the resultant optical rotation may be analyzed in terms of parallel and perpendicular components to the helix axis which depend differently on the structural parameters, that the perpendicular components are degenerate in zeroth order, and finally that the rotation may also be analyzed primarily in terms of the intrinsic rotation, if any, of the strong monomeric transitions and in terms of the rotation produced by the helical exciton interaction, the latter having a different frequency dependence than the former.

Up to this point, no emphasis has been put on the coupled oscillator nature of Moffitt's approach because it does not appear explicitly in the intensity formulations for absorption or chiral activity on which we have focused attention. It is, however, explicit in the description of the perturbation interaction that leads to the splitting of the energy levels for the parallel and perpendicular bands of the helix. As we saw in Chapter 4, it is explicit also in the intensity expressions in Tinoco's expansion of the Moffitt, Fitts, and Kirkwood theory (8), as it was indeed in Kirkwood's original theory for nondegenerate situations (21). For example, in Eq. 4.400, the expression for the rotational strength produced by coupling electric dipole transition momenta on two nearby ($R_2 - R_1$ is the distance between them) chromophoric groups contains the interaction potential $V_{10a,20b}$ between them. In Eq. 4.401 this potential is expressed in terms of the coupling of transition moments (oscillators) and in Eq. 4.402 in terms of the corresponding optical polarizabilities of the two chromophores. In attempting to formalize the relationship between the exciton and polarizability approaches, Moffitt, Kirkwood, and Fitts found what they believed to be an error in Moffitt's approach. The error was more imagined than real in the sense of its theoretical origin, but the analysis did result in disclosing a missing term that can make an important contribution to the optical activity.

Kirkwood's 1937 treatment of the polarizability theory of optical activity was for nondegenerate systems and therefore did not apply to polymers. By dropping this restriction, Moffitt, Fitts, and Kirkwood arrived at a formulation of the rotational strength that involved two, rather than one, terms with the frequency of the helical term $[M_{oh}]$ in Moffitt's expression and reproduced the rest of the expression as well. The new term, which, for example, for $\pi \rightarrow \pi^*$ transitions in polypeptides is opposite in sign to $[M_{oh}]$, was thought to appear because the polarizability approach required no restriction with respect to periodic boundary or reentry conditions. Moffitt had used this condition, as we have seen, to define the possible exciton states in a way that had been used successfully in explaining the optical properties of crystals. Since it is apparent that real helical polymers may not contain an exactly integral number of helical turns, this new term was thought of as an end effect. Let us examine this problem a little more closely both because of its intrinsic interest and because much recent work (2, 9, 10) has been devoted to demonstrating that these boundary conditions are, in fact, applicable to long helical polymers, that is, long compared to the wavelength of the light used to make the measurements.

The essence of the problem of periodic boundary conditions is the following: It is a characteristic of theories of optical activity that the derivatives of the amplitudes of the electromagnetic fields play an important role. In Chapter 2 (Eqs. 2.310 and 2.311), for example, we saw that $\partial H/\partial t$ and $\partial E/\partial t$ are necessary to fulfill the requirement for the induced electric and magnetic moments associated with chiral activity. Since H and E are represented by trigonometric functions over space, the derivatives are trigonometric functions and vary in amplitude in the direction of propagation. In the derivation of the Rosenfeld equation, the spatial extent of the chromophoric electron that interacts with the radiation field is therefore restricted to be much smaller than the wavelength of the light because the semiclassical assumption is made that the amplitudes of the E and H fields are given by sine or cosine functions that periodically every half wavelength go from a maximum value of $E = E_0$ and $H = H_0$ to zero. If the interaction of the chromophoric electron had to be treated at each point in this mutual space, not only would the complexity be considerably increased (22), but by having to account for the amplitude variation of the fields in addition, a method would have to be found to deal with the ıncertainty in the position of the electron resulting from the Heisenberg condition. To avoid this, the value of E and H must be assumed to be constant over the space occupied by the chromophoric electron, which means that this space must be very small compared to the wavelength of the light. For ordinary chromophores with an electron distribution that extends, at most, over a very few angstroms beyond the nuclei, wavelengths from 1800 to 10,000 Å (near ultraviolet and visible) easily meet this criterion. The wave nature of the radiation fields enters, then, only as a factor to determine the shape of the chiral absorption or refraction because the vibrational substates at each frequency give rise to different contributions to the oscillator and rotational strengths.

Now consider the problem of a helical polymer whose length is of the order of the wavelength; a DNA molecule of about one-half-million-gram molecular weight is about 2600 Å long, just the wavelength of the peak of its long-wavelength ultraviolet absorption band. If the electronic wavefunction is to be the polymer wavefunction, that is, cover the entire space of the polymer, the Rosenfeld equation could not be applied. However, another alternative is open. Let the polymer wavefunction consist of a product of monomer functions perturbed by interaction with nearby monomers. Then the Rosenfeld equation can be applied by summing over all the perturbed monomer functions. But this is a very formidable task. In the case of a helical polymer (or a crystal with finite unit cells), there is a very special way of getting around this because one can generate the polymer helix from many smaller helices by a symmetry operation, that is, by a combined rotation of one or more turns of the helix and a translation along it. Immediately we are met with still another problem; by definition this symmetry operation will work only if the entire helix is exactly an integral number of subhelices. (Note that it is not necessary that the subhelices contain an integral number of monomers.) In group-theoretical language, this means that the application of the full helical symmetry operation is equivalent to the identity operation. Otherwise the helical symmetry operation is not a true member of the group that controls the selection rules governing the allowedness of electric and magnetic dipole transitions. This requirement is termed a reentry condition. That is, after each application of the rotation-translation operation, every subhelix must superimpose exactly on another subhelix: *The boundary conditions must be periodic.* If, despite this principle, it is possible to apply the periodic boundary conditions to a real helical polymer regardless of its actual length, then the Rosenfeld equation can be applied. Is this possible? Or, as an alternative, is it possible to develop an equation that does not have the restriction with respect to the relation between the size of the chromophoric electron and the wavelength of the light, in which case periodic boundary conditions can be used with impunity? This, indeed, is what Stephan (23) and (later) Deutsche (11) did, using another procedure to develop rigorous extensions of the Rosenfeld equation that do not include this restriction. Thus, as we shall see in the next section, Tinoco's expression, which was based on Stephan's equation, bypassed the boundary condition problem. However, not until Loxsom (2) and Ando (10) demonstrated that the reentry boundary condition was a perfectly valid criterion to apply, was Stephan's equation found to be valid. An infinite helix can be defined as one long enough that the effect of imperfect ends is unimportant. The reason Tinoco's summation of the mechanism of optical activity is applicable to helical polymers is that his expression was based on the unrestricted quantum mechanical formulation of the rotational strength.

The wavelength dependence of the optical activity predicted on the basis of the exciton treatment by Moffitt and Moffitt and Yang (24) was not affected by the boundary condition problem and, as we shall see in Section 8.43, was immediately put to test. However, very few serious attempts (25, 26) were made in the next few years to calculate the magnitude of the optical activity of a helical polymer from its monomer optical properties. In 1960–1962, Tinoco published his theoretical elaboration based, in part, on the Moffitt, Fitts, and Kirkwood theory. The polymer is now treated as an extended asymmetric system; exciton effects result from the interactions between coupled oscillators.

8.3   THE POLYMER AS AN EXTENDED ASYMMETRIC SYSTEM

In the original work of Kirkwood and of Moffitt, as well as in their combined work, the wavefunctions used to obtain the necessary electric and magnetic transition moments were derived from zeroth-order wavefunctions of the monomer in Kirkwood's treatment or of the polymer when the exciton theory was developed. These are the wavefunctions given in Eqs. 4.200 and 4.201 and 8.207–8.209. The basis for this lies in the assumption, which has continued to be made (and for which more explicit wavefunctions for the polymer were developed) even in the treatment by Tinoco, that the monomer or at least the monomer chromophoric groups are sufficiently isolated so that there is negligible electron exchange between them. Kirkwood in his 1937 paper, justified this approximation by calling attention to the fact that "the (ultraviolet) spectrum of an organic radical retains to a large extent its identity in different compounds of which it is a substituent....The approximate additivity of atomic refractions and atomic volumes of organic compounds are related to this fact." (18) When it came to considering a polymer having degenerate excited states, that is, the exciton states of Eq. 8.208, which can be N-fold degenerate if all the N groups in the molecule are identical, it became apparent that such states could not each be a true description of the states of the system. A better description could be obtained by applying perturbation theory to obtain first-order polymer states from the zeroth-order polymer states (still using zeroth-order monomer states). These first-order ground and first-order polymer excited states are, respectively,

$$\psi_0' = \psi_0^0 - \sum_i \sum_{j \neq 1} \sum_a \frac{V_{i0a}\psi_{ia}^0}{\mathcal{E}_a - \mathcal{E}_0}$$

$$- \sum_i \sum_{j > i} \sum_a \sum_b \frac{\left(V_{i0a,j0b}\psi_{ia,jb}^0\right)}{(\mathcal{E}_a - \mathcal{E}_0) + (\mathcal{E}_b - \mathcal{E}_0)} \qquad (8.300)$$

$$\psi_{AK}' = \sum_i^N C_{iAK}\left[\psi_{ia}^0 + \sum_{j \neq 1} \frac{V_{i0a,j00}\psi_0^0}{\mathcal{E}_a - \mathcal{E}_0} \right.$$

$$- \sum_{j \neq 1} \sum_{b \neq a} \frac{V_{i0a,j0b}\psi_{jb}^0}{(\mathcal{E}_a - \mathcal{E}_0) + (\mathcal{E}_b - \mathcal{E}_0)}$$

$$\left. - \sum_{b \neq a} \frac{V_{iab,jo0}\psi_{ib}^0}{(\mathcal{E}_a - \mathcal{E}_0) + (\mathcal{E}_b - \mathcal{E}_0)} \right], \qquad (8.301)$$

where ψ_0^0 and ψ_{ia}^0 are the zeroth-order ground and excited states of the polymer and ψ_0' and ψ_{AK}' are the corresponding first-order states referred to but not described in Chapter 4. We shall not attempt to explain them in detail. Most of the symbols have been defined or are obvious from earlier discussions. There are, however, two important points to be made here. One is that in calculating *these* wavefunctions, as distinct from the zeroth-order functions, only nearby states need be considered because otherwise the energy denominators like $\mathcal{E}_a - \mathcal{E}_0$ get very large and the terms get insignificant. Second, the manner of formulating the interaction potentials makes a difference. In the Moffitt-Fitts-Kirkwood development, these potentials contained only the effects of the dynamic fields of surrounding groups and of optical transitions but not of the static fields. Tinoco took this one step further (12) and accounted for the fact that already in the ground state there are interactions between the chromophoric electrons in one group and the field produced by the permanent (time-averaged) distribution of charges in the surrounding groups; this accounted for the Condon one-electron mechanism.

The words "static" and "dynamic" as applied to the chiral perturbation of a symmetric charge distribution are literally descriptive of our most physically intuitive notions. Think, if you will, of a chromophore in terms of its transition dipole moment, which, for example, lies in a plane of symmetry *with respect to the chromophore*. This moment results from a charge distribution and is measured by the transition moment integral. Regardless of its measure, however, we conceive of some distribution of charge density—some accumulation of pluses and minuses that cluster so that the centers of gravity of the positive and negative charges are separated by a small but finite distance such that the line connecting the centers lies in the symmetry plane of the chromophore. We know that this situation (this transition dipole) will not contribute to optical activity. This chromophore, however, is sitting in an asymmetric environment; perhaps there are one or more groups (asymmetric carbon atoms, for example) that are so disposed that the system as a whole does not have a symmetry plane (or center). The nominally symmetric charge distribution of the chromophore is, of course, distorted by the fields of the nearby asymmetric group, and this distortion can come about in two ways. First, the charge distribution of the asymmetric group in its ground electronic state forms a field that polarizes the dipole so as to remove one or both centers of gravity from the symmetric plane. While it is true that we think of the electronic charge that provides this asymmetric field as being in constant motion, it is equally true that we consider it to have a time-average distribution, which we picture as stationary or *static*. So this field, which in other contexts we would say produces van der Waal's forces (sometimes separating these into dipole and polarizability components), represents the static perturbation. It enters into the Tinoco representation through terms such as $V_{i0a,j00}$, that is, the potential energy of interaction between the $0 \to a$ transition on the ith symmetric group and the ground state ($0 \leftrightarrow 0$) or static charge distribution on the perturbing jth group.

The second or *dynamic* perturbation comes from the transition dipoles (which are produced in the almost infinitely small time intervals of the interaction of the radiation with a chromophoric electron) of the asymmetric group. They enter into the Tinoco representation as the potentials $V_{i0a,j0b}$, etc., that is, as the potential energy of interaction between the $0 \to a$ transition dipole on the ith symmetric group and the field of the $0 \to b$ transition on the jth perturbing group. Both of these are produced by the time-dependent fields of the radiation and so are thought of as dynamic; the perturbation $V_{i0a,j0b}$ is a *dynamic* coupling phenomenon.

As a result of this treatment, the rotational strength in Eqs. 4.202-1 to 4.202-6 of Chapter 4, applicable to small molecules and in principle to nonrepeating polymers, contains both static and dynamic coupling terms. The expression is perfectly general and accounts for both electric- and magnetic-dipole-allowed transitions. The modifications necessary to account for a repeating polymer is to sum over the N monomeric groups. When the intrinsic monomer transition $0 \to a$ is either magnetic or electric dipole forbidden (μ_{i0a} or $M_{i0a} = 0$), then the first is null and the remaining terms may be reduced to their important components. Expressions for these special but common situations are easily derivable from Eqs. 4.202-1 to 4.202-6 and appear together with detailed discussions in a number of places (11, 27, 28). One point that is especially worth noting is that an electric-dipole-forbidden transition will not give rise to splitting of the exciton transition in the polymer. Recall that the splitting is dependent on the strength of the monomer transitions. Consequently, if a contribution to the rotation from a nominally electric-dipole-forbidden transition does appear through the asymmetric perturbation or vibronic coupling mechanisms, it is likely to appear as a single band at the positions of the monomer band. Recall also in this connection that both the exciton theory and the extended asymmetric molecule treatments are dependent on perturbation theory for descriptions of the wavefunctions of the polymeric system. In each case the perturbation energy consists of interacting dipoles or monopoles, or their equivalent polarizabilities. Since perturbation theory is restricted to situations in which the perturbing energies are small compared to the total energies, these treatments are better applicable (see Eq. 4.401) to weak transitions for which the transition dipole moments are small. However, as can also be observed from Eq. 4.401, the interaction energies are proportional to the inverse third power of the distance between the interacting units; in many repeating polymers this distance of more than 3 or 4 Å for nearest neighbors is barely sufficient to allow perturbation theory to be used even with moderately strong transitions (29). The use of a monopole formulation of the interaction energy (30),

$$V_{i0a,j0b} = - \frac{q_{it0a} \mathbf{R}_{it0a,j0b} \cdot \boldsymbol{\mu}_{j0b}}{|R_{it0a,j0b}|^3},$$

(8.302)

in place of the dipole formulation (Eq. 4.401) also helps for some systems, especially since the placement of equivalent monopoles at the atom positions reduces the sensitivity of the calculation to the accuracy of the positioning of the transition dipole. Here q_{it0a} and $\mathbf{R}_{i0ta}$ are, respectively, the charge and position of the monopole t in the ith group, the monopole usually being placed at an atom position, and $\mathbf{R}_{it0a,j0b} = \mathbf{R}_{j0b} - \mathbf{R}_{it0a}$. The polarizability corresponding to the higher transitions is then used to approximate the $\boldsymbol{\mu}_{j0b}$ transition dipole moments (using Eq. 2.403 in which the sum is over states $|b\rangle \neq |a\rangle$). In order to further avoid the problem of using perturbation theory for interactions at short distances, a number of other treatments have recently been developed: The linear response theory developed primarily by Rhodes and his co-workers in a series of papers starting in 1967 (3, 14, 15, 31); a modification of the coupled oscillator approach using a matrix formalism by Bailey, Nielsen, and Schellman (19) and by Madison and Schellman (17); and variations of the exciton treatment using scattering theory by Loxsom (9) and by Philpott (18). We have noted above the availability of the very lucid discussion of the latter approach by Philpott (29).

While the matrix formalism does avoid the large perturbation problem, it is restricted in practice to the consideration of polymer molecules containing only a small number of units, dimers to decimers, perhaps, or slightly longer. The reason for this involves the magnitude of the computing time as the number of interacting subunits or transitions becomes very large. This restriction is not too limiting because most polymer optical characteristics, as distinct from their other physical properties, are usually fixed by the time the polymer consists of ten or more repeating units. The principal advantage of the method is that it makes it possible to avoid having to calculate transition moments from wavefunctions, which must first be obtained by perturbation theory. Instead, empirical parameters, especially transition energies and moments that are obtainable from spectral data, may be used to describe the interaction between groups and between transitions.

Very succinct summaries of the method have been given (17, 27). Rather than repeat or expand on these, let us try to examine the special features by which it intrinsically differs from the perturbation methods. The aim, as always, is to calculate the rotational strength using the Rosenfeld approximation; the desired quantities are the electric and magnetic dipole transition moments of the system. In the perturbation methods, the transition moment matrix elements $\mu_{ab} = \langle a|\mu|b \rangle$ and $\mathbf{m}_{ba} = \langle b|\mathbf{m}|a \rangle$ for *each* of the interacting transitions are calculated from the perturbed wavefunctions, and these are used to calculate the interaction energies by the Tinoco formalism of Eq. 4.202 and its polymer counterpart. In the matrix formalism the electric and magnetic dipole transition moments for the *entire* system are found from

$$\mathbf{U} = \mathbf{U}^0\mathbf{C} \tag{8.303}$$

$$\mathbf{M} = \mathbf{M}^0\mathbf{C} \tag{8.304}$$

where $\mathbf{U}^0$ and $\mathbf{M}^0$ are row vectors containing an entry for each subunit (integers) and each transition (Greek letters) considered: $\mathbf{U}^0 = (\mathbf{U}_{1\alpha}, \mathbf{U}_{1\beta}, \ldots, \mathbf{U}_{2\alpha}, \mathbf{U}_{2\beta}, \ldots, \text{etc.})$, $\mathbf{M}^0 = (\mathbf{M}_{1\alpha}, \mathbf{M}_{1\beta}, \ldots, \mathbf{M}_{2\alpha}, \mathbf{M}_{2\beta}, \ldots, \text{etc.})$, and $\mathbf{C}$ are the coefficients of a matrix by means of which the allowed energy states of the system have been found by diagonalizing the Hamiltonian for interaction between transition moments and the subunits of the system, a process symbolized by

$$\mathcal{H} = \mathbf{C}^{-1}\mathcal{H}^0\mathbf{C} \tag{8.305}$$

It is this process that avoids the use of the perturbation methods because the Hamiltonian energy can be specified in terms of experimental parameters.

The process of diagonalization of a matrix equation of the form in Eq. 8.305 is equivalent to the solution of a differential equation such as the Schrödinger equation by the ordinary methods of differential calculus. In the case of the Schrödinger relation, the solution yields the values of the energy of the allowed quantum states and the wavefunctions describing the charge distribution in those states. The matrix diagonalization in the Schellman method yields the energy of the allowed states and the coefficients C, relating the values of the transition moments for the states of the system to the "known" transition moments of the isolated unit chromophore.

Let us consider, for example, the simplest interaction possible that will still involve all the mechanisms described in Chapter 4 for producing chiral activity from a nominally symmetric chromophore, defined here as one for which either the transitions are electric dipole forbidden and/or the magnetic dipole and electric dipole transitions are perpendicular to each other. The molecule consists of two identical groups, and we consider only transitions to two lower excited states a and b as important, since the interaction energy with higher states is small because of the large energy denominator. In this case, the Hamiltonian consists of the following possible interactions: (i) transition $a \rightarrow b$ on group 1 with the ground state static field of group 2, and vice versa, and (ii) the interaction of the $0 \rightarrow b$ (allowed) transition on one unit with that on the second, and vice versa. The entire Hamiltonian is given by

$$
\mathcal{H} = \left[\begin{pmatrix} \mathcal{E}_a = h\nu_a & V_{1ab,200} \\ V_{1ba,200} & \mathcal{E}_b = h\nu_b \end{pmatrix} \begin{pmatrix} V_{10a,20a} & V_{10a,20b} \\ V_{10b,20a} & V_{10b,20b} \end{pmatrix} \\ \begin{pmatrix} V_{20a,10a} & V_{200,10b} \\ V_{20b,10a} & V_{2b,10b} \end{pmatrix} \begin{pmatrix} \mathcal{E} = h\nu_a & V_{2ba,100} \\ V_{20b,100} & \mathcal{E} = h\nu_b \end{pmatrix} \right]. \tag{8.306}
$$

With the restrictions of electric dipole forbiddeness for $0 \rightarrow a$ and/or its perpendicularity with the magnetic dipole, several of the elements are virtually zero, for example, $V_{10a,20a}$. In any case, the V's may be specified in either the dipole or monopole approximation using experimental values obtained from spectral data, that is, transition moments obtained from polarized absorption measurements and polarizabilities from refractive index anisotropies. The ν_a and ν_b are the frequencies (at maximum absorbance for wide bands) of the $0 \rightarrow a$ and $0 \rightarrow b$ absorption bands. We are, of course, making the reasonable assumption here that although the $0 \rightarrow a$ transition is electric dipole forbidden, it appears weakly in absorption. With the elements of C obtained in this way, the various components of U and M, and therefore of the rotational strength, are found. In the next section we will see how successful this method has been in treating molecular polymers.

Finally, but by no means because of its secondary importance, we return to the "classical" theory of DeVoe (16). This too is a "polarizability" theory, but since the nomenclature has been preempted for the original Kirkwood theory, we will refer to it, as DeVoe did, as a classical theory; and classical theory it is in the sense that it makes direct use of Maxwell's electromagnetic theory and does not depend on quantum mechanical perturbation theory. No wavefunctions are directly involved, although some of the parameters may be obtained from quantum mechanical calculations if direct measurements are not available. It does, however, make use of one important nonclassical assumption, namely, that the absorption of radiation is polarized in the direction of the quantum mechanical transition moments μ_{ab} and $\mathbf{m}_{ab}$. As a consequence, the unit vectors, $\mathbf{e}_i$ and $\mathbf{e}'_i$, polarized in these directions, appear in the theory. The procedure develops the classical polarizabilities arising from the interaction of the electromagnetic fields with the electric and magnetic oscillators (chromophores) of the monomers, substituting these appropriately in expressions for the induced electric and magnetic moments (Eqs. 2.310 and 2.311). In order to account for the polarization directions, however, the directional parts ($\mathbf{e}_i$ and $\mathbf{e}'_i$) and frequency-dependent parts (α_i and β_i) are separated. The dipoles induced by the effective electric field $\mathbf{D}_i$ and magnetic field $\mathbf{B}_i$ are given by

$$\mu_i = \alpha_i(\mathbf{D}_i \cdot \mathbf{e}_i)\mathbf{e}_i - \frac{\beta_i(\dot{\mathbf{B}}_i \cdot \mathbf{e}'_i)\mathbf{e}_i}{c}, \tag{8.307}$$

$$\mathbf{m}_i = \frac{\beta_i(\dot{\mathbf{D}}_i \cdot \mathbf{e}_i)\mathbf{e}'_i}{c}, \tag{8.308}$$

where $\dot{\mathbf{B}}_i$ and $\dot{\mathbf{D}}_i$ are the time derivatives of B_i and D_i. It is the fact that $\mathbf{D}_i$ and $\mathbf{B}_i$ are Maxwell's *local* electric and magnetic fields experienced by the oscillator (chromophoric electron), which accounts for the dependence of the induced moments on the environment of the chromophore. The local field is the applied field $\mathbf{E}$, including the field induced in the medium by the applied field (and this is given in the usual way by Eq. 2.015), minus the field resulting from the interaction of the transition dipoles held in fixed relationship to each other, as they would be in a crystal or a stereoregular polymer:

$$\mathbf{D}_i = \mathbf{E} + \left(\frac{4\pi}{3}\right)\mathbf{P} - \sum_j \mathsf{T}_{ij} \cdot \mu_j, \tag{8.309}$$

where T_{ij}, a tensor specifying the geometric relationship between the monomers is given by

$$T_{ij} - r_{ij}^{-3}(I - 3e_{ij}e_{ij}).\tag{8.310}$$

e_{ij} is a unit vector along the direction connecting the dipole oscillators on monomer units i and j, and $I = ii + jj + kk$. If the oscillators on the surrounding units are randomly arrayed around the ith unit or monomer, then $\sum_j T_{ij} \cdot \mu_j = 0$ because the dipole interactions cancel. The local field is then just the Lorentz field $E + (4\pi/3)P$, as it is for the monomer in a randomly oriented solvent. However, if the polymer has a preferred conformation, then the sum is greater than zero and the effective field is dependent on the geometry of the monomers with respect to each in an axial system defined in the polymer. In Chapter 2, we saw that the dipoles induced by the electromagnetic fields, in this case D_i and B_i, are responsible for the polarizability α (see Eq. 2.019). Substituting Eq. 8.309 for the local effective field, the polarizability α_i in Eq. 8.308, associated with the transitions due to the interaction of the oscillators with the external field and *with each other*, is similarly determined:

$$\alpha_i = \frac{\mu_i \cdot e_i}{\left[E_i + (4\pi/3)P\right] \cdot e_i - \sum_j T_{ij}\mu_j \cdot e_j} - \frac{\beta_i(\dot{B}_i \cdot e_i')e_i e_j}{c}\tag{8.311}$$

The frequency dependence can be explicit by noting that the E, P, and B have the frequency dependence defined in Chapter 2 with $\dot{B}_i = i\omega B$, and that the "magnetic polarizability" β_i is dependent on the frequency difference between the measuring frequency ω and the oscillator or transition frequency ω_i (see, for example, Eq. 3.423).

This is all that is required to understand the origin of the theory because once we see that the optical polarizability depends in part on the interaction between transition dipoles on different monomers, the remainder of the derivation of the final expressions is formally explicit and clearly given in DeVoe's paper (16). Basically, for an aggregate or polymer the induced dipoles and, therefore, the polarizabilities α_i are summed over all the oscillators, and the macroscopic refractive index calculated from the resulting polarizabilities (recall Eq. 2.021). The frequency-dependent parts are obtained empirically from the spectrum of the monomers in dilute isotropic solution and the interaction parts by a matrix method, the latter, of course, being limited in practice by the computing facilities to some reasonable number of transitions having polarization e_i on the monomeric units i and j reasonably close to each other (r_{ij} small). The difference in the refractive indices for right and left circularly polarized light for optically active aggregates or polymers then yields the molar rotation

$$[\phi] = \sum_{ij} C_{ij} \operatorname{Re} A_{ij} \qquad (8.312)$$

or the ellipticity

$$[\theta] = \sum_{ij} C_{ij} \operatorname{Im} A_{ij}, \qquad (8.313)$$

where

$$C_{ij} = \left(\frac{6\omega^2 N_0}{c^2}\right)\left[(e_i \times e_j)\cdot(r_j - r_i) - 4b_j e_i \cdot e_j'\right],$$

$$b_j = \frac{\beta_j}{\alpha_j},$$

$$A_{ij} = \left(\frac{\delta_{ij}}{\alpha_i} + G_{ij}\right)^{-1}, \qquad \delta_{ij} = \begin{cases} 0 & \text{for} \quad i \neq j \\ 1 & \text{for} \quad i = j \end{cases}.$$

The required quantities are, therefore, the geometry of the polymer (to obtain r_i, r_j), the isotropic spectrum of the monomers to obtain the frequency dependence of α_i, and either quantum calculations or polarization measurements on oriented or crystalline samples of the monomer to determine the directions e_i and e_i' of the electric and magnetic dipole transition moments. Of course, e_i' is, not directly measurable because of the weakness of magnetic transitions, but may be inferred from symmetry arguments and the direction of e_i or from quantum mechanical calculations. When the monomers have a high degree of symmetry, both e_i and e_i' may also be inferred from group-theoretical considerations, for example, long- and short-axis in-plane polarizations for e_i in monosubstituted aromatic rings.

Recently, Levin and Tinoco (32) derived a theory for the optical rotation and circular dichroism of helical polymers based on this nonperturbation treatment of the optical absorptive and refractive properties of a polymer. In doing so, they extended the treatment to include all intermonomer interactions. Their treatment differs from DeVoe's original treatment (just as Moffitt's did) in that periodic boundary conditions are introduced; consequently, the condition that the molecular dimensions be small compared to the wavelength of the light can be relaxed. Not unnaturally, the application of their theory to a helical polymer containing only one repeating unit arrives at the same final result for the optical rotatory dispersion as did Moffitt, Fitts, and Kirkwood (6). However, the method is, in principle, capable of treating polymers containing more than one repeating unit and incorporates all the optical transitions of the units in terms of the monomer optical transitions. It has not yet been explicitly applied.

8.4   APPLICATIONS

We turn now from theory to observations and applications. Here we can examine the variety of phenomena to which optical activity measurements have been applied, as we did in Chapter 7, namely, the questions of structure, absolute configurations, conformation, and conformational changes. The extension of these phenomena to polymers is, however, not unique. Intuitively and observationally, we find that only to the extent that the theory of a molecule containing repeating units is unique, are the applications different. Let us, therefore, take a slightly different approach and divide the discussion into the following areas: (i) the optical activity of nonhelical polymers; (ii) the ability of the theory to distinguish between the contribution of the helical structure to the chiral activity and other contributions; (iii) the ability of the theory to predict helical parameters, for example, the screw sense of the helix and the degree of helicity in a repeating polymer or protein; and (iv) the significance of different types of optical transitions in the chiral activity of helical polymers. These by no means exhaust the range of applications of the theories or of problems associated with the chiral activity of polymers. It is hoped, however, that they will be sufficiently illustrative and informative in achieving our objective of demonstrating relationships between the electronic and nuclear structure in the stationary energy states of polymers and the chiral activity arising from optical transitions between these states.

8.41   Optical Activity of Nonhelical Polymers

Except for the biologically important proteins, many of which contain only partially helical polypeptide chains, most nonhelical polypeptides and other synthetic polymers do have organized structures. They are stereoregular, frequently with flat or folded-sheet conformations that endow them with chiral optical properties if they have been synthesized from chiral monomers. Their chirality arises not from helical conformations, but from the intrinsic chirality of the monomers and from the asymmetric but nonhelical interactions between monomer units. Excitonlike interactions are not, in principle, excluded, but since in nonhelical polymers the chromophoric groups tend to be more distant from each other, and since the exciton interaction energy that produces the splitting (Eqs. 8.214 and 8.215) varies as the inverse third power of the distance between the chromophore transition dipoles, the positive and negative exciton components are almost degenerate in frequency, with a resultant nil contribution to optical activity. Even in the stereoregular poly-α-olefins, which do take on helical conformations, no clear example of exciton activity has been found, in part because the strong transitions in these compounds are generally at higher energies than measurements can conveniently be made and possibly also because the geometric relationships between repeating monomers is unfavorable (33). In the nonhelical β structures of polypeptides, however, exciton splittings are demonstrably significant (34).

Stereoregular polymers having an aliphatic hydrocarbon backbone form the large bulk of synthetic polymers and fall into four main classes of which one, the stereo block polymer, is either optically inactive or else may be considered as a special combination of the other types. The principal types then are the atactic, syndiotactic, and isotactic polymers (Fig. 8.03) that result from steric hindrances or strong conformational energy minima. Measurements (35) of the optical activity of poly-α-olefins in these confor-

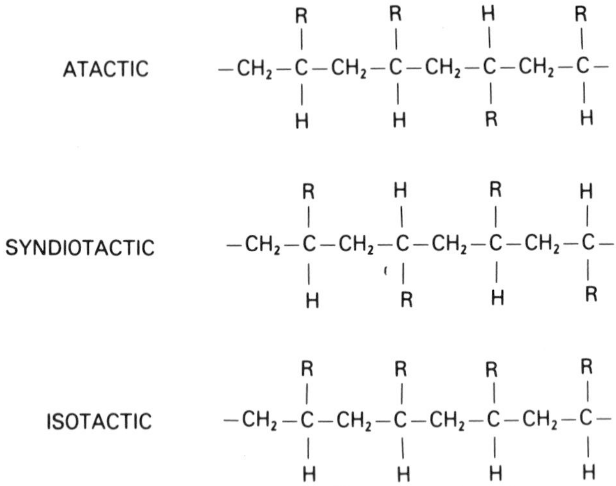

ATACTIC

$$-CH_2-\underset{\underset{H}{|}}{\overset{\overset{R}{|}}{C}}-CH_2-\underset{\underset{H}{|}}{\overset{\overset{R}{|}}{C}}-CH_2-\underset{\underset{R}{|}}{\overset{\overset{H}{|}}{C}}-CH_2-\underset{\underset{H}{|}}{\overset{\overset{R}{|}}{C}}-$$

SYNDIOTACTIC

$$-CH_2-\underset{\underset{H}{|}}{\overset{\overset{R}{|}}{C}}-CH_2-\underset{\underset{R}{|}}{\overset{\overset{H}{|}}{C}}-CH_2-\underset{\underset{H}{|}}{\overset{\overset{R}{|}}{C}}-CH_2-\underset{\underset{R}{|}}{\overset{\overset{H}{|}}{C}}-$$

ISOTACTIC

$$-CH_2-\underset{\underset{H}{|}}{\overset{\overset{R}{|}}{C}}-CH_2-\underset{\underset{H}{|}}{\overset{\overset{R}{|}}{C}}-CH_2-\underset{\underset{H}{|}}{\overset{\overset{R}{|}}{C}}-CH_2-\underset{\underset{H}{|}}{\overset{\overset{R}{|}}{C}}-$$

Fig. 8.03 Stereoregular hydrocarbon polymers. The asymmetric carbon atoms are shown in full Fischer projection. If the R groups are bulky enough to hinder or restrict rotation about the bonds between the methylene carbon and the asymmetric carbon atoms, then each of these types of polymers may become optically active through the formation of helical or other stereoregular structures. The isotactic polymer will be optically active independently of the formation of any secondary structure because all the asymmetric centers have the same absolute configuration.

mations have helped establish features of the monomeric structures that lead to different types of polymer structure. The stereochemical aspects of poly-α-olefins, as well as other synthetic macromolecules as reflected in their crystalline structures, has been concisely but excellently reviewed by Natta and Allegra (36). We wish only to point out here that there are many indications that although statistical fluctuations do take place in solution as a result of limited barrier heights between different conformations, these structural parameters in the crystals are heavily reflected in the solution conformations. Because of the fluctuations, however, the chiral activity of these molecules in solution is very temperature and, in some cases, solvent dependent. Consequently, all conjectures about their conformation derived from measurement of the circular dichroism or optical rotation must take this into account. Also, because of the uncertain predictability of the conformational equilibria, very little progress has been made in applying the theoretical considerations of this chapter to predictions of the optical activity, both of the nonhelical and helical polymers of this type. In the case of those poly-α-olefins, where bulky groups on the carbon atoms α to the main chain cause one conformation of the polymer to predominate strongly over the others, classical polarizability theory has been reasonably successful in explaining the observed activity in at least one case (37), that of a helical polymer with an aromatic side chain. Another informative case is that of poly-(S)-3-methylpentyne, where despite the moderately strong transition of the olefin ($\varepsilon \approx 2400$ at $\lambda \approx 280$ nm) and the almost certain skew tendency of the conjugated systems, the absence of any evidence of exciton splitting in CD and ORD curves (Fig. 8.04) indicates that in solution the polymer probably does not exist with long sequences of helical conformation. It undoubtedly does, however, have a predominance of one skew sense over the other in short sequences of the conformationally mobile polymer. Few nonhelical systems have been treated by any of the methods; but as a result of the discovery of extensive β-pleated-sheet structures in a number of globular proteins (38–40), several attempts have been made to calculate the optical activity of both the parallel and antiparallel forms of polypeptides with these structures (Fig. 8.05). These almost planar forms of polypeptides have proved rather refractory to the application of theory, but as the nature of the optical transitions in the polypeptides have become increasingly better understood, results indicative of the origin of the chiral activity, and almost quantitatively accurate, have been achieved. Except for the most recent of these by Ronish and Krimm (41), the subject has been thoroughly reviewed (27). It is interesting and perhaps not too surprising that both the matrix formalism (15) and the general perturbation method (41) have produced quantitatively similar results for the β structures of polypeptides, although the same procedures applied to the related

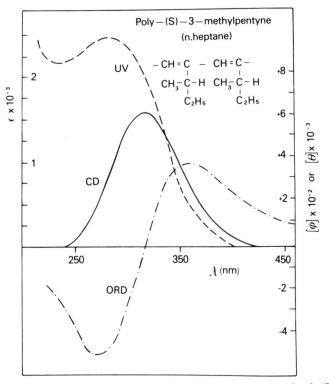

Fig. 8.04 Ultraviolet, circular dichroism, and optical activity spectra of poly-(S)-3-methyl-pentyne. [From preprint, O. Pieroni and F. Ciardelli, *International Symposium on Macromolecules, Helsinki, July* 2–7, 1972, Vol. 3, with the permission of the authors.]

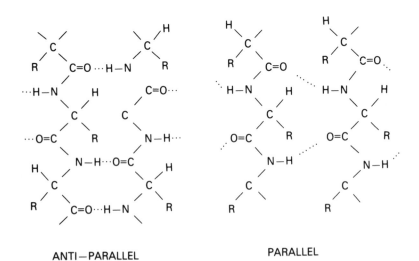

ANTI–PARALLEL   PARALLEL

Fig. 8.05 Structural relationships between the polypeptide chains in parallel and antiparallel β-pleated-sheet conformation.

274

polyproline structures and to α-helices do not necessarily agree. In part, this is due to different estimates of the effect of the far ultraviolet transitions on the near ultraviolet optical activity. These interactions, it will be recalled, enter directly in both formalisms. In the matrix method, it enters through the number (and magnitude) of entries in the row vectors defining U^0 and M^0 in Eq. 8.303 and 8.304. In the Tinoco form of the perturbation method, it enters through the number b of transitions over which the sums in Eq. 4.202 are performed. Because of the different geometry of the helices and the β structures, the interactions with higher energy transitions are quite different, probably tending to be somewhat smaller in the more extended β sheet structures than in the helical structures. In both cases exciton interactions between the *same* transitions on adjacent groups are sufficient to split the amide 190-nm $\pi \rightarrow \pi^*$ band into two or more components, but not the weaker band of the $n \rightarrow \pi^*$ transitions. We will have more to say about this in Section 8.44, where the nature of the optical transitions is discussed in detail.

8.42   Distinguishing between Helical and Nonhelical Contributions to the Chiral Activity

In 1955, Carolyn Cohen called attention to the fact that while the specific optical rotation $[\alpha_D]$ of proteins is negative at the sodium D line, as is the specific optical rotation of the L-amino acids of which the proteins are composed, the magnitude of the rotations of the native proteins is much less negative than that of the denatured proteins (42); the values of $[\alpha_D]$ (native) are approximately -20 to -70 degrees compared to values of $[\alpha_D]$ (denatured) of approximately -80 to -120 degrees (43). In keeping with the conventional wisdom that the native proteins, unlike the denatured ones, have an orderly structure, she suggested, therefore, that the backbone structure of the native proteins, namely, the polypeptide linkage, must contribute a positive optical activity at the sodium D line. This suggestion set in action an area of research that to this day contributes a very substantial fraction of the literature on chiral optical effects. It was natural, of course, to turn to the α-helix (44,45) as the structure that must be present in at least many of the proteins to account for this behavior. The immediate result of the suggestion was a rash of experimental and theoretical attempts to understand the optical activity of polypeptides in terms of their α-helical structure. We have already discussed some of the theoretical work, and the experimental research has been reviewed in so many places that there is no point in repeating it here. We wish only to call attention to some salient features of both, which can help illuminate how the helical structures (not only in polypeptides and proteins, but also in polynucleotides and natural or synthetic nucleic acids) contribute to the chiral activity, thereby enabling the latter to be used as a probe of the structure and behavior of these macromolecules. At the beginning of this chapter, attention was called to the way in which the rotational strength per mole of residue varies with the number of residues in polymers of adenylic acid, and also to the correlation of the highly increased rotational strength associated with the low pH form of poly(A), which is known to form a double-stranded helix. We will be examining in this section those elements of the theory and observations that interrelatedly enable us to understand how the helical array of molecular residues produces marked changes of the rotational strength from that of the simple sum of the monomeric rotational strengths and determines the shape of the circular dichroism and optical rotatory dispersion.

Consider the shape first. Moffitt predicted that in a helix the interaction of strong monomer transitions $(0 \rightarrow \sigma)$ would lead to the splitting into components parallel and perpendicular to the helix (exciton components) separated in frequency by $2\zeta\sigma K$ symmetrically disposed at higher and lower frequencies about the monomer transition frequency ν_o. This prediction has been amply borne out by experiment and is a critical diagnostic feature in circular dichroism experiments. For a variety of reasons, one of the components of the split band may not appear in the spectrum, and this itself is evidence for the existence of interactions that have to be accounted for. We will see that there is a long history involved in the analysis of polypeptide chiral activity concerned with just such an effect. If both components appear, then the band shapes in both CD and ORD depend on the magnitude of the splitting $2\zeta\sigma K$, on the rotational strength of the $K = M$ and $K = (N - M) + K = M$ components $R_N, R_{M, N-M}$, and on the screw sense of the helix. We are, of course, ignoring the intrinsic shape of the band vibrational envelope which also contributes to the band shape (Section 3.6) and are considering here only the helical determinants of the overall dispersion of the circular dichroism and optical rotation.

Since the Moffitt equations were developed to describe the results of the helical exciton interactions at frequencies far from the resonance absorption, we turn instead to the general perturbation theory (Tinoco; references 12, 13, 46) expressions to examine these same effects in absorbing regions. Note, however, that the linear response theories, the matrix formalism, and the classical theory of DeVoe, all yield similar predictions. Equations 4.201–4.206 of general perturbation theory may be divided rather naturally into two parts to confront separately *electric-dipole-allowed* and *electric-dipole-forbidden* transitions (28). We have already called attention to the fact that electric-dipole-forbidden transitions are too weak to give rise to strong coupled oscillator or exciton interactions and, of course, do not contribute to the intrinsic activity of the dissymmetric chromophore term of Eq. 4.201-1. Consequently, only the one-electron terms of Eqs. 4.202-3 and 4.202-4 and the very weak term of Eq. 4.202-5 will contribute in this case. Furthermore, in a helix with twofold or greater symmetry, even these terms are likely to be small or null because interactions with the static field of the nearby groups tend to be equal and opposite in sign. Since the symmetry in the helix cannot be truly spherical or planar, weak or nominally forbidden transitions do derive, however, some rotational strength from these terms. Since no exciton terms appear, the CD spectrum for these transitions has the shape of the corresponding monomer isotropic absorption of CD bands, while the ORD appears with the typical shape of an isolated monomeric transition illustrated in Fig. 2.12. The longest wavelength chirally active band of the polypeptides (one of the two main classes of helical polymers that have been studied by chiral-optical methods) arises from just such weak or forbidden transitions (see Section 8.44). This band is not particularly easy to see because of the overlap from the strong $\pi \rightarrow \pi^*$ transition at slightly shorter wavelengths, which does show exciton splitting in the α-helix conformation, but not in the pleated-sheet conformation (Fig. 8.06). No such *weak* transitions have been found to give observable CD in the polynucleotides, although they have been identified as responsible for some weak bands in the isotropic or linearly polarized absorption

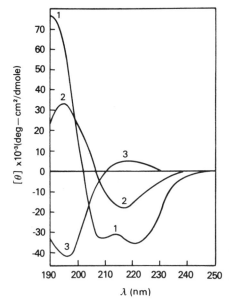

Fig. 8.06 The CD bands of poly-L-lysine in the nonhelical disordered (2) and β-pleated-sheet (3) forms showing the bands near 212 and 198 nm (of opposite sign in each of the two forms), which correspond to the bands appearing in the isotropic ultraviolet absorption spectrum. By comparison, the α-helical form (1) is seen to have a CD spectrum in which the strong 198-nm $\pi \rightarrow \pi^*$ band is exciton split into bands at approximately 208 and 192 nm; but the weaker 212-nm $n \rightarrow \pi^*$ remains unsplit with an apparent shift to a slightly longer wavelength. [From N. Greenfield and A. D. Fasman, *Biochem.*, **8**, 4108 (1969), reproduced with permission.]

spectra. These weak transitions are thus not traceable to the helical symmetry, and consequently while they may only occur or be enhanced upon helix formation, they are not intrinsically diagnostic of the special dissymmetry of the helix.

The situation with regard to allowed electronic transitions is quite different. While the one-electron terms can make weak contributions to the chiral activity of allowed transitions, the significant rotational strength comes either or both from the terms in the $\sum \mathrm{Im}\, \mu_{i0a} \cdot \mathbf{m}_{i0a}$ (Eq. 4.202-1) and the coupled oscillator terms in Eqs. 4.202-2 and 4.202-6 which correspond to the Moffitt exciton contributions. The intrinsic dissymmetry term is just the sum of the monomer contributions, and the resulting rotational strength normalized to the equivalent monomer concentration will be identical with that of the monomer. The conclusion to be drawn about a polymer exhibiting CD or ORD of this type is that the polymer is either completely disordered (no correlation exists between successive monomer subunits) or that the particular arrangement of the subunits is such as to reduce the coupled oscillator term to zero. Since the only geometry that would completely prevent any exciton coupling is a collinear arrangement of the transition moments in successive subunits along the helix axis (a geometry not likely to be found in a helical polymer), all strong transitions will lead to helix-induced coupled oscillator rotational strength. Let us consider, then, more explicitly the dispersion resulting from the monomer dissymmetry and the intermonomer exciton coupling (terms 1, 2, and 6 of Eq. 4.202):

$$
\begin{aligned}
R_{OA} =\ & \sum_i \mathrm{Im}\, \mu_{i0a} \cdot \mathbf{m}_{ia0} \\[2mm]
& + 2 \sum_{j\neq i} \sum_{b\neq a} \frac{\mathrm{Im}\, V_{i0a,j0b}(\mu_{i0a} \cdot \mathbf{m}_{jb0}\nu_a + \mu_{j0b} \cdot \mathbf{m}_{ia0}\nu_b)}{h(\nu_b^2 - \nu_a^2)} \\[2mm]
& - \frac{2\pi}{c} \sum_{j\neq i} \sum_{b\neq a} \frac{V_{i0a,j0b}\nu_a\nu_b(\mathbf{R}_j - \mathbf{R}_i)\cdot(\mu_{j0b} \times \mu_{i0a})}{h(\nu_b^2 - \nu_a^2)}.
\end{aligned}
\qquad (8.400)
$$

Rather severe restrictions can be imposed on this general expression in order to simplify an examination of the shape of the circular dichroism and optical rotatory dispersion curves. The first term, of course, has the frequency dependence of the intrinsically dissymmetric chromophore, and its contribution to the CD and ORD, if present, will be of the type we have examined for monomers earlier. Since the frequency dependence of the two coupled oscillator terms are the same, conditions can be chosen to examine just one of them. For example, it is possible to choose a polymer with subunits whose magnetic dipole transition moments are small or interact negligibly with the electric dipole transition moments on neighboring monomers. In this case the second term is negligibly small. A polymer consisting of a single repeating monomer, all of whose strongly allowed electric dipole transitions lie in a symmetry plane more or less perpendicular to the helix axis, is just such a case. Because the strong 260 nm transition moments of the nucleic acid bases are essentially in the molecular plane and the bases planes are approximately perpendicular to the helix axis, the helical polynucleotides are real representatives of this situation. The second term is zero or negligible because the interacting magnetic transition moments are perpendicular to the base planes or nearly so, and therefore to the electric dipole transition moments on adjacent monomers. In fact, it is not necessary to be quite so restrictive because with perpendicular in-plane transitions in the helical geometry, only the third term can contribute even if the monomeric units are not identical.

For the repeating monomer, the situation is the same as that of a dimer —all transitions are identical (degenerate). Considering any particular transition, we have seen (Eq. 4.400) that the rotational strength associated with each dimeric unit ($i = 1$, $j = 2$) consists of a positive and negative component separated in frequency by the extent of the exciton splitting. As the interactions between nonnearest neighbors, trimers, and higher polymers, ($i = 1, j = 3, 4, \ldots$), enter, the exciton splitting and rotational strength are different for each nonnearest neighbor pair so that the shape of the circular dichroism obtained by summing over the N groups becomes somewhat distorted for very small (less than one or two helical turns) polymers. But as the polymer reaches sizes where further interactions between more distant monomers becomes negligible, the shape of the CD and ORD spectra attributable to a strong polymer transition becomes characteristic of that of the coupled oscillator illustrated in Fig. 8.07.

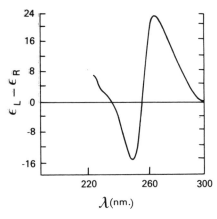

Fig. 8.07 The CD spectrum of the 258-nm transition of single-stranded poly(A) showing the positive and negative components produced by the strong exciton splitting. (62)

For the nondegenerate situation in which the dimeric units of the polymer consist of two or more different groups, the transitions in neighboring groups are no longer identical. Nevertheless, if there are strong allowed transitions in both groups sufficiently close in energy so that the factor $\nu_a\nu_b/(\nu_b^2 - \nu_a^2)$ is not negligible, then the rotational strengths associated with their interactions will exhibit the same behavior.

The rotational strength of exciton components of interacting transitions is strongly dependent on the energy difference between them. The dependence is illustrated in Fig. 8.08, where the ratio of the frequencies for two monomer transitions, $0 \to a$ and $0 \to b$, is plotted against the frequency factor in the third term of Eq. 8.400. As the two frequencies approach true degeneracy, $(\nu_A/\nu_B)=1$, the relationship appears to lose its meaning since the equal and opposite rotational strength of the components leads to a cancellation of the chiral activity, while we know that degenerate bands split into separate exciton components. This is just a

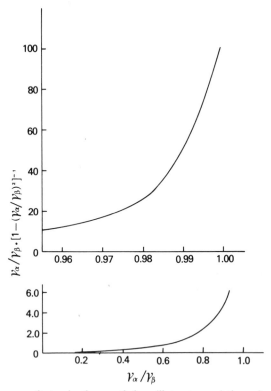

Fig. 8.08 The frequency factor in the coupled oscillator term of the polymer rotational strength (Eq. 8.400) showing the very strong dependence of the rotational strength on the separation between the transition energies in nondegenerate coupled oscillators.

result of the approximations made in developing the frequency dependence in these equations. Nevertheless, the strong dependence on the energy separation is quite clear. For example, the relative magnitudes of the frequency factors of the exciton rotational strengths for two interacting transitions at 250 and 244 nm (1000-cm^{-1} separation) is 70 times that of transitions with the same dipole strength at 250 and 200 nm (10,000-cm^{-1} separation).

For the purpose of simplicity, let us limit consideration to only two strong electric dipole transitions, $0 \rightarrow a$ and $0 \rightarrow b$ (in local symmetry planes perpendicular to the helix so that terms 1 and 2 of Eq. 8.400 are small or null). The rotational strength of one exciton component is given by

$$R_{OA} = - \sum_{j \neq 1} \frac{\pi}{c} \left[\frac{V_{i0a,j0b} \nu_a \nu_b}{h(\nu_b^2 - \nu_a^2)} (\mathbf{R}_j - \mathbf{R}_i) \cdot (\boldsymbol{\mu}_{j0a} \times \boldsymbol{\mu}_{i0a}) \right.$$

$$\left. + \frac{V_{i0b,j0a} \nu_a \nu_b}{h(\nu_b^2 - \nu_a^2)} (\mathbf{R}_j - \mathbf{R}_i) \cdot (\boldsymbol{\mu}_{j0b} \times \boldsymbol{\mu}_{i0a}) \right]. \qquad (8.401)$$

The rotational strength for the other component, R_{OB}, is obtained by permuting the subscripts a and b. The magnitude of all terms are the same, and while it is true that the cross products ($\boldsymbol{\mu}_{i0a} \times \boldsymbol{\mu}_{j0b}$) may change sign on interchanging subscripts a and b, each term appears as the negative of the other in both R_{OA} and R_{OB}; so the net effect of the interchange on this factor is only to interchange terms. However, $\nu_b^2 - \nu_a^2$ appear identically in both terms of the equation, so that interchanging subscripts for this factor has the effect of changing the sign of the rotational strength. The net result is that the two components of the nondegenerate helical exciton interaction for the in-plane transitions in a helical polymer give rise to the same kind of circular dichroism as does the degenerate exciton interaction. Bush and Brahms (47) have termed this type of CD dispersion "conservative" because the sum of the rotational strengths of the two components is zero. The appearance of a nonconservative or distorted conservative spectrum is evidence, therefore, that rotational strength is appearing from sources other than that of the coupled oscillator terms of Eqs. 4.202 and 8.400. The frequency dependence of the polymer chiral activity, therefore, can take characteristically conservative or nonconservative shapes, or some combination of the two. Figure 8.09 illustrates just such combined behavior for both the circular dichroism and the optical rotation. The magnitude of each of the contributions when both types of chiral activity are present strongly affects the shape. We have already seen how the frequency separation of neighboring transitions affects the magnitude of the coupled oscillator terms. The geometry of the helix or the relative polarization

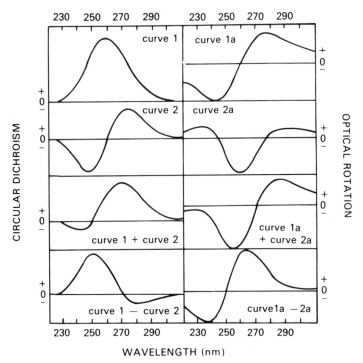

Fig. 8.09 A comparison of the intrinsic and coupled oscillator (exciton) contributions to the CD spectra on ORD of a helical polymer. The sum of the contributions when the integrated chiral activity of the contributions are about the same is shown for each case. Obviously, many variations of the composite contributions are possible. (Modified from Fig. 2 of reference 48.)

directions, of course, also exert profound effects on the magnitude of these terms through the interaction potential and through the $\mu \times \mu$ factors. We saw this in the case of the dimer in Chapter 4. Figure 8.10, taken from a 1968 paper by Tinoco (48), gives the dependence of the circular dichroism of each of the exciton components of a model helical polynucleotide polymer on the orientation of the transition moment in the plane of the perpendicular bases (curves 1-4), and also shows the effect of tilting the transition moment by 15 degrees in the direction of the positive helix axis. Almost any change in the helix geometry, for example in the distance of the chromophores from the helix axis, perturbs the interchromophoric distance $R_j - R_i$. Thus, it affects the magnitude of the exciton chiral strength (49). This same result, as well as the conclusion that the conservative shape of the strong transitions in the helical polymer CD and ORD is almost identical with that of a dimer (one nearest neighbor only), has been

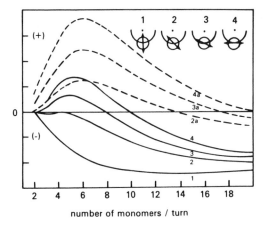

Fig. 8.10 The exciton contribution to the circular dichroism of transitions perpendicular to (——) and tilted at 15 degrees to (----), the axis of a right-handed helical polymer. The separate curves for each show the effect of the angle that the transition moment makes with respect to the perimeter surface of the helix. Note that for helices with about 10 monomers per turn (as in DNA and many polynucleotides), transitions radial to the helix axis make strong contributions to the rotational strength, whereas, transitions tangent to the perimeter contribute very little (for tilted monomer the situation is reverse). In the case of helical polymers of benzenoid-type compounds, where the first and second in-plane transitions are almost always nearly perpendicular to each other, this virtually guarantees that the exciton contributions to the CD will be large for one but not the other. If the monomers are intrinsically optically active, then both situations lead to strong nonexciton contributions, but of opposite sign. (48)

reached without involving the formalism of exciton theory. For example, the classical theory discussed in Section 8.3 and the time-dependent Hartree theory, as specifically developed for chiral problems by McLachlan, Ball, and Harris (50, 51), also predict the appearance of conservative components in the chiral optical activity of helical polymers. Using the strong transition of adenosine near 260 nm as a model, Schneider and Harris (52) have calculated the dependence of the molar rotation on the 10 nearest neighbor interactions along a single-stranded helical polymer chain (Table 8.1). It is clear from the result (which accounts for only a single interaction with a neighbor and thus has to be doubled in an "infinite" polymer) that except when the geometry produces a collinear or coplanar arrangement of the transition polarizations (180 degrees), not only does the magnitude of the contributions fall off as expected with distance, but also that contributions can be both positive and negative for the same helical sense.

Table 8.1   First 10-Neighbor Contributions to the 250-nm Molar Rotation for a Right-Handed Helix

Neighbor Pair	Stacking Angle (degrees)	Intermonomer Distance D (Å)	$[\Phi]_{260} \times 10^{-4}$
1st	36	3.4	−5.64
2nd	72	6.8	−1.27
3rd	108	10.2	+0.373
4th	144	13.6	+0.485
5th	180	17.0	0
6th	216	20.4	−0.302
7th	252	23.8	−0.151
8th	288	27.2	+0.135
9th	324	30.6	+0.175
10th	360	34.0	0

$$\sum_{i=1}^{10} [\Phi]_{260} = -6.20$$

8.43   The Prediction of Helical Parameters

A helix is characterized by its pitch, its sense of skew, and its diameter. The detailed molecular architecture of a helical polymer which determines the values of these parameters, depends, in part, on the number of monomeric units per turn of the helix, which, of course, is determined by the various intra- and intermonomer bond lengths and bond angles. Values of all of these are obtainable in principle from the comparison of theoretically calculated chiral activity with the experimental observations. In practice, no such detailed structural information is obtainable with the degree of accuracy available from x-ray diffraction of fibers and crystals, but by invoking known structural details of the monomer units and their electronic transitions, the helical parameters can be explored using chiroptical information. We have seen examples of the way the rotational strength and extent of splitting of the exciton bands are dependent on these parameters. Let us examine some other aspects of the effect of the structural details of the molecular helix.

Theoretical predictions do not always achieve such a ready reception and rousing enthusiasm as did those of Moffitt and Moffitt and Yang, when in respective papers (1, 24) they predicted the exciton splitting of the strong amide ($\pi \rightarrow \pi^*$) transitions near 190 nm in an α-helical polypeptide and demonstrated the frequency dependence of the optical rotation of a helical polypeptide in long-wavelength regions far from its absorption band. It matters little that shortcomings in these predictions and practical problems in their precise quantitative application have subsequently been found, or that the development of equipment permitting measurements of circular dichroism in the higher energy region of these strong transitions have demonstrated new complexities that must be accounted for (53). The basic physical phenomena had been sighted, the right questions asked, the underlying physical theory developed and was ready to be tested and applied. In the course of the next five years (to 1961), and in the decade and a half since then, these predictions have borne fruit in an ever-increasing knowledge of the structure and properties of helical polymers. The shortcomings and complexities have served only to stimulate further research. Let us consider separately the two predictions and their influence on the determination of helical parameters. The *frequency* dependence of the terms of the Moffitt equations (Eqs. 8.218 and 8.219) were expressed by Moffitt and Yang (24) in the form of an empirically testable relationship. The specific rotation as a function of the wavelength λ of the incident radiation is

$$[\alpha]_\lambda = \frac{100}{M}\left[(a_0)\frac{\lambda_0^2}{\lambda^2 - \lambda_0^2} + (b_0)\frac{\lambda_0^4}{\left(\lambda^2 - \lambda_0^2\right)^2}\right], \qquad (8.402)$$

where M is the molecular weight per monomer residue and a_0, b_0, and λ_0 are characteristic values related to the intrinsic and helical but nonexciton (a_0) and the helical exciton (b_0) contributions to the optical rotation of a strong optical transition or near $\lambda_0 = \nu_0^{-1}$. These are bold assumptions in simplifying what must be a more complex situation, but not unreasonable ones. The general wavelength dependence of optical rotation for simple molecules far from absorption bands had long been known (since 1900!) to follow the Drude equation (54), which is just the first term of Eq. 8.402. Also, the helical contribution of strong transitions according to Moffitt's exciton treatment has a different frequency (wavelength in this form of the equation) dependence given by the second term. Finally, we have seen how the exciton interaction falls off sharply with increasing energy separation between the transitions. So it is reasonable to expect that only one transition at λ_0, the nearest in energy to the measuring wavelength λ unless it is particularly weak, will be dominant in determining the rotation at large values of $\lambda^2 - \lambda_0^2$, consequently, a single characteristic, λ_0, should suffice, albeit its value might not be exactly coincident with that of the true position of the longest-wavelength exciton band. Granting these assumptions, we could (Moffitt and Yang did) predict that a_0 and b_0 have values characteristic of a *particular* conformation (and thus of specified helical parameters) of the helical polymer. It is important to remember that the nonexciton part may also be due to the existence of the helix through the one-electron perturbations, as well as those of the intrinsic dissymmetry of the monomers, but that it is only the exciton interaction that gives rise to the distinct dispersion we have described as resulting from the coupled oscillator interaction. Although some confusion arose and persisted for a long time because of the real and supposed errors in Moffitt's theoretical treatment, it is quite clear that the sign of b_0, and possibly of a_0, should reflect the sense of helix rotation, since the relative signs of the exciton component rotational strengths are exact reflections of the screw sense.

The *magnitude* of b_0 reflects the rotational strength of the strong transition centered at or near λ_0. The rotational strength in turn is determined by the combined effects of the geometry (the pitch and diameter of the helix determine the polarization direction of the transitions) and of the intrinsic strength of the transition. Unscrambling these latter parameters requires additional information, but if the assumption is made that the transition moments for a given chromophore are only slightly affected by the environment, comparisons of values of the constants should yield information on relative geometries of different but related polymers or different conformations of the same polymer. Not all of this has yet been realized; partly because more recent developments have provided additional tools with which to examine these questions. However, very shortly after the Moffitt and Yang work was reported, a series of papers appeared sparked by these predictions, and by Cohen's 1955 observation (42). It was quickly established that the rotatory dispersion of helical polypeptides does, indeed, follow the course predicted by the Moffitt-Yang relation, and that the sign of b_0 reflects the relative screw sense of the polypeptides, but no unique answer could be derived from the chiral observations alone, as to which sign resulted from which helical sense. Since the qualitative shape of the rotatory dispersion at long wavelengths (but not at wavelengths close to λ_0) is very similar for the one-term Drude equation (Fig. 8.11) and the two-term Moffitt-Yang relation, it became imperative to establish whether the dispersion was still adequately described by the latter if the natural optical activity of the monomer units contributed little or nothing to the optical activity of the polymer. Since the monomers in this case are optically active amino acids, a method had to be devised to eliminate their intrinsic contribution. To do this, Doty and Lundberg (55) measured the optical activity of a polymer that consisted of a racemic mixture of the D and L forms of the monomer prepared by initiating the additional polymerization of the D-form with an L-form polymer so as to retain the

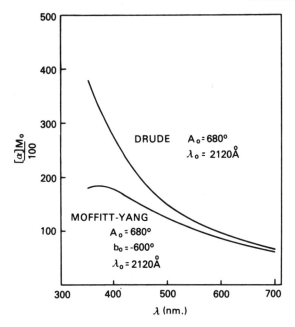

Fig. 8.11 Plots of the Drude and Moffitt dispersion curves. Note the similar shapes and not too dissimilar magnitudes at wavelengths far removed from the band center.

screw sense of the L-helix. They observed that the sign of b_0 and its magnitude, as well as the magnitudes of a_0 and λ_0, as determined by fitting the data to the Moffitt-Yang equation, did not change appreciably from those of the L-polymer alone. This was convincing evidence that the chiral power arose from the helical character of the polymer and the resulting exciton bands of the strong amide transitions.

The dispersion relations and the characteristic values of the Moffitt-Yang constants for the α-helix and other polypeptide structures have been reviewed many times (56). We should add only that equations of this type can be formulated from other considerations, (57, 58), but that this does not vitiate the intrinsic importance of their predictive power which derives from the fact that they are a consequence of quantum mechanical exciton theory; their coherence with physical observations, therefore, can be interpreted in terms of the theory.

At numerous points in this chapter, reference has been made to the determinants of the sign of the rotational strength, or to the sign of the exciton components of the rotational strength; but, at no point has it been suggested that a definitive or unique relationship exists between the sign and helical sense of a polymer. By now it should be apparent that the reason for this is that no such relationship exists because the sign is dependent both on the absolute sense of the helix and on the polarization directions of the transition moments. Thus, every polymer is unique to itself, or at least to a class of polymers containing the same or very similar chromophores. So the polypeptides of a given conformation in which the amide is the only significant chromophore form a class, and the polynucleotides have a set of subclasses; not even all polypurines or all polypyrimidines belong to a single class. We can see rather easily why the skew sense alone is insufficient to determine the sign of the chiral activity. If we restrict our attention to allowed transitions and so to Eq. 8.400, the first term for the contribution of the intrinsic dissymmetry of the monomer transitions is patently independent of the helical sense, since no interaction between the monomers is involved. The sign is completely dependent on the polarization directions of the intrinsic monomer electric and magnetic transition moments. The first of the coupled oscillator terms (the second term of Eq. 8.400) does involve the direct interaction between different groups. Note, however, that while the optical factor $\boldsymbol{\mu} \cdot \boldsymbol{m}$ is a pseudoscalar and therefore signed, the sign is determined by the cosine of the angle between the moments:

$$\boldsymbol{\mu}_i \cdot \mathbf{m}_j = |\mu_i| |m_j| \cos \phi_{ij},$$

where the angle between $\boldsymbol{\mu}_i$ and $\mathbf{m}_j$ is defined for the i, j monomer pair for translation along a helix axis. It will be positive or negative, depending on the magnitude of ϕ_{ij}. The only difference between the right- and left-handed helix is that if ϕ_{ij} is positive for one, it is negative for the other. Since the cosine function is the same absolute value for the same subtended angle in different quadrants, $\cos(-\phi_{ij}) = \cos\phi_{ij}$, the sign of the second term is also independent of the helix sense. This is not so with respect to the third term of Eq. 8.400, the major exciton term. Here, instead, the contribution is proportional to the cross product of two vectors, $\boldsymbol{\mu}_i \times \boldsymbol{\mu}_j$. The angular dependence is a sine function:

$$\boldsymbol{\mu}_i \times \mathbf{m}_j = |\mu_i| |\mathrm{m}_j| \sin \theta_{ij},$$

where θ_{ij} is the angle between $\boldsymbol{\mu}_i$ and $\mathbf{m}_j$ corresponding exactly to ψ_{ij} between $\boldsymbol{\mu}_i$ and $\mathbf{m}_j$. Since $\sin(-\theta_{ij}) = -\sin(\theta_{ij})$, it follows that a reversal of the helix sense reverses this component completely. If this is the only contribution to the optical activity, the curves will be related as enantiomorphs, that is, mirror images.

To further illustrate how the sign of the rotational strength depends on transition polarization directions as distinct from the helical sense, we need only refer back to Table 8.1, where the rotational contributions for the different neighbor pairs in a *polynucleotide helix of a single sense*, right-handed in this case, are observed to go from negative to positive and back again as the angle that the nth monomer makes with the first one goes through the four quadrants, corresponding to the behavior of the sine function. This very pronounced behavior is dependent on the existence of the exciton interaction as the most prominent source of rotational strength. In a helix the only other major source is likely to be the intrinsic activity from a dissymmetric chromophore in the monomer, the one-electron perturbations generally being small because of the nearly symmetric nature of these perturbations.

In this section we have tried to show that the structure of a helical polymer is reflected in observed chiral-optical parameters in rather particular ways—in its characteristic wavelength dispersion and in its sign especially, as well as in its magnitude. Starting in 1962 (excluding the original theoretical work) with an effort by Schellman and Oriel (59) on polypeptides and continuing in 1966 with the work by Bush and Tinoco (60) on polynucleotides, the more detailed architecture of biological and biologically related polymers has been continuously investigated by these methods. Considerable progress has been made. The most recent research has centered on the details of the effects of the tilting of the nucleic acid bases and of the helix expansion in polynucleotides (49, 61–63) and on the differences between the polyproline I and polyproline II helices (41, 64). In the next section we consider, in slightly more detail, the nature of the molecular electronic transitions that enter into the chiral activity of the polypeptide helices.

8.44   Helical Polypeptides

It should be possible to test Moffitt's prediction (Eq. 8.210) that inter-monomer exciton interaction of strong electronic transitions in a helical polymer leads to two chirally active spectral bands—one at lower frequency polarized parallel to the helix axis and the other a degenerate perpendicular band at higher frequency—by measuring the circular dichroism of α-helical polypeptides. In these helical molecules a strong amide $\pi \to \pi^*$ transition lies in the plane of the peptide unit (Section 7.33); the plane is almost parallel to the helix axis, but the transition moment makes an angle that has been variously estimated from 18 to 45 degrees to the N–C bond in this plane. Taking into account the orientation of the

N–C bonds in the helix, the linear moment $(\mathbf{p}_\sigma = \mathbf{p}_{\pi \to \pi^*})$ associated with this transition is directed predominantly, but not entirely, in the "vertical" v direction, that is, in the direction of the helix axis. Thus it has very substantial components both parallel and perpendicular to the axis, with the parallel component expected to be the stronger. As a result, both of the Moffitt exciton contributions to the rotational strength (Eqs. 8.214 and 8.215) should be finite, with the rotational strength of the perpendicular band, $R_{\sigma, M} + R_{\sigma, N-M}$ at higher frequency. This band is degenerate in first order but should itself be split by exciton interaction. Consequently, three bands that have their origin in the amide $\pi \to \pi^*$ transitions are predicted to appear in the CD spectrum of α-helical polypeptides. If we add a fourth band arising from the $n \to \pi^*$ amide transition, which is too weak [oscillator strength for the $n \to \pi^*$ is about 0.01 or less compared to 0.27 for the $\pi \to \pi^*$ (65)] to produce observable exciton bands, then the CD spectrum above 160 nm should contain four bands: (1) a weak $n \to \pi^*$ band probably at lowest energy; (2) a strong band polarized parallel to the helix axis, the polarization of which could be discerned by measurements on oriented systems in which the perpendicular bands contribute little, if at all; (3) and (4) two perpendicular bands of opposite sign slightly split by exciton interaction, the sum of whose rotational strength should be approximately zero. The observed circular dichroism of an unoriented sample of α-helix polyglutamic acid is shown in Fig. 8.12 (66). Four bands are distinctly discernable between 170 and 250 nm, but the two highest energy bands,

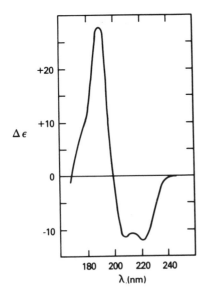

$\Delta \epsilon$

λ (nm)

Fig. 8.12 The circular dichroism of α-helical polyglutanic acid (Adapted from Fig. 1 of reference 66).

the shoulder at about 175 nm and peaks near 195 nm, appear to be of the same rather than opposite sign. This unexpected result had, in fact, been predicted twice on the basis of experimental measurements of the ORD and CD at longer wavelengths (67,68). Two reasons were advanced for the absence of the predicted strong negative component of the perpendicular band, inadequacy of the theory and/or the presence of a canceling band of opposite sign. The evidence is still not conclusive (69,70), but the existence of the negative component of the band, much weaker than it was predicted to be, has been shown in a study of the CD of a helical polypeptide oriented by an electric field (71).

In the discussion of the amide chromophores in Chapter 7, we noted that a number of other transitions have been observed or predicted in this spectral region (72,73). Because the exciton polarization directions in a molecule of helical symmetry are constrained to be parallel or perpendicular to the helix axis, both linear and circular dichroism additional measurements on oriented systems would be helpful. Some measurements have been made (71; see also Chapter 9), but attention has been focused more recently on other polypeptide structures (41,64). A very thorough summary of the work in this field up to 1973 has been given by Sears and Baychock (27). Experimental work since then has concentrated largely on poly-nucleotides, for which the most recent theoretical work [Cech, Hug, and Tinoco (63)] reduces to Moffitt's result when applied to helical polymers containing a single repeating unit. The work of Goodman and his colleagues on the effect of chain length on the optical rotation of peptide oligomers is especially interesting from the point of view of the interpretation of helical effects discussed in this chapter (74). Reviews by Blout, Yang, and Urry and by Imahori and Nicola of the experimental information are also available (75–78). Extensive reviews and critical analyses of the general theoretical work and of the experimental work on biopolymers has been made by Woody (79) and by Balasubramian and Kumar (80).

8.5 SUMMARY

We have discussed the facts that polymers have some of the characteristics of crystals, principally that the electronic transitions of the repeating units strongly interact to form exciton bands, and that the special symmetry of the helix leads to the strong selection rules governing these bands. The discussion of the application of theory of these two effects has been limited to a very few informative problems in polynucleotide and polypeptide chirality. An extensive literature that covers these and many other problems in polymer chirality is available.

REFERENCES AND NOTES

1. W. Moffitt, *Chem. Phys.* **25**, 467 (1956).

2. F. M. Loxsom, *Phys. Rev.*, **B1**, 858 (1970).

3. W. Rhodes and M. Chase, *Rev. Mod. Phys.*, **39**, 348 (1967).

4. J. Brahms, A. M. Michelson, and K. E. van Holde, *J. Mol. Biol.*, **15**, 467 (1966).

5. A. Rich, D. R. Davies, F. H. C. Crick, and J. D. Watson, *J. Mol. Biol.*, **3**, 71 (1961).

6. J. Frenkel, *Phys. Z. Sovjetunion*, **9**, 158 (1936).

7. A. S. Davydov, *Theory of Molecular Editions*, M. Kasha and M. Oppenheimer, Jr., (Transls.), McGraw-Hill Book Co., New York; 1962, A. S. Davydov, *J. Exp. Theor. Phys.*, **18**, 210 (1948); *ibid.*, **21**, 673 (1951).

8. W. Moffitt, D. D. Fitts, and J. C. Kirkwood, *Proc. Natl. Acad. Sci. (U.S.A.)* **43**, 723 (1957).

9. F. M. Loxsom, *Int. J. Quant. Chem.*, **3**, 147 (1969); F. M. Loxsom, *J. Chem. Phys.*, **51**, 4899 (1969).

10. T. Ando, *Prog. Theor. Phys.*, **40**, 471 (1968).

11. C. W. Deutsche, *J. Chem. Phys.*, **52**, 3650 (1970).

12. I. Tinoco, Jr., *J. Chem. Phys.*, **33**, 1332 (1960); *ibid.*, **34**, 1067 (1961). See also the short but interesting comments in the paper by P. J. Stiles, *Mol. Phys.*, **22**, 73 (1971).

13. I. Tinoco, Jr., *Adv. Chem. Phys.*, **4**, 113 (1962).

14. D. G. Barnes and W. Rhodes, *J. Chem. Phys.*, **48**, 817 (1966).

15. W. Rhodes and D. G. Barnes, *J. Chim. Phys.*, **65**, 78 (1968).

16. H. DeVoe, *J. Chem. Phys.*, **43**, 3199 (1965); *ibid.*, **41**, 393 (1964).

17. V. Madison and J. A. Schellman, *Biopolym.*, **11**, 104 (1972); J. A. Schellman and E. B. Nielsen, *Conformation of Biopolymers*, Academic Press, London and New York, 1967, pp. 109–122.

18. M. R. Philpott, *J. Chem. Phys.*, **56**, 683 (1972).

19. P. M. Bailey, E. B. Nielsen, and J. A. Schellman, *J. Phys. Chem.*, **73**, 228 (1969).

20. Note that in Eq. 8.216 as in many others throughout this book, we have omitted a factor of the order of $(\eta^2 + 2)/3$, which appears in the original paper, where η is the index of refraction of the solvent at the frequency of measurement. We have long felt that including this factor can be misleading. For a recent discussion, see J. A. Schellman, *Chem. Rev.*, **75**, 323 (1975).

21. J. G. Kirkwood, *J. Chem. Phys.*, **5**, 479 (1937).

22. See, for example, the analysis of this problem by L. J. Oosterhoff, in *Modern Quantum Chemistry*, Pt. 3, "Action of Light and Organic Crystals," O. Sinanglu (Ed.), Academic Press, Inc., New York, 1965.

23. M. J. Stephan, *Proc. Cambridge Philos. Soc.*, **54**, 81 (1958).

24. W. Moffitt and J. T. Yang, *Proc. Natl. Acad. Sci.* (U.S.A.), **42**, 596 (1956).

25. H. Murakami, *J. Chem. Phys.*, **27**, 1231 (1957).

26. I. Tinoco, Jr., and R. W. Woody, *J. Chem. Phys.*, **32**, 461 (1960).

27. D. W. Sears and S. Baychock, *Physical Principles and Techniques of Protein Chemistry*, Pt. C, S. J. Leach (Ed.), Academic Press, New York, 1973, Chapter 13.

28. V. A. Bloomfield, D. M. Crothers, and I. Tinoco, Jr., *Physical Chemistry of Nucleic Acids*, Harper and Row, New York, 1974.

29. A particularly lucid description of this aspect of the problem and of the application of modern exciton theory to molecular crystals and biopolymers is to be found in the review by M. R. Philpott, in *Advances in Chemical Physics*, Vol. 23, I. Prigogine and S. Rice (Eds.), John Wiley & Sons, New York, 1973. See also Table 8.1 in the text where the variation of rotational strength with neighbor distance is tabulated for a particular model.

30. R. W. Woody and I. Tinoco, Jr., *J. Chem. Phys.*, **46**, 4927 (1967).

31. W. Rhodes, *J. Chem. Phys.*, **53**, 3650 (1970).

32. A. I. Levin and I. Tinoco, Jr., *J. Chem. Phys.*, **66**, 3491 (1976).

33. P. Pino, F. Ciardelli, and N. Zandomeneghin, *Ann. Rev. Phys. Chem.*, **21**, 561 (1970).

34. K. Rosenheck and B. Sommers, *J. Chem. Phys.*, **46**, 532 (1967).

35. P. Pino, *Adv. Polym. Sci.*, **4**, 393 (1965).

36. G. Natta and G. Allegra, *Tetrahedron*, **30**, 1987 (1974).

37. W. Hug, F. Ciardelli, and I. Tinoco, Jr., *J. Am. Chem. Soc.*, **96**, 3407 (1974).

38. R. D. B. Frazer and T. P. MacRae, *J. Mol. Biol.*, **5**, 457 (1962).

39. S. Arnott, S. D. Dover, and A. Elliot, *J. Mol. Biol.*, **30**, 201 (1967).

40. L. Pauling and R. B. Corey, *Proc. Natl. Acad. Sci. (U.S.A.)*, **39**, 253 (1953).

41. E. W. Ronish and S. Krimm, *Biopolym.*, **13**, 1635 (1974).

42. C. Cohen, *Nat. (London)*, **175**, 129 (1955).

43. J. T. Yang, *Tetrahedron*, **13**, 143 (1961).

44. L. Pauling and R. B. Corey, *J. Am. Chem. Soc.*, **72**, 5349 (1950).

45. L. Pauling, R. B. Corey, and H. R. Branson, *Proc. Natl. Acad. Sci. (U.S.A.)*, **37**, 205 (1951).

46. I. Tinoco, Jr., R. W. Woody, and D. F. Bradley, *J. Chem. Phys.*, **38**, 1317 (1963).

47. C. A. Bush and J. Brahms, *J. Chem. Phys.*, **46**, 79 (1967).

48. I. Tinoco, Jr., *J. Chim. Phys.*, **65**, 91 (1968).

49. D. S. Moore and T. E. Wagner, *Biopolym.*, **12**, 201 (1973).

50. A. D. McLachlan and M. A. Ball, *Mol. Phys.*, **8**, 581 (1964).

51. R. A. Harris, *J. Chem. Phys.*, **43**, 959 (1965).

52. A. S. Schneider and R. A. Harris, *J. Chem. Phys.*, **50**, 5204 (1969).

53. W. B. Gratzer and D. A. Cowburn, *Nat. (London)*, **222**, 426 (1969).

54. P. Drude, *Lehrbuch des Optik*, S. Hirzel, Leipzig, 1900; translated into English as *The Theory of Optics*, by C. R. Mann and R. A. Millikan, Longmans, Green & Co., New York, 1929.

55. P. Doty and R. D. Lundberg, *Proc. Natl. Acad. Sci. (U.S.A.)*, **43**, 213 (1957).

56. See B. Jirgensons, *Optical Activity of Proteins and Other Macromolecules*, 2nd ed., Springer-Verlag, New York, Heidelberg, Berlin, 1973, and references contained therein.

57. W. Kauzmann, *Ann. Rev. Phys. Chem.*, **8**, 413 (1957).

58. C. Schellman and J. A. Schellman, *C. R. Trav. Lab. Carlsberg Ser. Chim.*, **30**, 463 (1958).

59. J. A. Schellman and P. Oriel, *J. Chem. Phys.*, **37**, 2114 (1962).

60. C. A. Bush and I. Tinoco, Jr., *J. Mol. Biol.*, **23**, 601 (1967).

61. D. S. Studdert and R. C. Davis, *Biopolym.*, **7**, 1377, 1391, 1405 (1974).

62. C. L. Cech, Ph.D. Thesis, University of California, Berkeley, 1975. The data of figure 8.07 are attributed to the unpublished results of D. Ghee.

63. C. L. Cech and I. Tinoco, Jr., *Nucl. Acids Res.*, **3**, 399 (1976); C. L. Cech, W. Hug, and I. Tinoco, Jr., *Biopolym.*, **15**, 131 (1976).

64. E. S. Pysh, *Biopolym.*, **13**, 1563 (1974).

65. H. D. Hunt and W. T. Simpson, *J. Am. Chem. Soc.*, **75**, 4540 (1953).

66. W. C. Johnson, Jr., and I. Tinoco, Jr., *J. Am. Chem. Soc.*, **94**, 4389 (1972).

67. J. P. Carver, E. Schecter, and E. R. Blout, *J. Am. Chem. Soc.*, **88**, 2250 (1966).

68. J. Y. Cassim and J. T. Yang, *Biopolym.*, **9**, 1475 (1970).

69. M. A. Young and E. S. Pysh, *Macromol.*, **6**, 790 (1973).

70. I. Tinoco, Jr., in *Fundamental Aspects and Recent Developments in Optical Rotatory Dispersion and Circular Dichroism, Proceedings of NATO Advanced Study Institute*, F. Ciardelli and P. Salvadori (Eds.), Heyden & Son, London, 1971, Chapter 2.4.

71. R. Mandel and G. Holzwarth, *J. Chem. Phys.*, **57**, 3469 (1972).

72. V. A. Zubkov and M. V. Vol'kenshtein, *Mol. Biol.*, **4**, 483 (1970).

73. H. Basch, M. B. Robin, and N. A. Kuebler, *J. Chem. Phys.*, **47**, 1201 (1967).

74. M. Goodman, E. E. Schmitt, and D. Yphantis, *J. Am. Chem. Soc.*, **82**, 3483 (1960); M. Goodman, F. Boardman, and I. Listowsky, *J. Am. Chem. Soc.*, **85**, 2491 (1963); M. Goodman, A. S. Verdini, C. Toniolo, W. D. Phillips, and F. A. Bovey, *Proc. Natl. Acad. Sci. (U.S.A.)*, **64**, 444 (1969).

75. E. Blout, in *Fundamental Aspects and Recent Developments in Optical Rotatory Dispersion and Circular Dichroisms, Proceedings of NATO Advanced Study Institute*, F. Ciardelli and P. Salvadori (Eds.), Heyden & Son, London, 1971, Chapter 4.5.

76. J. T. Yang, *Conformation of Biopolymers*, G. N. Ramachandran (Ed.), Academic Press, New York, 1967, p. 157 ff.

77. D. W. Urry, *Spectroscopic Approaches to Biomolecular Conformation*, American Medical Association, Chicago, 1970, p. 33 ff. See also the chapter by Urry in *Annual Reviews of Physical Chemistry*, H. Eyering, C. J. Christensen and H. S. Johnston, (Eds.) Annual Reviews, Inc. Palo Alto Cal. 1968, vol. 19.

78. K. Imahori and N. A. Nicola, *Physical Principles and Techniques of Protein Chemistry*, Pt. C, S. J. Leach (Ed.), Academic Press, New York and London, 1973, Chapter 22.

79. R. W. Woody, *J. Polym. Sci. Macromol. Rev.*, **12**, 181 (1977).

80. D. Balasubramian and C. Kumar, *Applied Spectros. Rev.*, **11**, 222 (1976).

CHAPTER NINE

ORIENTED SYSTEMS

9.1 INTRODUCTION TO DIRECTIONAL EFFECTS

At this point it comes as no surprise that chiral-optical effects are highly dependent on the propagation direction of the measuring light with respect to the orientation of the molecular axes of the optically active material. However, except for brief prior references, we have restricted the discussion to the optical activity of randomly oriented (isotropic) systems of molecules. The theoretical treatments, both the perturbation theories and the direct quantum calculation methods, depend on the original formulations by Kirkwood, Stephan, and Rosenfeld in which the activity associated with directional interactions are averaged over all possible orientations in order to relate theory to the observations made on isotropic solutions. We turn our attention now to chiral measurements of oriented systems where the potential for obtaining more detailed information is considerable. This should already be apparent, for example, from the discussion of the Moffitt analysis of the optical rotation of a helical polymer for which rotational strengths are predicted to be different for light propagating perpendicular and parallel to the helix axes (see Eqs. 8.214 and 8.215). Measurements of isotropic solutions detect only the orientationally averaged sum of these; the relative magnitudes of the exciton and nonexciton components are much more distinctly observed in the axial components. Since the frequency dependence of these components may be different along principal molecular axes, observations on oriented polymers help to assign the chiral effects to their origin in structural parameters and spectroscopic states.

The ultimate oriented system is, of course, a crystal. Any theory developed for general orientation should be capable of eliciting the special cases of crystals. Crystals were treated earlier than partially oriented systems, but the complexity precluded exact solutions (1, 2). The difference between crystals and oriented systems of polymers depends on the method and degree of orientation and on the crystal symmetry. If the orientation is uniaxial such as that achieved in electric or cylindrical flow fields, and if it is complete in the uniaxial direction, then the system is exactly analogous with that of a uniaxial crystal. Otherwise the only correspondence between crystals and partially oriented molecules is the tensor with a minimum of nine components (some of which may be zero) required to describe the optical activity of both. The tensor relates the physical property (rotational strength) to the coordinate axes in the system and to the direction of propagation of the light with respect to these axes. Rosenfeld's rotatory parameter β and the rotational strength derived from it are tensors whose orientationally averaged values we have been discussing up to this point (see Section 3.4 and Eqs. 3.428–3.430). Now we will be interested in the components along specified molecular directions and their projections in space-fixed directions.

If crystals were the only oriented systems, we could ignore the question of how partial orientation affects the observations. Because the CD or optical activity of partially oriented systems are also informative, we will have to consider the orientation question as well. For this purpose in the discussion that follows, the assumption will be made that the molecules are fixed and rigid, regardless of size. This is not too restrictive because it is possible to incorporate the effects of molecular flexibility and conformational distributions, but these problems require separate treatment. They have been investigated with some success with respect to the measurement of physical properties other than chiroptical activity, most notably with respect to hydrodynamic properties (3). It is an important subject, one to which measurements of circular dichroism and optical rotatory dispersion on oriented systems could contribute. For small molecules with conformational isomers, the methods are reasonably straightforward (4) and have been applied to isotropic chiral measurements. For large flexible polymers, the effects of the orienting field on the size and shape of the polymer molecule are highly complex, with the result that the calculation of meaningful values for the activity associated with fixed axes in the molecular framework is subject to considerable uncertainty. The linear dichroism of partially oriented flexible systems, in which only the electric dipole transition moment components are involved, is still not unequivocally analyzed (5,6). For circular dichroism, where both the polarizability tensor and the rotatory parameter tensor (or the components of the matrix elements of both the electric dipole and quadrupole and the magnetic dipole transition moments) must be specified in terms of the coordinates of the molecule, no definitive solutions are available. Either the assumption is made that the molecule, the most prominent example being a helical polymer, has a rigid rodlike structure of known conformation and configuration (7–15), or it is assumed to be an ellipsoid of revolution whose components of optical activity can be empirically examined and interpreted in terms of the structural parameters if the structure itself is assumed (16–21). It should be interesting to measure the chiral activity of long flexible polymers, which fit into this category, oriented by flow (16, 17, 22) or electric fields (7, 20) or stretched films (23, 24), where they are constrained to a plane, thereby reducing the three-dimensional problems to a two-dimensional one. Small molecules may be treated this way as well.

9.2   THE ORIGIN OF ORIENTATION EFFECTS

Simply put, the origin of orientation effects in chiral phenomena lies in the vector properties of the transition moments. The product of the vector components of the orientationally averaged rotational strength,

$$\boldsymbol{\mu} \cdot \mathbf{m} = \mathbf{i} \cdot \mathbf{i} \mu_x m_x + \mathbf{i} \cdot \mathbf{j} \, \mu_x m_y + \mathbf{i} \cdot \mathbf{k} \mu_x m_z + \mathbf{j} \cdot \mathbf{i} \mu_y m_x + \mathbf{j} \cdot \mathbf{k} \mu_y m_z + \mathbf{k} \cdot \mathbf{i} \mu_z m_x + \mathbf{k} \cdot \mathbf{j} \, \mu_z m_y$$

$+ \mathbf{k} \cdot \mathbf{k} \mu_z m_z$, which specifies the chiral activity of isotropic systems, has only three nonzero components, $\mu_x m_x + \mu_y m_y + \mu_z m_z$, because of the orthogonality of the unit vectors, $\mathbf{i}$, $\mathbf{j}$, $\mathbf{k}$. While each of the three components, and therefore the sum, may have different signs, the direction of propagation of the light need not be specified in order to make the measurement meaningful. For oriented systems, however, the measured rotation is the dot product of two vectors and a pseudoscalar tensor. Snir and Schellman (25), for example, have pointed out that the circular dichroism associated with a transition $0 \rightarrow a$ is given by (before orientational averaging):

$$\Delta \epsilon_k = \frac{32 \pi^2 N \nu_{a0}}{2302 hc} \mathbf{k} \cdot \mathbf{R} \cdot \mathbf{k} \tag{9.201}$$

where $\mathbf{R}$ is the rotatory strength tensor and $\mathbf{k}$ the unit vector in the direction of propagation of the radiation. [In the Snir and Schellman treatment, a factor $\phi(\nu)$, which describes band shape, is included, but the factor is normalized to a value of unity.] Note that now it is the propagation direction that enters the mathematical formalism, not the direction of the electric and magnetic fields, although it is the latter that induces the moments that result in the optical rotation. This can be demonstrated rigorously, but it should be intuitively apparent that since the circularly polarized electromagnetic vectors of the radiation change direction continuously in traversing the medium, incorporating explicit directional effects associated with the electromagnetic field directions, would complicate the formalism enormously. The product $\mathbf{k \cdot R \cdot k}$ is a second-rank tensor (26). The rotatory parameter $\underline{\underline{\beta}}_{ij} = \mathbf{k \cdot R}_{ij} \cdot \mathbf{k}$ is, in fact, a rather particular kind of second-rank tensor, whose properties conform to the requirements of optical activity. Let us see if we can discern some of these properties before attempting a more mathematical discussion of oriented systems. We know, for example, that the optical activity must be independent of a 180 degree reflection in the axis system; if the radiation propagates through the system in the direction of $-x$, the measured optical activity must be the same in sign and magnitude as if it were propagated in the $+x$ direction. (Remember that the convention for the handedness of the axes system does have to be fixed in order to assign any sign to the rotation.) The tensor, therefore, must be invariant to reflection. Thus, if the rotation ϕ_k of the polarization plane of the incident radiation by the chiral system (or its CD, $\Delta\epsilon_k$) is proportional to $\underline{\underline{\beta}}_{ij}$, the relationship

$$\phi_k = K\underline{\underline{\beta}}_{ij} = K\underline{\underline{\beta}}_{ji} \qquad (9.202)$$

must be obeyed; the interchange of indices, i,j, specifies a 180 degree reflection. As a consequence, if $\underline{\underline{\beta}}_{ij} = -\underline{\underline{\beta}}_{ji}$, the only value it can have is zero, which means that it is describing an optically inactive material. Note, of course, that not all nine components of the tensor $\underline{\underline{\beta}}_{ij}$ must be identically zero, only the sum. Still another property of the tensor results from the requirement that changing the axis system from left-handed to right-handed requires that the sign of the optical activity change. So the tensor must be describable in terms of the axis system and be sensitive to its handedness. (A tensor with these two properties is called an axial second-rank tensor.) For example, the tensor $\underline{\underline{\beta}}_{mn} = \Sigma_{m,n} \mathbf{B}_{mn} l_m l_n$ describes the rotatory property in terms of its values in the planes of the axes m and n and of the cosines l_m and l_n, which the wave normal (the propagation axis) of the light makes with the axes fixed in the molecular system. m and n are running indices $1,2,3$, where if the axis system is Cartesian, $1,2,3$ are the set of right-handed axes usually designated x,y,z. The indices $1,2,3$, are frequently used to describe orientation phenomena and crystal axes because they are easy to keep track of in lengthy derivations, and because by convention if the system is uniaxial, the 3 axis is used to represent the unique axis ($l_1 = l_2 \neq l_3$ for light propagating along the unique axis). For light propagating in a cubic system perpendicular to the faces of the cube or along principle diagonals, $l_1 = l_2 = l_3$. An ordered array of long helical polymers constitutes a uniaxial system with the unique axes parallel to the helix axes. In Section 9.3, the relation of these tensors to crystal symmetry and to the chiral activity of crystals will be examined more closely. The reason for considering them here is that they are the most general form of the Rosenfeld rotatory parameter before orientational averaging. These tensors or related mathematical quantities specify the rotational strength of oriented molecules for light propagating along specific axes fixed in the molecules.

Starting with Gibbs in 1882 (27), there has been a succession of theoretical treatments of the chiral activity of oriented systems (1, 7–9, 28–30). These treatments, especially the modern ones, have almost universally described the interaction of the radiation with the molecular moment, using as parameters the vector potentials of the radiation and the linear moments associated with the optical transitions—a description we have generally eschewed in this monograph (except to call attention to it in Chapter 4 and again in Moffitt's treatment of polymer excitons in Chapter 8). Although the discussion that follows is concerned mainly with the consequences of these treatments, it is cast in terms of the vector potentials. The reader may wish to review Sections 3.4 and 8.2 before going on. The most recent theoretical treatment is that of Kuball, Karstens, and Schonhofer (30). They follows Gō's earlier analysis but develop the linear momentum transition moments in terms of the electric and magnetic dipole and the electric quadrupole components. A number of authors have pointed out that the early theories of optical activity ignored the effect of quadrupole moments. No difficulties were encountered because the direct interaction of the radiation with the molecular quadrupole moments sums to zero when the chiral activity is averaged over all orientations of the molecules (31). Since the theoretical treatments were designed to explain isotropic solution measurements, this was a perfectly valid approximation. We can see it both intuitively and in a more formal way. Referring to Fig. 9.00, we find that the interaction of incident polarized electric and magnetic field vectors, $\mathbf{E}$ and $\mathbf{H}$, with the electric dipole transition moment of an oriented molecule polarized along the x axis depends on which Cartesian axis the field vector is parallel to. (Recall that the linearly polarized electric and magnetic fields of the radiation may be considered to consist of two equivalent left and right circularly polarized fields.) Consequently, the interaction averaged over the three Cartesian directions is not zero. On the other hand, the interaction of the incident radiation with a quadrupole field fixed in the x,y plane is equal and opposite for the radiation field polarized parallel to the x and y axes and zero for the radiation field polarized along the z axis, resulting in a null average over the three orientations. This relatively naive conception can be seen more formally. Referring to Eq. 9.201, we can write rotatory strength R as the sum of contributions from electric quadrupoles R_Q and magnetic dipoles R_m, each

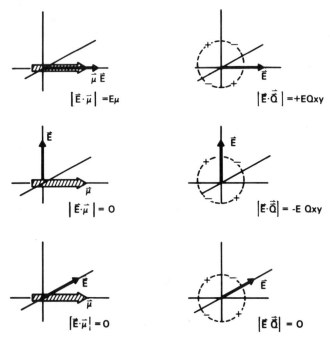

Fig. 9.00 A model of the interaction of radiation field vectors with quadrupole (right) and dipole (left) charge distributions. The charge distributions are constrained to the xy plane and oriented so that at the instant of interaction, the field vectors are parallel and/or perpendicular to an axis of the charge distribution.

of which may have nine tensor components. The orientational averaging is equivalent to taking one-third the sum of the diagonal elements (the trace) of these tensors; the magnetic dipole and electric quadrupole rotatory strengths, respectively, are given by

$$R_m = \tfrac{1}{3}(R_{m,11} + R_{m,22} + R_{m,33}),\tag{9.203}$$

$$R_Q = \tfrac{1}{3}(R_{Q,11} + R_{Q,22} + R_{Q,33}).\tag{9.204}$$

The magnetic dipole components are the cyclic permutants of

$$R_{m,11} = \tfrac{3}{2}\mathrm{Im}\left[\boldsymbol{\mu}_{0a}\boldsymbol{\cdot}\mathbf{m}_{a0} - (\mu_1)_{0a}(m_1)_{a0}\right],\tag{9.205}$$

and in linear momentum description, the quadrupole moments are the cyclic permutants of

$$R_{Q,11} = \frac{3e}{4mc} \text{Im}\left[(\mu_2)_{0a}(p_3 r_1 + r_3 p_1) - (\mu_3)_{0a}(p_2 r_1 + r_2 p_1)_{a0} \right] \quad (9.206)$$

The orientationally averaged magnetic dipole part of the rotatory strength tensor is, therefore,

$$R_m = \left(\tfrac{1}{3}\right)\left(\tfrac{3}{2}\right)\text{Im}\left[\boldsymbol{\mu}_{0a} \cdot \mathbf{m}_{a0} - (\mu_1)_{0a}(m_1)_{a0} + \boldsymbol{\mu}_{0a} \cdot \mathbf{m}_{a0} \right.$$
$$\left. - (\mu_2)_{0a}(m_2)_{a0} + \boldsymbol{\mu}_{0a} \cdot \mathbf{m}_{a0} - (\mu_3)_{0a}(m_3)_{a0} \right], \quad (9.207)$$

$$= \tfrac{1}{2}\text{Im}(3\boldsymbol{\mu}_{0a} \cdot \mathbf{m}_{a0} - \boldsymbol{\mu}_{0a} \cdot m_{a0}), \quad (9.208)$$

$$= \text{Im}\, \boldsymbol{\mu}_{0a} \cdot \mathbf{m}_{a0}. \quad (9.209)$$

The orientationally averaged quadrupole part is

$$R_Q = \tfrac{1}{3}\left(R_{Q,11} + R_{Q,22} + R_{Q,33} \right) = 0, \quad (9.210)$$

a fact that may be verified by writing out the sum of the similar cyclic permutations of $R_{Q,11}$. The total rotatory strength, R_{0a} for $0 \rightarrow a$ transition of the isotropic system is the sum of Eqs. 9.209 and 9.210, which is just the familiar Rosenfeld expression

$$R_{0a} = \text{Im}\, \boldsymbol{\mu}_{0a} \cdot \mathbf{m}_{a0}. \quad (9.211)$$

It contains no quadrupole contributions.

For oriented systems of rigid molecules, the situation is quite different. Each of the tensor components of both the magnetic and the quadrupole moments may contribute. Furthermore, each of these components contributes to the extent of its projection on the direction of propagation of the incident radiation. If the orientation is complete, then the projection cosines are unity or zero depending on the relationship between the molecular axes and the propagation axis. If the molecular axes are not defined by symmetry axes or if the molecular system is only partially oriented, then any or all of the components have finite projections. The analysis in this case is much more complex.

Kuball, Karstens, and Schonhofer have defined a few particular situations where symmetry considerations make the analysis of experimental observations more tractable (30). The experimental problem is a separate and difficult one, but has been carefully considered (21, 32, 33, 34). The special cases considered by Kuball, Karstens, and Schonhofer include: (1)

a molecular system in which the molecules are in complete parallel alignment, the equivalent of a biaxial crystal; (2) a system with rotational symmetry about a fixed axis, the equivalent of a uniaxial crystal; (3) the special case of a uniaxial system in which the molecules are D_2 symmetry; (4) a partially oriented system in which the model is that of a fraction ρ of molecules uniaxially oriented as in (2) and a fraction $1-\rho$ of isotropically distributed molecules. Their treatment results in an expression for the circular dichroism associated with the Born-Oppenheimer electronic transition $0 \to a$, which incorporates in place of the propagation direction $\mathbf{k}$, the tensor form of the orientation function g_{klij}:

$$\Delta \epsilon = \frac{32\pi^2 N \nu_{0a}}{hc \cdot 2303} g_{klij} \sum R_{0a}^{kl}, \tag{9.212}$$

$$g_{klij} = \int f(\alpha, \beta, \gamma) a_{ki} a_{lj} \sin \beta \, d\alpha \, d\beta \, d\gamma. \tag{9.213}$$

The a_{mn} are matrices that transform the molecule-fixed Cartesian coordinates x_1, x_2, x_3 on to the space-fixed Cartesian axes in terms of the angles α, β, γ, which define their respective orientations. For values of the indices $ij = 33$, the light propagates along the "3" space-fixed axis. Equation 9.212 is the general expression for the circular dichroism associated with the $0 \to a$ transition of a molecular system in terms of the projections of the kl molecule-fixed tensor components of the rotatory strength, R_{0a}^{kl}, and the tensor defining the orientation. If the molecular system is isotropic, $f(\alpha, \beta, \gamma) = 1$ and $g_{kl33} = \frac{1}{3}$; Eq. 9.212 reduces to

$$\Delta \epsilon = \frac{32\pi^3 N \nu_{a0}}{3hc \cdot 2303} R_{0a}, \tag{9.214}$$

which is identical with Eq. 9.202 after orientational averaging for which $\mathbf{k} \cdot \mathbf{R} \cdot \mathbf{k} = \frac{1}{3} R$; only the overall rotational strength can be measured. The special cases then become

(1) Completely parallel alignment of the molecules (equivalent to a biaxial crystal). In this case

$$R_{0a} = R_{0a}^{33} + \sin^2 \beta \left[R_{0a}^{11} - R^{33} + \sin^2 \gamma \left(R_{0a}^{22} - R_{0a}^{33} \right) \right.$$

$$\left. - \sin 2\gamma R_{0a}^{12} \right] - \sin 2\beta \left(\cos \gamma R_{0a}^{13} - \sin \gamma R_{0a}^{23} \right)]. \tag{9.215}$$

By suitable choices of the propagation direction with respect to the oriented systems, $\gamma, \beta, 2\gamma$, or 2β can be separately reduced to zero. Therefore, it is possible to obtain each of the rotational strength components. In a biaxial crystal this requires measurements with the light propagating along each of the six unique principal axes. Because of the interference of linear dichroism and/or birefringence effects, such measurements are extremely difficult to accomplish, and no example is known to the author. We should note that by the requirement in Section 9.1, the symmetric nondiagonal elements of the rotatory strength tensor must be equal to each other, $R_{0a}^{ij} = R_{0a}^{ji}$, or equal to zero. Therefore, the six measurements should suffice to obtain all nine tensor components.

(2) A system with rotational symmetry about a unique axis: There are several theoretical calculations and numerous measurements of such systems and we will discuss a few in Section 9.3. The measurement with the propagation direction along the uniaxial direction, $kl_{ij} = 3333$, gives a true measure of the circular dichroism because no linear effects are present; the system is linearly isotropic to the incident polarized light. Measurement perpendicular to the unique axis, $kl_{ij} = 1133 = 2233$, give the values of the rotational strength perpendicular to that axis, but only if linear effects are absent or can be separated.

$$R_{0a} = R_{0a}^{33} = \tfrac{1}{2} \sin^2 \beta \left(R_{0a}^{11} + R_{0a}^{22} - 2R_{0a}^{33} \right). \tag{9.216}$$

(Note that off-diagonal elements, if present, cannot be measured with $\beta = 0$ and $\beta = \pi/2$—only the diagonal elements, $k1 = 11$, 22, and 33.)

(3) The special case of molecules of D_2 symmetry oriented uniaxially: D_2 symmetry is not especially common in the electronic structure of molecules. It requires a molecule (or chromophore) with three twofold rotational axes, but no center or plane of symmetry. Ethylene twisted axially about its double bond is a typical example. It is, however, an interesting case and one that other systems sometimes acquire or approximate. Examples are hydrogen-bonded dimers or uniaxial crystals with two appropriately disposed molecules per unit cell. Kuball et al. have tabulated the values of the rotatory strength tensor components for transitions of different symmetry in these systems in terms of the electric, magnetic, and quadrupole contributions (30). In Table 9.1, we show only the existence or nonexistence of chiral components.

In accord with the general result for uniaxial systems given above, only two measurements are required to obtain all the chiroptical information, but because of the special symmetry, one of these can be the isotropic circular dichroism, which is just one-third the sum of absolute values of the diagonal components. Table 9.1 shows that the measurement with light propagating along the unique axis gives null chiral activity for the $A \rightarrow B_1$ transition and two-thirds the measured isotropic activity for $A \rightarrow B_2$ and

Table 9.1   The Chiral Activity of Uniaxially Oriented D_2 Systems[a]

Transition Symmetry	R_{0a}^{11}	R_{0a}^{22}	R_{0a}^{33}
$A \to A$	0	0	0
$A \to B_1$	$-\frac{1}{2}R_{0B_1}^{11}$	$\frac{1}{2}R_{0B_1}^{22}$	0
$A \to B_2$	$\frac{1}{2}R_{0B_2}^{11}$	0	$-\frac{1}{2}R_{0B_2}^{33}$
$A \to B_3$	0	$-\frac{1}{2}R_{0B_3}^{22}$	$\frac{1}{2}R_{0B_3}^{33}$

[a]B_1, B_2, B_3 used as subscripts to the rotational strength components refer to states of corresponding symmetry.

$A \to B_3$ transitions. The assignment of axes and therefore the transition symmetry is arbitrary for strict D_2 symmetry because the C_2 axes are indistinguishable; but as the applicable chromophores generally derive from subclasses of higher symmetry, the transitions are naturally separable into the separate classifications. It is interesting to notice, for example, that D_2 systems may be only slight distortions of D_{2h} and D_{2d} systems; the former contains no chirally active transitions since all transitions to magnetically allowed states are electric dipole forbidden, and vice versa. In the latter systems, the doubly degenerate E transitions can exhibit chiral activity only by the intervention of an asymmetric crystal field. It is expected, therefore, that the chiral activity of uniaxially oriented D_2 chromophores may be rather small, but since no off-diagonal components of the rotational strength tensor exist, only two uncomplicated measurements are required to obtain detailed values if the components are large enough. These systems, therefore, offer a sharp target for theoretical approaches. [Although we will discuss in Section 9.4 some aspects of symmetry as it affects the chiral properties of crystals, it is beyond the scope of this discussion to examine the details of crystal interactions and symmetry (35, 36).]

(4) Partially oriented systems: It is not clear that the same approximation that has been proposed for the linear dichroism of partially oriented systems (37) is valid for chiroptical measurements, although its use has been suggested by several authors (7–9, 30) and it may be valid for interesting uniaxial systems such as helical polymers. The linear dichroism tensor is totally symmetric, while the circular dichroism tensor is pseudoaxial with components of arbitrary sign; off-diagonal elements of nonuniaxial systems, therefore, do not necessarily add to zero for the isotropic unoriented fraction. The model does lead to a relatively simple result: For the purpose of computing chiral strength tensor components, a partially oriented system may be treated as a mixture containing a fraction, ρ, of fully oriented molecules and the fraction $1 - \rho$ of disoriented molecules. If

ρ is known, then the measurement of the isotropic optical activity of the completely disoriented system and of the optical activity parallel to the optic axis of the partially oriented system, suffices to obtain the sum $R_{0a}^{11} + R_{0a}^{22}$, and the parallel component, R_{0a}^{33}, of the rotational strength tensor as well. Initial attempts to apply this result to some ketosteroids do not justify its further uncritical application (30). For molecules with true cylindrical or pallindromic helical symmetry, all strong transitions are confined to polarizations parallel and perpendicular to the symmetry axis. In this case the model may produce a more valid comparison with experimental results, since the principal tensor components of the rotational strength are, in any event, only the diagonal elements. There have been a number of attempts to use the model to compute the circular dichroism of polynucleic acids from observation or uniaxially flow-oriented systems in order to compare the results with theoretical calculation (17, 18, 21). In particular, Holzwarth and his colleagues (17, 18) have calculated the circular dichroism tensor of DNA based on the theory of Johnson and Tinoco (38). The calculated magnitudes, derived from the flow orientation measurements (18, 39), are not in particularly good accord with the predictions for the separate exciton and one-electron contributions parallel and perpendicular to the helix axis. However, they are qualitatively in agreement if the assumption of the theoretical treatment includes the strict perpendicularity of the DNA bases to the helix axis and the absence of any interactions of the nucleotide bases with the phosphate-sugar backbone. In Table 9.2, for example, adapted from the paper by Wooley and Holzwarth (17), we see that within the experimental error there is agreement with the theoretical predictions for the perpendicular dichroism of both the 275- and 245-nm bands. For the former, the calculated parallel circular dichroism $\Delta\varepsilon_{33}$ is also in good quantitive agreement with the theoretical prediction.

Table 9.2   Components of the Oriented Circular Dichroism of DNA

Wavelength[a] (nm)	$\Delta\varepsilon_{11}$	$\Delta\varepsilon_{22}$	$\Delta\varepsilon_{33}$	$\Delta\varepsilon$
275	-0.2 ∓ 0.8	-0.2 ∓ 0.8	7.7 ± 1.6	2.46
	0	0	7.8	2.6
245	1.2 ± 1.1	1.2 ± 1.1	-11.0 ∓ 2.2	-2.84
	0	0	-4.5	-1.5

[a]The upper row at each wavelength refers to the values calculated from the experimental data and the lower row to the theoretical values computed from a perturbation theory treatment by Johnson and Tinoco (38).

Subsequent calculations (see Section 8.12) have shown that the optical activity is very sensitive to choices made with respect to helical parameters; so it is not known how to proportion the discrepancies between deficiencies in the "model" and in the "theory."

Measurements of partially oriented helical polypeptides preceded those on DNA. They were obvious candidates for such measurements in view of the Moffitt prediction of different values for the rotational strengths of strong exciton bands perpendicular and parallel to the helix axis, given by Eqs. 8.214 and 8.215, respectively. While a weak $n \rightarrow \pi^*$ band is interposed between the predicted $\pi \rightarrow \pi^*$ exciton bands and the long wavelengths at which the early measurements were made, the rotation should still be dominated by the $\pi \rightarrow \pi^*$ bands. The first measurements on oriented polypeptides (7,30) were only semiquantitative but were sufficient to show that the long-wavelength rotatory strength of polybenzyl-L-glutamate parallel to the helix axis is large and positive, but that perpendicular, it is large and negative. In Section 8.44 we saw that the $n \rightarrow \pi^*$ transitions of polypeptides, polarized almost perpendicular to the helix axis (40), are assigned to absorption near 222 nm, which is responsible for the trough near 233 nm in the ORD spectrum. The chiral activity of this transition, as expected, is only moderately intense. Several calculations (10,41) have predicted an isotropic rotational strength of only about -0.03 DBM compared to a predicted value of -0.11 DBM for the nearby $\pi \rightarrow \pi^*$ exciton band. While these values are somewhat smaller than observed, it is significant that for oriented molecules, the parallel component of the rotational strength was predicted to be much smaller than the perpendicular, only about -0.01 DBM compared to -0.04 DBM for the perpendicular component. Blout and Schecter (42) measured the optical activity of polypeptides in this region on oriented films and indeed found no detectable Cotton effect at 233 nm for light parallel to the helix axis but found a band about four times the observed isotropic strength for light perpendicular to the axis. The situation with respect to the $\pi \rightarrow \pi^*$ exciton bands is much less clear. Significant reviews and comments appear in several places (21,34,43—45).

Kuball has recently (46) pointed out that the measurements of the CD of oriented systems can also be used to distinguish different vibronic contributions, since these develop along different directions in chromophores of high or moderately high symmetry. Thus a strong electronic transition may gain most of its intensity from promotions to totally symmetric vibrations that develop along the symmetry axis appropriate to the electronic state, but may also couple weakly to antisymmetric vibrations along different directions. In the isotropic achiral absorption spectrum, as well as in the isotropic CD, the latter are likely to be completely unobservable because of the dominance of the transitions to symmetric vibrational states. However, as is well known from linearly polarized absorption spectra, the weak transitions can appear prominently in directions forbidden to the strong ones. In a weak electronic transition, almost all the intensity may come from vibronic coupling to transitions along different axes, and these should appear separately in the oriented CD components.

9.3   THE OPTICAL ACTIVITY OF CRYSTALS

The discovery of optical activity was made during experiments by Biot (47) on crystal quartz cut perpendicular to its unique axis. In 1812–1813, he discovered that polarized light would be rotated by the crystal in proportion to its thickness and that the rotation is inversely related to the wavelength of the light utilized in the measurement; in fact, he anticipated Drude's analysis of the dispersion by showing that the dependence on wavelength was approximately inversely quadratic. He also discovered that different quartz crystals, if cut to the same thickness, rotated polarized light equally and oppositely (enantiomorphs). In view of the fact that this discovery was made more than a century and a half ago, it seems almost surprising that crystals have not been more intensively investigated for their chiroptical properties than they have been. One of the difficulties has been in measuring the rotation in the presence of linear effects, although between 1904 and 1928 techniques were evolved for doing so (48) and a modest number of crystalline substances have been measured. The earliest "theoretical" considerations were those of Fresnel in 1824 and Pasteur in 1848. Both suggested that the particles or molecules in crystals having rotatory power must be arranged in spirals or helices (49). More modern theoretical analyses were attempted by Born (1) and Sommerfeld (2). Since the molecules of crystals exist in the strong fields of the surrounding molecules, a problem modern quantum theory can deal with only for relatively simple cases or with severe approximations, the theory remains substantially at the structural rather than the electronic level. At this level, symmetry considerations, as expected, are a great help. Most of the applications of modern symmetry considerations to the optical properties of crystals stem from the work of Jahn (50) and of Bhagavantam (51). The direct application of group theory to chiroptical properties is that of Bhagavantam and Suryanarayana (52). The most recent phenomenological treatment and application of group theory to the optical activity of crystals appears to be that of Jerphagnon and Chermia (53). Going back to the roots of the electromagnetic theory of optical activity (see Section 2.3 and especially Eqs. 2.314 and 2.316 defining the rotatory parameters g), they arrive at the result that the optical activity of crystals can be expressed as the sum of contributions from (1) the intrinsic activity, if any, of the molecules that make up the crystal (we have seen that the tensor value of the intrinsic molecular chiral activity can be expressed as a pseudoscalar); (2) a very small longitudinal component resulting from light polarized along the optic axis (this component occurs in 3 of the 32 crystal classes and can be expressed as a vector); and (3) a component that can be large and, in some cases, the only contribution to the optical activity (it arises from the chiral structure of the crystal and can be expressed as a pseudo-deviator, defined with respect to its tensor properties). With the help of group theory, Jerphagnon and Chermia are able to identify the contributions associated with the 21 noncentrosymmetric crystal symmetries classified into 12 subclasses. [Crystal symmetry is more often classified according to Hermann-Mauguin notation than Schoenflies's notation (54). Both are given in Table 9.3 to facilitate reference to the literature. Because the Schoenflies notation is used throughout the rest of this book and is the notation used in point symmetry classifications, we retain it in this discussion.] One of the 12 subclasses, although noncentrosymmetric, is optically inactive (see Table 9.3). Another, consisting of crystals with symmetries C_{3v}, C_{4v}, and C_{6v}, contains only the very small vector contributions. No optical activity has been observed in crystals of this class.

There are 32 different ways in which crystal axes can be assembled so that the symmetry operations of reflection, inversion, rotation, and rotation-reflection (or improper rotation) can be performed when the definition of these symmetry operations is governed by point group theory. Of the 32, 11 include the inversion operation; that is, they are centrosymmetric and must, perforce, refer to symmetry classes characteristic of nonoptically active crystals. Selection rules for optical activity are not, however, identical for point groups and crystal classes, so that the S_4 rotation-reflection axis of D_{2d}, S_4, T, and O symmetry classes do not preclude optical activity in crystals characterized by these symmetry representations. Similarly, the reflection planes of C_s and C_{2v} do not prevent crystals with this symmetry from being optically active. Of these 6 classes, the C_s, C_{2v}, D_{2d}, and S_4 form a separate interesting class of crystals for which no enantiomorphs exist, but which are nevertheless optically active. For a long time there was some question as to whether these crystals would exhibit true optical activity. In the absence of enantiomorphs and in the presence of interfering linear effects along all the axes of D_{2d} and S_4, and along the axes perpendicular to the unique axis for C_s and C_{2v} (and no optical activity parallel to the unique axis), there was no hard evidence for its existence. Bhagavantam and Suryanarayana (52), however, demonstrated from theoretical arguments that these crystals should be active. In

Table 9.3 Irreducible Components of the Optical Activity for Noncentrosymmetric Crystal Symmetry Classes

Class		Pseudoscalar	Vector	Pseudodeviator
Schoenflies	Hermann-Mauguin			
T_d	$\bar{4}3mm$	0	0	0
D_{3h}	$\bar{6}m2$			
C_{3h}	$\bar{6}$			
O	432	+	0	0
T	23			
D_3	32	+	0	+
D_4	422			
D_6	622			
C_{3v}	$3m$	0	+	0
C_{4v}	$4mm$			
C_{6v}	$6mm$			
C_3	3	+	+	+
C_4	4			
C_6	6			
D_{2d}	$\bar{4}2m$	0	0	+
S_4	$\bar{4}$	0	0	+
C_{2v}	$mm2$	0	+	+
D_2	222	+	0	+
C_2	2	+	+	+
C_3	m	0	+	+
C_1	1	+	+	+

1967, Hobden (55) found that the linear birefringence of the D_{2d} crystal of AgGaS$_2$ changes sign at about 4970 Å. Thus its linear birefringence is essentially zero at that wavelength, making it possible to measure the optical activity without interference from linear optical effects (it is also transparent at 4970 Å). The 21 noncentrosymmetric crystal symmetry classes are tabulated in Table 9.3. The entries show the existence (+) or nonexistence (0) of an irreducible component of the optical activity tensor g. The table is adapted from reference 53, where the values of the tensor components are given.

There are a number of interesting observations that can be made with respect to the relationship between the group classifications and the optical activity. Strong optical activity from the crystal structure is possible in all the optically active classes except those containing O and T representations and C_{3v}, C_{4v}, and C_{6v}. Examination of the point group character tables (Appendix I) reveals that for the former all electric and magnetic moments belong to the same degenerate species, which cannot be distinguished. Therefore, the only optical activity possible must arise from the intrinsic activity of the molecules and not from the organization of the molecules in the crystal lattice, a fact consistent with the existence of the pseudoscular component and no others. For the latter, one component each of the electric and magnetic moments are in different symmetry representations, and the other two are degenerate and indistinguishable. (They can be shown to give rise to activity equal and opposite in sign.) Consequently, again no activity is possible from the lattice per se, but a weak induced component can arise along the direction containing the third component of either the electric or magnetic moment. Similar observations could be made with respect to the nonenantiomorphic classes, D_{2d}, S_4, C_{2v} and C_s, and with respect to all the optically active crystal symmetries. A possible source of confusion is the relationships of crystalline form to the symmetry classifications of optical activity because they do not necessarily correspond to intuitive notions. For example, cubic crystals that are isotropic to linear optical effects may be chirally active. The symmetry classifications available to cubic crystals are T, C_{3v}, O, T_d, and O_h. The O_h group is centrosymmetric and totally inactive, as is T_d for other reasons; but as can be observed from Table 9.3, the other groups can exhibit optical activity. Similar nonintuitive observations can be made about other crystalline forms (54).

Although it is difficult, there have been a number of more or less successful attempts to analyze crystal chiroptical activity directly in terms of the molecular transitions. Examples include the uniaxial crystals of CO^{3+} and Rh^{3+} complexes (56), the divalent hexaaquo coordinated metal ions (57), and an interesting organic crystal of D_3 symmetry. The organic crystal of D_3 symmetry was investigated by Chadhuri and El-Sayed (58).

This crystal contains three molecules of benzil per unit cell disposed helically about the threefold axis. Although the free molecules in solution are optically inactive because of free rotation about the bond connecting the two benzoyl groups, they become optically active by fixing the conformation of the two groups into dissymmetric right- or left-handed skew positions when the crystal is formed. Limitations of the experimental method and complexity of the spectra of conjugated bicarbonyl precluded a quantitative analysis of the contributions of all the oriented components, but symmetry analysis provided a useful tool for gathering information on the bicarbonyl $n \rightarrow \pi^*$ transitions. It is interesting that α-quartz, on which the first optical activity measurements were made, is in some ways analogous to benzil, in that the SIO_4 unit of the crystal is a tetrahedron with no intrinsic optical activity. Nevertheless, the crystal symmetry has the D_3 point group classification which requires that at least some of the optical activity come from the intrinsic activity of the molecular units. The SIO_4 tetrahedra have, in fact, been found to be slightly distorted in the crystal (59). A number of other applications to trigonal D_3 and C_3 systems have been described (60–63).

Strickland and Richardson (64) have pointed out that the theories of Gō (8,9) and Tinoco (7,10) for oriented systems are "oriented gas" theories. This is also true of Kuball's treatment (30). They are based on internally generated chiroptical activity of the molecules with radiation propagating along particular axes; no specific account is taken of intermolecular interactions. The activity generated in crystals by the molecular interactions that give rise to exciton bands has been investigated (65–67), but even these do not account for activity due to interactions between the crystal (site) symmetry and the intrinsic molecular dissymmetry. A treatment of Deutsche (13), which does not include the local interactions, has been applied by Strickland and Richardson to a crystal of D_4 symmetry, $\alpha NiSO_4 \cdot 6H_2O$ (68–70). The exciton but not the crystal site interactions may be neglected in this system because the ligand field transitions of the Ni^{2+} ion in the distorted symmetry of the crystal are weak. (It will be recalled that observable exciton effects have their origin in strong molecular optical transitions.) Using crystal field theory of the spectra of transition metal ions and CNDO calculations to obtain charge distributions, they calculated the rotational strength associated with the transitions along the unique axis in this crystal with the following results (64):

Band	Wavelength (cm^{-1})	Transition Symmetry	R(calc'd.)(cgs)	(exp.)(cgs)(70)
I	8,000–11,000	$^3A_{2g} \rightarrow {}^3T_{2g}$	1.7×10^{-40}	1.7×10^{-40}
II	13,000–18,000	$^3A_{2g} \rightarrow {}^3T_{1g}$	0.4	0.25
III	22,000–24,000	$^3A_{2g} \rightarrow {}^1T_{2g}$		0.005
IV	24,000–28,000	$^3A_{2g} \rightarrow {}^3T_{1g}$	0.05	0.07

This excellent agreement has been encouraging and is rather better than has been generally expected or obtained. Some recent progress has been made in applying vibronic intensity borrowing theory to crystal chiroptical effects (71), but very little experimental work has been done in this field.

9.4   SUMMARY

In this brief examination of oriented systems, we have seen that chiroptical effects develop along different molecular and crystal axes in specific and different ways. The informational content is, therefore, more detailed for directional than for isotropic measurements. For oriented polymers the discussion has pointed to the origin of the effects in spectroscopic states. Crystal transitions that give rise to ordinary absorption have been treated in considerable detail. The theory of exciton activity in crystals is well developed (72). Specific applications of crystal field theory to the optical activity of crystals of different symmetry have also been developed, largely with respect to the optical transition of inorganic complexes. Apparently because of measurement difficulties, less work has been done in relating the theoretical predictions to experimental observations. This field remains ripe for exploration.

REFERENCES AND NOTES

1. M. Born, *Optik*, Julius Springer-Verlag, Berlin, 1933, p. 413.

2. A. Sommerfeld, "Optik," *Voslessungen über Theoretische Physik*, Vol. 4, Dietrichsche Verlagbuchandlussg, Wiesbaden, 1950, p. 161.

3. See for example, V. N. Tsvetkov, in *Newer Methods of Polymer Characterization*, B. Ke (Ed.), Interscience, New York, 1954, and V. A. Bloomfield, D. M. Crothers, and I. Tinoco, Jr., *Physical Chemistry of Nucleic Acids*, Harper and Row, New York, 1974.

4. A. J. Hopfinger, *Conformational Properties of Macromolecules*, Academic Press, New York and London, 1973.

5. J. Hofrichter and W. A. Eaton, in *Annual Reviews of Biophysics and Bioengineering*, Vol. 5, Annual Reviews, Inc., Palo Alto, 1976.

6. R. W. Wilson and J. A. Schellman, *Biopolym.*, **16**, 2143 (1977).

7. I. Tinoco, Jr., and W. G. Hammerle, *J. Phys. Chem.*, **60**, 1619 (1956); I. Tinoco, Jr., *J. Am. Chem. Soc.*, **79**, 4248 (1957); *ibid.*, **81**, 1540 (1959).

8. N. Gō, *J. Chem. Phys.*, **43**, 1275 (1965).

9. N. Gō, *J. Phys. Soc. Jpn*, **21**, 1579 (1966).

10. I. Tinoco, Jr., and R. W. Woody, *J. Chem. Phys.*, **32**, 461 (1960); R. W. Woody and I. Tinoco, Jr., *J. Chem. Phys.*, **46**, 4927 (1967).

11. J. Snir and J. Schellman, *J. Phys. Chem.*, **77**, 1653 (1973).

12. M. L. Tiffany and S. Krimm, *Biopolym.*, **8**, 347 (1969).

13. C. W. Deutsche, *J. Chem. Phys.*, **52**, 3703 (1970).

14. E. S. Pysh, *J. Chem. Phys.*, **52**, 4723 (1970).

15 A. W. Levin and I. Tinoco, Jr., *J. Chem. Phys.* **66**, 3491 (1977).

16. R. Mandel and G. Holzwarth, *Biopolym.*, **12**, 655 (1973).

17. S. Y. Wooley and G. Holzwarth, *J. Am. Chem. Soc.*, **93**, 4066 (1971).

18. S-Y. Chung and G. Holzwarth, *Biopolym.*, **14**, 1531 (1975).

19. S. F. Mason and A. J. McCafferey, *Nat.*, **204**, 468 (1964).

20. K. Y. Yamaoka, Ph.D. Thesis, University of California, Berkeley, 1964.

21. H. J. Hofrichter, Ph.D. Thesis, University of Oregon, 1971.

22. A. Wada, *Appl. Spectrosc. Rev.*, **6**, 1 (1972).

23. M. J. B. Tunis-Schneider and M. F. Maestre, *J. Mol. Biol.*, **52**, 521 (1970).

24. A. Yogev, L. Margulies, D. Amar, and Y. Mazur, *J. Am. Chem. Soc.*, **91**, 4558, (1969). A. Yogev, J. Riboid, J. Marero, and Y. Mazur, *J. Am. Chem. Soc.*, **91**, 4559 (1969).

25. J. Snir and J. Schellman, *J. Phys. Chem.*, **77**, 1653 (1973).

26. See, for example, G. G. Hall, *Matrices and Tensors*, Pergamon Press, Oxford, 1963. For a description of how second-rank tensors enter into the description of the optical activity of crystals, see, for example, J. F. Nye, *Physical Properties of Crystals*, Oxford University Press, England, 1957.

27. J. W. Gibbs, *Am. J. Sci.*, **23**, 460 (1882).

28. Y. N. Chiu, *J. Chem. Phys.*, **52**, 1042 (1970).

29. A. D. Buckingham and M. B. Dunn, *J. Chem. Soc.*, A, 1988 (1971).

30. H. G. Kuball, T. Karstens, and A. Schonhofer, *Chem. Phys.*, **12**, 1 (1976).

31. The indirect coupling of quadrupole moments pertinent to the **μ-m** mechanism in perturbation theory treatments described in Section 4.5 is a second-order effect arising from interaction between charge distributions on vicinal groups.

32. Oriented systems of chiral molecules (with the exception of cubic crystals) are not only circularly dichroic and circularly birefringent, but are also linearly dichroic and birefringent. Separating chiral effects in the measurement can be exceedingly difficult. The problem has been examined by Disch and Sverdlich (33) and Hofrichter (21) and recognized by many others. In the discussion we assume that the chiral activity is separable from the linear effects, either because the measurements are made with the light propagating along the unique axis of a uniaxially oriented system, in which case the system is isotropic to linearly polarized light, or because the linear effects are insignificantly small relative to the chiral effects in the spectral region of interest.

33. R. L. Disch and D. E. Sverdlich, *J. Chem. Phys.*, **41**, 2137 (1967).

34. R. L. Disch and D. E. Sverdlich, *Anal. Chem.*, **4**, 82 (1969).

35. R. M. Hochstrasser, *Molecular Aspects of Symmetry*, W. A. Benjamin, New York, Amsterdam, 1966.

36. J. C. Decius and R. M. Hexter, *Molecular Vibrations in Crystals*, McGraw-Hill Book Co., New York, 1977.

37. R. D. B. Fraser, *J. Chem. Phys.*, **21**, 1511 (1953); *ibid.*, **24**, 89 (1956).

38. W. C. Johnson and I. Tinoco, Jr., *Biopolym.*, **7**, 727 (1969).

39. S-Y. Chung and G. Holzwarth, *J. Mol. Biol.*, **92**, 449 (1975).

40. G. Holzwarth and P. Doty, *J. Am. Chem. Soc.*, **87**, 218 (1965).

41. J. Schellman and P. Oriel, *J. Chem. Phys.*, **37**, 2114 (1962).

42. E. R. Blout and E. Schecter, *Biopolym.*, **1**, 568 (1963).

43. J. Y. Cassim and J. T. Yang, *Biopolym.*, **9**, 1475 (1970).

44. W. C. Johnson and I. Tinoco, Jr., *J. Am. Chem. Soc.*, **94**, 4389 (1972).

45. M. A. Young and E. S. Pysh, *Macromol.*, **6**, 790 (1973).

46. H. G. Kuball, presentation at the Eighteenth European Spectroscopy Congress, Wroclaw, Poland, 1977.

47. The observation was first made by L. Arago [*Mem. Inst. Fr.*, **1**, 93-134 (1811)], who did not discriminate between the effects of optical rotation and linear birefringence, and then by J. B. Biot [*Mem. Inst. Fr.*, 1–372 (1812)], who made the measurement without interference from linear birefringence by measuring with light propagating along the uniaxial direction. Both references are from T. M. Lowry, *Optical Rotatory Power*, Green and Co., London, 1935.

48. See the reference to the work by Dufet, Wallerant, and Longchambon in the book by Lowry, (47, p. 340).

49. A Fresnel, *Oeuvres*, **2**, 477 (1824) and L. Pasteur, *Oeuvres*, **66** (1848); both quoted in Lowry (47).

50. H. A. Jahn, *Z. Krystallogr.*, **98**, 191 (1937).

51. S. Bhagavantam, *Proc. Indian Acad. Sci.*, **16**, 359 (1942).

52. S. Bhagavantam and D. Suryanarayana, *Nat. (London)*, **160**, 750 (1947); S. Bhagavantam and D. Suryanarayana, *Acta Cryst.*, **2**, 21 (1949).

53. J. Jerphagnon and D. S. Chermia, *J. Chem. Phys.*, **65**, 1522 (1976).

54. See, for example, J. F. Nye, *Physical Properties of Crystals*, Oxford University Press, Oxford, England, 1967, and the paper by J. Donahue in which the relationship between the symmetry notations are spelled out. Nye's monograph also has an excellently concise chapter (14) in which the phenomenological aspects of the optical activity of crystals are discussed and appendix (B) in which the symmetry elements and conventions for the choice of axes in the 32 crystal classes, together with their relationship to crystal forms, are illustrated.

55. M. V. Hobden, *Nat. (London)*, **216**, 678 (1967).

56. Galsboel, F., Steeboel, P., Soerensen, B., Soendergaard, F. *Acta Chem. Scand.* **26**, 3605 (1972).

57. K. D. Gailey, H. F. Giles, Jr. and R. A. Palmer, *Chem. Phys. Lett.*, **19**, 561 (1973).

58. N. K. Chadhuri and M. A. El-Sayed, *J. Chem. Phys.*, **47**, 1133 (1967).

59. R. A. Young and B. Post, *Acta Crystallogr.*, **15**, 337 (1962).

60. W. Moffitt, *J. Chem. Phys.*, **25**, 1189 (1956).

61. H. Poulet, *J. Chem. Phys.*, **59**, 584 (1962).

62. A. J. McCafferey and S. F. Mason, *Mol. Phys.*, **6**, 359 (1962).

63. F. S. Richardson, D. Caliga, G. Hilmes and J. J. Jenkins, *Mol. Phys.*, **30**, 257 (1975), and references therein.

64. R. Strickland and F. S. Richardson, *J. Chem. Phys.*, **57**, 589 (1972).

65. V. L. Ginzburg, *Zh. Eksp. Teor. Fiz.*, **34**, 1593 (1958). [*Soviet Phys. JETP*, **7**, 1096 (1958).]

66. V. M. Agronovich and A. A. Rukhadza, *Zh. Eksp. Teor. Fiz.*, **35**, 982 (1958). [*Soviet Phys. JETP*, **8**, 685 (1959).]

67. Yu A. Tsvirko, *Zh. Eksp. Teor. Fiz.*, **38**, 1615 (1960). [*Soviet Phys. JETP*, **11**, 1163 (1961).]

68. P. L. Merideth and R. A. Palmer, *Chem. Commun.*, **1337**, (1969).

69. R. Grinter, M. J. Harding and S. F. Mason, *J. Chem. Soc.*, A, 667 (1970).

70. F. Costano, *Spectrochim. Acta*, **25A**, 401 (1969).

71. F. S. Richardson, G. Hilmes and J. J. Jenkins, *Theor. Chim. Acta*, **39**, 75 (1975).

72. A. S. Davydov, *Theory of the Absorption of Light in Molecular Crystals*, Works of the Institute of Physics, Ukranian Academy of Sciences, Kiev, U.S.S.R., 1951. Available in English translation as *Theory of Molecular Exciton*, M. Kasha and M. Oppenheimer, Jr. (Transls.) McGraw-Hill Book Co., New York, 1962.

CHAPTER TEN

VIBRATIONAL AND MAGNETICALLY INDUCED CHIRAL ACTIVITY: CIRCULARLY POLARIZED LUMINESCENCE

10.1 INTRODUCTION

In this chapter we discuss three separate chiroptical phenomena: (1) chiral activity of entirely vibrational origin that gives rise to CD (or ORD) in absorption and Raman scattering spectra in the infrared region; (2) circularly polarized luminescence (CPL); and (3) magnetically induced circular dichroism (MCD) and optical rotation (MORD). All three are given limited treatment. In the case of MCD and MORD, this results from a decision to limit this monograph to chiral-optical phenomena involving *natural* optical activity, which has its origin only in the interaction of light with matter. That we treat MCD (or MORD) at all reflects the fact that it elicits the chiral activity of degenerate naturally optically active, but otherwise chirally nonobservable, transitions and is thus related to our earlier discussions of symmetry and optical activity. MCD and MORD are of wide current interest. The sparse discussion of the other two phenomena reflects the limited specialized interest shown in them up to the present, rather than their potential utility.

10.2 VIBRATIONAL CHIROPTICAL EFFECTS

As early as 1930, Kastler, who predicted a number of optical phenomena without formalizing the theories, suggested that the vibrational Raman spectra of optically active molecules would be different if the incident radiation were right and left circularly polarized rather than unpolarized (1). Small wonder, however, that he was unable to detect optical activity, since two factors, the mass of the particles and the low frequency of the optical transitions, cooperate to reduce the magnitude of chiral-optical effects in the infrared region where vibrational spectra are manifest. We have already called attention to the fact that the magnetic transition moments are inversely proportional to the mass of the particles. In this case the transitions between vibrations of the nuclei, produce moments several orders of magnitude smaller than those in magnetically allowed electronic transitions. The frequency factor in the optical activity relations (see, for example, Eq. 2.404) not only operates in the same direction relative to electronic transitions, but is significant within the infrared region. Thus, no optical activity has yet been observed in the spectra of low-frequency bending modes, notwithstanding the fact that these may develop higher intrinsic rotational strengths than the high-frequency stretching modes for which infrared circular dichroism has been observed.

Note that in this discussion of vibrational effects, which refers generally to observations at or near room temperature in solution or in pure liquids, we attribute no special effects to the transitions between different rotational states of the molecules which contribute to the width of each vibration band. The thermal energy at room temperature corresponds to about 200 cm^{-1}. Since rotational energies are typically less than 10 cm^{-1}, the Boltzmann factor, $\exp(\Delta E/kT)$, ensures that many molecular rotational energy transitions are active and contributing to the vibrational bandwidth. The situation is similar to that assumed for vibrations in the treatment of allowed electronic transitions.

Vibrational spectra of fundamental modes are in the region from approximately 3000 to 10 cm^{-1}, covering the range of high-frequency stretching vibrations (of the carbon-hydrogen bonds, for example) to the bending vibrations of heavy atoms. In the ground electronic state, and in some cases in excited staes, they are usually measured as isotropic (or linearly polarized) infrared absorption spectra or as Stokes and anti-Stokes Raman scattering (2). Frequencies in the visible region, where most of the molecules containing the chromophores discussed in Chapter 7 are transparent, are most frequently used to excite vibrational Raman bands. The availability of a variety of good radiation sources and the fact that at the low energy of these frequencies the radiation only rarely produces photochemical decomposition or reaction make this region particularly favorable for Raman excitation. In this discussion we ignore resonance Raman effects that are produced when the exciting light is at or near the same frequency as an electronic absorption band. The Raman spectra consist of emission bands displaced from the frequency of the exciting radiation by a frequency difference corresponding exactly to the frequency of the infrared absorption band, if both are allowed by the symmetry selection rules, or sometimes at a slightly different frequency if interaction with another nearby allowed band is possible in either the infrared or Raman spectra, but not in both. The well-known phenomenon of Fermi resonance, which causes two close interacting bands of the same symmetry to move to higher and lower frequencies, is the most frequent cause of this displacement from exact correspondence with the infrared frequency. Since the symmetry selection rules for infrared absorption and achiral Raman scattering are different from those for rotational strength, it should not be surprising that chiral vibration-rotation bands will be found at slightly different frequencies than those of the achiral phenomena. The observation of such spectral shifts would be useful in the analysis of vibrational spectra, but have not yet been observed because of the limited instrumental sensitivity.

The theory of vibrational chiral activity has been developed in stages reminiscent of the postquantal development of the theories of electronic chiral effects. First, rather specific models were examined. Thus the earliest theoretical treatment of pure vibrational optical activity was specifically directed to the vibrational modes of helical polymers (3). The important general point was made, however, that only the normal displacement coordinates of the vibrational modes would be needed in order to obtain the required vibrational electric and magnetic dipole transition moments. In the case of the electric dipole moment, for example, the total moment associated with the nuclear framework of the molecule is formally equivalent to the corresponding electric moment

$$\mu = \sum_j e_j \mathbf{q}_j, \tag{10.200}$$

where e_j is the charge of the jth particle and $\mathbf{q}_j$, its coordinate ($q_j = x, y, z$ for Cartesian coordinates). The value of the coordinate for the vibrating particle can be expressed as the sum of its equilibrium value $\mathbf{q}_j^0$ and its displacement from equilibrium $\Delta\mathbf{q}_j$:

$$\mu = \sum_j e_j(\mathbf{q}_j^0 + \Delta\mathbf{q}_j). \tag{10.201}$$

In these terms, the quantum mechanical transition moment for the radiation-induced vibrational transition is, therefore,

$$\sum_j \langle v|e_j\mathbf{q}_j^0|v'\rangle = \sum_j \left[\langle v|e_j\mathbf{q}_j^0|v'\rangle + \langle v|e_j\Delta\mathbf{q}_j|v'\rangle \right], \tag{10.202}$$

where, however, the first of the two terms is zero, since the equilibrium value of the coordinate is independent of the coordinates involved in the wavefunctions for the vibrational states $\langle v|$ and $\langle v'|$ of the jth vibrational mode. Thus the electric dipole transition moment contains only the displacement coordinates Δq_j or normal coordinates Q_j (see Section 3.6) associated with the vibrations

$$\mu = \sum_j \langle v|e_j\Delta q_j|v'\rangle. \tag{10.203}$$

Since the vibrational wavefunctions are well known (4), it is the determination of the normal coordinates that represents the calculational problem in determining the transition moments. The Deutsche and Moscowitz treatment (3) was specifically directed toward developing these coordinates for helical polymers.

In 1973 Schellman developed a more general treatment for chiral effects in vibrational modes (5). This theory still retains some aspects of a specific model in that it is based on the assumption that the vibrational motion of the atoms in the molecule does not perturb the charges on the atoms themselves. This assumption, which has been used by Lifson, Warshel, and Levitt (6) in the analysis of the conformations of amides and polypeptides, is sufficient to provide very reasonable agreement between the predicted and observed values of frequencies and intensities of ordinary achiral infrared absorption bands. The theoretical values of the intensities, proportional to $|\langle v|\mu|v'\rangle|^2$, are obtained by developing the dipole moment operator in terms of the normal coordinates, as in Eq. 10.203. (It is in the definition of the normal coordinates that the fixed-charge assumption enters.) A similar calculation can be made of the magnetic vibrational transition moment. The final expression obtained by Schellman for the transition of the jth normal mode from the ground to the first vibrational state is

$$R_{01}^j = \mu_{01}^j \cdot m_{01}^j = \frac{h}{4c} \sum_{i<k} e_i e_k (\mathbf{R}_{ki} \cdot \mathbf{S}_k^j \times \mathbf{S}_i^j), \tag{10.204}$$

where $\mathbf{R}_{ki} = \mathbf{R}_k - \mathbf{R}_i$ are the vector magnitudes of the equilibrium internuclear distance between atoms k and i, and $\mathbf{S}_k^j$ and $\mathbf{S}_i^j$ are the vector components associated with the matrix that transforms normal coordinate displacements to internal coordinate displacements (along and perpendicular to bond directions). The charges on the atoms are e_i, e_k, and the summation is over the atoms and transformation vectors associated with the jth vibrational normal coordinate. It is interesting to examine properties of the magnetic transition moment, m_{01}^j, because effects of isotopic mass in vibrational chiral activity result directly from these properties. The operator $\mathbf{m}$ in

$$\mathbf{m}_{01}^{j} = \langle \nu_0 | \mathbf{m} | \nu_1 \rangle \qquad (10.205)$$

differs slightly but importantly from the corresponding operator in the electronic magnetic transition moment, because where the mass of the electron is the same for all electrons, the mass of the nuclei involved in the different normal modes is not. Thus the value of the ratio of the charge to the mass, e_j / m_j, in

$$\mathbf{m} = \sum_{j} \frac{e_j}{m_j} (\mathbf{R}_j \times \mathbf{P}_j) \qquad (10.206)$$

is not the same for all j and the $\sum_j$ does not vanish, despite the varying signs of $\mathbf{R}_j \times \mathbf{P}_j$, which ensures that the *electronic* magnetic moment operator vanishes when summed over all transitions. As a result, vibrational optical activity should be sensitive to isotopic substitution both as to magnitude and to frequency, the latter because of the dependence of the energy of the vibrational modes on the mass.

Using the Lifson-Warshel procedure to calculate the values of $\mathbf{R}_{ki}$ and $\mathbf{S}_k^j, \mathbf{S}_i^j, \ldots$, Schellman made a provisional assignment of the 42 normal modes of 3, 4, and 5-methyl-substituted pyrrolidone (see Fig. 7.33) and calculated the electric and magnetic dipole strengths associated with one quantum transition (i.e., the fundamentals) of these vibrational modes. Let us select a few of the more interesting predictions to examine. Consider, for example, a band that is predicted to be very weak in the ordinary achiral infrared spectrum; the 907-cm^{-1} band with a dipole strength of only 5.8×10^{-42} (esu-cm)^2. (For comparison, the torsional mode at 334 cm^{-1} associated with the rotation of the methyl group about the bond to which it is attached to the ring is predicted to have a dipole strength almost three orders of magnitude larger.) The rotational strength is also predicted to be small, only -5.1×10^{-45} cgs unit, but the resulting anisotropy factor, $g = \Delta\epsilon/\epsilon$, is -3.5×10^{-3}, which is large enough to permit its observation as a CD band because the weak absorption permits the use of long path lengths of sample.

Carbon-hydrogen stretching vibrations offer another informative example. If only one or two isolated groups exist in the molecule, the anisotropy factors of their fundamental modes is likely to be small because only very small changes in magnetic moments can be expected from the linear change in effective mass when the bond between a carbon and a hydrogen atom lengthens. When as in pyrrolidone, however, there are many C–H bonds whose vibrational energies are almost identical, the normal modes can couple, and because of the overall molecular asymmetry, the resultant rotational strength can be quite large. In the case of 5-methyl pyrrolidone,

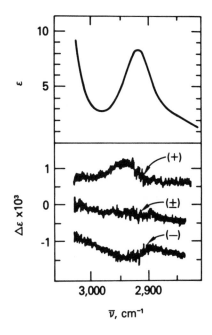

Fig. 10.00 Infrared absorption (upper) and CD spectra (lower) of 2, 2, 2-trifluoro-1-phenyl-ethanol. The CD + and − curves have been displaced from the ± curve for visual clarity; the actual curves overlap at 3030 cm^{-1}. (7)

the C–H stretching vibration at 2908 cm^{-1} is calculated to have an anisotropy factor of -3.8×10^{-4}, not large but probably observable. The first authenticated measurement of a vibrational CD band on liquid samples was, in fact, that of a C–H stretching vibration with an anisotropy factor of only 6.5×10^{-5} (Fig. 10.00) (7). In this case the degenerate C–H stretching vibrations of the phenyl group of optically active trifluorophenylethanols couple into an optically *inactive* band because of the planar symmetry, while the C–H stretch of the α-carbon couples with the C–O–H angle bend of the ethanol moiety to produce a significantly large CD band, which was successfully predicted (7). The success of this type of calculation is, however, dependent on the accuracy of the description of the nuclear vibrations. There is a vast literature on this subject. In the calculations on trifluorophenylethanol, two different descriptions were used (8, 9), one considerably more successful than the other.

Vibrational Raman scattering is of relatively low intensity. It is only the recent introduction of laser light sources and highly sensitive photon detection systems that has permitted the observation of circularly polarized Raman scattering. The subject has recently been discussed in detail in a review by Barron and Buckingham (10) from which the discussion that follows has been radically condensed. The term circular intensity differential (CID) has been coined to distinguish chiral Raman scattering from the circular dichroism observed in absorption. The theory was developed by the same authors in 1971 (11), and the first authenticated observations reported two years later (12). The polarization effects in Raman scattering arise primarily from that part of the incident light scattered from a molecular system, the intensity of which is proportional to the product of the ordinary molecular polarizability α, and the rotatory response parameter β. It will be recalled that ordinary achiral absorption intensity is proportional to the square of the electric dipole transition moment (Chapters 2 and 3) to which α is proportional, while chiral strength, which determines ordinary CD and ORD amplitudes, is proportional to the product β of an electric and magnetic dipole transition moment. The same interactions with the electromagnetic field that produce CD and ORD, therefore, produce CID in the Raman effect. A factor that depends on quadrupole moments may also be present, but as in the case of CD, it tends to be small or null except in oriented samples. The molecular parameters that control achiral and CID Raman intensities are contained in the following expressions which were deduced, respectively, by Placzek (13) and by Barron and Buckingham (11).

Ordinary (achiral) Raman:

$$\langle v'|\alpha_{mn}|v\rangle = \alpha_{mn}\delta_{v'v} + \sum_j \frac{\partial \alpha_{mn}}{\partial Q_j}\langle v'|Q_j|v\rangle. \qquad (10.207)$$

Chiral Raman (CID):

$$\langle v'|\alpha_{mn}|v\rangle\langle v|\beta_{mn}|v'\rangle = \sum_j \left(\frac{\partial \alpha_{mn}}{\partial Q_j}\right)\left(\frac{\partial \beta_{mn}}{\partial Q_j}\right)|\langle v'|Q_j v\rangle|^2. \qquad (10.208)$$

In these equations $\langle v'|$, $\langle v|$, and Q_j are, as before, the vibrational states and the normal coordinate of the jth vibrational mode. The operators α_{mn} and β_{mn} are defined in the same way as for electronic transitions (see Eqs. 2.402 and 2.403) except that the states are vibrational rather than electronic or vibronic. Since Raman scattering results from optical transitions between different vibrational states, $\langle v'|\neq\langle v|$ and the first term in Eq. 10.207 drops out ($\delta_{v'v}=0$). We see, then, that an important selction rule for both ordinary Raman and CID Raman activity is that the polarizability tensor α_{mn} varies with a change in the normal coordinate, that is, as bonds stretch, bend, or rotate. In chiral molecules all or almost all vibrational coordinates give values of $\partial \alpha_{mn}/\partial Q_j$ and $\partial \beta_{mn}/\partial Q_j$ greater than (or less than) zero, so that virtually all bands should exhibit Raman CID; the magnitude of many of them may be too small to be observable. Details of the experimental method have been published (10). At best the experiments are difficult to perform because the true Raman CID signals are frequently no larger than artifacts produced by very small birefringence in the incident beam. Recently, Boucher, Brocki, Moskovitz, and Bosnich (14) have reported the CID Raman scattering from a series of chiral sulfoxides including one measurement made on a new instrument that uses multichannel photon counting. The observation of moderately intense CID in the deformation modes of the carbon-sulfur-oxygen bonds, which deform the molecules about their chiral center (the sulfur atom), and little or no CID associated with the corresponding bond-stretching vibrations is compatible with the idea that the low-frequency bending modes should tend to develop high intrinsic rotational strength.

10.3   MAGNETICALLY INDUCED CHIROPTICAL ACTIVITY

MORD and MCD are manifestations of the Faraday (15) and Cotton-Scherer (16) effects, which were discovered in 1846 and 1932, almost a century apart. Just as for natural chiral activity, there have been many attempts to explain these effects, which are observed only when the sample is subjected to a longitudinal magnetic field. A good concise history of these developments is available in volume 4 of Partington's *Treatise on Physical Chemistry* (17). In this section, these phenomena are treated very briefly, only to indicate sufficiently some of the implications of the more modern analysis. We will focus on the relation of MCD and MORD to spectroscopic states and especially to degenerate states.

In Chapter 6 a division was made of the symmetry of chromophores into three classes. Chromophores belonging to one of the two classes of symmetry groups that are optically inactive can become active if perturbations split degeneracies. When the perturbation is a magnetic field, this statement needs expansion and modification. All optical transitions are optically active to a greater or lesser extent in a magnetic field. Transitions to degenerate states of the optically active point groups (class 1 in Chapter 6 and Appendix I) should exhibit optical activity that has its origin in both the natural chirality of the chromophore and in the splitting of the magnetic degeneracy. These may not be easy to sort out experimentally, although the magnitude of the component arising from the natural chirality should be independent of the magnitude of the magnetic field, provided only that the field is sufficient to split the bands by more than their natural bandwidth and that spectral resolutions of the measuring instrument are sufficient to resolve them. Transitions to degenerate states of the nonoptically active point groups of class 2 may or may not fall into the same category. If they contain symmetry elements for improper rotations S_n, then the magnetically induced chiral activity of their zero-field degenerate transitions will have no contributions from natural chirality. All of the point groups in class 3 have S_n elements (recall that the operation of reflection σ corresponds to S_1 and the operation of inversion i to S_2); for these, no natural chiral activity is possible under any perturbation.

MCD and MORD of degenerate states are associated with the Zeeman (magnetic field) splitting of the magnetic degeneracies. In achiral Zeeman spectra when unpolarized light is used, only the splitting of the absorption bands into doublets or higher multiplets is observed. If the spectra are taken with polarized light, a rotation of the plane of polarization may be observed just as with natural CD or ORD. High-resolution spectra of this type have been called magnetic rotation spectra (18). Kramers (19), Serber (20), Hougen (21), and Hameka (22), as well as others, have contributed to the development of the theory. The most modern treatment, which includes discussion of the effects for both small and large (compared to the bandwidth) Zeeman splitting, is due to Stephens (23) and Buckingham and Stephens (24).

Magnetically induced chiral activity is by no means limited to spectroscopic transitions to degenerate states. In an elegantly detailed paper (23) Stephens laid out the modern theory of magnetic circular dichroism based on the semiclassical theory of radiation absorption, but including now the effect of a longitudinal static perturbing magnetic field. We quote here only enough of the result to permit a brief discussion of the relation between the observations and the molecular parameters. The circular dichroism arising from the electronic transition between the states $\langle 0|$ and $\langle j|$ in the magnetic field of strength H is given by (25)

$$\Delta \varepsilon_H = \frac{8\pi^3 NH}{6909\,ch} \left\{ \left[A_{0j} \right] f(\omega) + \left[\frac{B_{0j} + C_{0j}}{kT} \right] f'(\omega) \right\}, \quad (10.300)$$

where the three terms represent, respectively, A_{0j} contributions from the degenerate upper or lower states, B_{0j} contributions from nondegenerate states, and C_{0j} contributions from changes in the population of ground degenerate states resulting from the interaction with the magnetic field— the latter two terms are temperature dependent as can be observed from the inverse kT part of the terms. As functions of molecular parameters, the three contributions are

$$A_{0j} = 2\Sigma \left[\langle j|\mathbf{m}|j\rangle - \langle 0|\mathbf{m}|0\rangle \right] \cdot \mathrm{Im}\{\langle 0|\boldsymbol{\mu}|j\rangle \times \langle j|\boldsymbol{\mu}|0\rangle\}, \quad (10.301)$$

$$B_{0j} = \Sigma \,\mathrm{Im} \left\{ \sum_{k\neq 0} \frac{\langle k|\mathbf{m}|0\rangle}{\mathcal{E}_k - \mathcal{E}_0} \cdot \langle 0|\boldsymbol{\mu}|j\rangle \times \langle j|\boldsymbol{\mu}|k\rangle \right.$$

$$\left. + \sum_{k\neq j} \frac{\langle j|\mathbf{m}|k\rangle}{\mathcal{E}_k - \mathcal{E}_j} \cdot \langle 0|\boldsymbol{\mu}|j\rangle \times \langle k|\boldsymbol{\mu}|0\rangle \right\}, \quad (10.302)$$

$$C_{0j} = 2\Sigma \langle 0|\mathbf{m}|0\rangle \cdot \mathrm{Im}\{\langle 0|\boldsymbol{\mu}|j\rangle \times \langle j|\boldsymbol{\mu}|0\rangle\}. \quad (10.303)$$

A number of interesting conclusions or predictions can be drawn from these expressions. Since the transition moments between identical states must be zero, the dependence of the terms A_{0j} and C_{0j} on the splitting of degeneracies comes from the necessity for splitting the states $\langle 0|$ and $\langle j|$ into $\langle 0_\pm|$ and $\langle j_\pm|$, which will allow matrix elements like $\langle 0_+|\mathbf{m}|0_-\rangle$ to be finite. Although it is not immediately obvious, these terms can be finite even if the states are already slightly split by some internal magnetic anisotropy and will contribute because of their additional dependence on the external magnetic field. The B_{0j} terms have no matrix elements of the type $\langle 0|\mathbf{m}|0\rangle$. They are, therefore, independent of the existence of degenerate states of the chromophore and finite for all allowed electromagnetic transitions. Further, there are very interesting implications when we consider the application to enantiomers. Enantiomers are mirror images that require at least one of the coordinates of each atom that does not lie in the reflecting plane to change sign between enantiomeric pairs. Since the electric dipole transition moment is a real quantity and depends on the vector connecting an origin fixed in the molecule to the coordinates of the electrons on the atoms, it must change sign unless it lies precisely in the reflecting plane. The products of components that lie in the plane, therefore, will not change sign, but that is equally true of the product of any two components that do not lie in the plane because both change sign. Therefore, $\boldsymbol{\mu}\times\boldsymbol{\mu}$ is invariant between enantiomeric forms, but so is $\mathbf{m}$ because the magnetic transition moment does not change sign on inversion of the axes. As a consequence, the MCD (MORD) of enantiomers have the same sign, and as a further consequence, there is no need to obtain optical isomers to measure these effects (26). In fact, since natural optical activity will be measured in the same experiment (27), it is easier to analyze the MCD spectra of racemic mixtures and of optically inactive molecules. Since all electronic transitions are magnetically active, even if only through vibronic interaction (28), MCD measurements hold out promise of utility comparable to that of ordinary absorption measurements; in many cases they can be even more useful. A strong geometric factor, just as in natural CD, arises from the presence of the cross product $\boldsymbol{\mu}\times\boldsymbol{\mu}$. Consequently, the dot product $\mathbf{m}\cdot(\boldsymbol{\mu}\times\boldsymbol{\mu})$ produces a much larger dependence on structure and conformation than does ordinary achiral absorption.

One other feature worth noting is the frequency dependence. In Eq. 10.300, different frequency dependencies are ascribed to the A term and to $B+C$ terms. Since most organic molecules do not have degenerate ground states, the C term is usually absent, and even in inorganic or complex systems that do have degenerate ground states, the C term can be distinguished by its temperature dependence. The different frequency dependencies can be used, therefore, to distinguish the presence of A and B

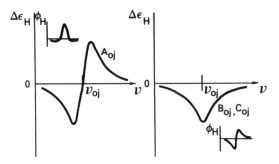

Fig. 10.01 The frequency dependence of MCD spectra of transition of degenerate states (A terms), of nondegenerate states (B terms), and of magnetically induced changes in the population of degenerate ground states (C terms). C terms are temperature dependent. The corresponding MORD spectra are shown in the inserts.

terms. Figure 10.01 illustrates the two different types of frequency dependence for MCD.

The MCD spectra (Fig. 10.02) of a metal porphyrin with D_{4h} symmetry reveal two strong transitions corresponding in shape to those designated for A terms in Fig. 10.01 (29). These arise from the splitting of degenerate ^1E_u excited states, which are responsible for the undifferentiated achiral absorption bands at about 17,340 and 18,430 cm^{-1}. The origin of these A terms has been discussed in detail (29,30). We note here only that ex-

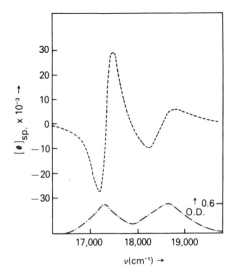

Fig. 10.02 MCD (----) and absorption spectrum (-·-·-) of zinc hematoporphyrin. (Adapted from Fig. 3 of reference 29)

amination of the D_{4h} symmetry table (Appendix I) is sufficient to rationalize these A-term MCD spectra. The A terms come from the triple product $\mathbf{m} \cdot \boldsymbol{\mu} \times \boldsymbol{\mu}$ (Eq. 10.301). Note that the degenerate ^1E_u transitions are allowed in the x and y directions. These will be allowed components of the type

$$\langle o|\boldsymbol{\mu}_x|j_\pm \rangle \times \langle j_\pm |\boldsymbol{\mu}_y|o \rangle = \langle o|\boldsymbol{\mu}_z|j_\pm \rangle,$$

where $\langle j_\pm |$ are degenerate ^1E_u states. Similarly, there are allowed magnetic dipole transitions to the degenerate ^1E_g states. The dot product, $\langle j_\pm |\mathbf{m}_z|j_\pm \rangle \cdot \langle o|\boldsymbol{\mu}_z|j_\pm \rangle$, is therefore finite, and $A_{0j} > 0$. B terms also appear in these spectra from the mixing of these and other states of suitable symmetry. The B terms are ubiquitous, contributing to the MCD spectra of all molecules. Interestingly enough, benzene, a D_{6h} molecule with electric- and magnetic-dipole-allowed transitions to the degenerate E_{1u} and E_{1g} states, respectively, has substantial MCD of B-term origin. In this case, much of the intensity comes from vibronic mixing after the fashion of Herzberg-Teller mixing in achiral absorption spectra (31). The vibronic mechanism is associated with induced B-term chiral effects, just as it is associated in natural optical activity with the weak electric-dipole-forbidden transitions in many organic molecules (32). Extensive calculations of MCD using the approximate quantum mechanical methods discussed in Chapter 7 have been done on a number of systems, notably the condensed aromatic hydrocarbons (33). Inorganic complexes have also been favored subjects for MCD investigations (34), as have the nucleic acid bases and polynucleotides (35–37).

10.4 CIRCULARLY POLARIZED LUMINESCENCE AND FLUORESCENCE-DETECTED CIRCULAR DICHROISM.

An electronically excited molecule that emits radiation without any change of spin multiplicity is said to fluoresce. Most molecules fluoresce following electronic excitation, but because there are other pathways for returning to the ground state, for example, phosphorescence with a change in spin multiplicity or radiationless decay in condensed phases, the fluorescence is extremely weak and cannot easily be detected in many cases. Strong fluorescence is concomitant with strong electric-dipole-allowed transitions. Circularly polarized fluorescence is no different in this respect. Chiral-optical effects involving fluorescence have been observed in at least two ways: One is the detection of the difference in intensities of the luminescence resulting from left and right circularly polarized incident beams. This method is just another technique for measuring circular dichroism, but it is capable of high sensitivity when the quantum yield for fluorescence is high. It can also be used to elicit information not easily available from ordinary CD measurements. For example, circular dichroism measured in absorbance is the weighted average of the CD of all the chiral conformers present, while fluorescence-detected circular dichroism (FDCD) is the weighted average of the chiral activity of the fluorescent chiral conformers. As a consequence, new information on conformational equilibrium may be obtained if the conformers fluoresce at different energies and with different intensities. The theory of FDCD from isotropic samples has been developed by Tinoco and Turner (38) and from oriented samples, by Tinoco, Ehrenberg, and Steinberg (39). Since fluorescence can be induced by energy transfer between chromophores or molecules, FDCD may be used in the same way that ordinary fluorescence measurements are used: as a structural probe by taking advantage of the fact that the excitation transfer between chromophores is highly dependent on their relative orientations.

The second chiral phenomenon utilizing fluorescence, circularly polarized luminescence (CPL), is entirely different because it is a measure of the chiroptical activity of the electronically excited state rather than that of the ground state. Thus, while FDCD is directly proportional to the CD and therefore in the usual way to the electric and magnetic transition moments associated with the excitation process from the ground state, CPL (or CPE, circularly polarized emission) is proportional to transition moments associated with the reverse process—the spontaneous deexcitation by radiation from an excited state (40). In the experimental measurement of CPL, the exciting radiation is unpolarized, but the luminescence is partially circularly dichroic with the circular dichroism luminescence intensity ΔI. ΔI is the difference in the intensities of the left and right circularly polarized radiation in the emitted radiation (41, 42). Possibly the most significant thing about CPL measurements arises from the recognition that molecules in the excited state may have physical and chemical properties that are markedly different from those in the ground state. We note that this follows the relationship between ordinary isotropic absorption and emission, recognizing of course, that differences exist because of the requirement of magnetic dipole transitions for chiral activity. The theory is given explicitly in the paper by Riehl and Richardson (42). Here we will discuss the phenomenon more qualitatively in terms of the concepts developed in earlier chapters.

The fact that the activity is governed by the excited state in CPL has several implications for the relationship between the band shapes and intensities of CD and CPL. Consider two different cases: (1) The structure of the molecule in the excited state from which it luminesces is substantially the same as that of the excited state; (2) the structures are different in the ground and excited state. (In these cases by structure we mean the geometric conformational parameters of bond angles and bond lengths— not bonding arrangements.) A molecule optically inactive in the ground state may become optically active in the excited state by a change in its structure; the net effect, however, in CPL is the same as in CD; no optical activity is observed because the unpolarized exciting light produces equal concentrations of optically active enantiomers, that is, a racemic mixture of excited state molecules. (As in ordinary CD, the application of a further perturbation—a magnetic field—results in observable MCPL; but no CPL is observed.) If, however, the molecule belongs to an optically active point group in its ground state and the sample measured is a pure optical isomer or contains a preponderance of one optical isomer, then the excited state structure is different, but also chiral, and gives rise to observable CPL. The resulting band shape can be significantly different from that of the CD.

There are at least two sources of this difference. The absorption (CD) is characterized by the vibrational structure of the excited state because electronic (vibronic) excitation normally takes place from the lowest vibrational state of the ground electronic state into the allowed vibrational manifold of the excited state. The shape of the CD band is dominated, therefore, by the electronic excited state vibrational manifold, and when spectral fine structure is resolved, by detailed allowedness specified in Section 3.6, similar considerations enter into the CPL band shapes. However, in CPL the electronic ground state vibrational manifold determines the shape, since emission is principally from the ground vibrational (vibronic) state of the excited electronic state into the manifold of excited vibrations of the *ground* electronic state.

The other principal source of the difference in band shapes between the CD and CPL of a chiral transition lies in the relative rotational strengths. While the pure electric and magnetic dipole transition moments for absorption and spontaneous emission are, at least in principle, the same and governed by the considerations of Chapters 3 and 4, their scalar product (the rotational strengths) depends on their relative orientations. Changes in structure between states can alter these directions markedly. The magnitude of the chiral activity arising from any of the coupling or intrinsic dissymmetry mechanisms discussed in Chapter 4 is, as we have seen, strongly affected by these orientations. All of these considerations probably apply to the CD and CPL of $(+)$-*trans*-β-hydroindanone reported by Emeis and Oosterhoff (41) and shown in Fig. 7.09.

If the structures are not substantially different in the two states, then the mechanism for induction of the chiral activity differs only slightly, primarily in the allowedness of the different vibrational components. If we define the asymmetry factor for CPL as

$$g_{CPL} = \frac{4R}{D}, \tag{10.400}$$

comparable to the Kuhn asymmetry factor

$$g_{CD} = \frac{4R}{D}, \tag{10.401}$$

where R and D are rotational and dipole strengths for the pure electronic parts of the spontaneous emission and absorption, then the difference between the *observed* values of g_{CPL} and g_{CD},

$$g_{CPL} = \frac{\Delta I}{I} \tag{10.402}$$

and

$$g_{CD} = \frac{\Delta \varepsilon}{\varepsilon}, \tag{10.403}$$

will be indicative of the extent to which vibronic allowedness and especially excited state geometry are factors in the differences between CPL and CD. In Eq. 10.402, ΔI and I, respectively, are the difference and half the sum of the luminescence intensity of left and right circularly polarized luminescence. In Eq. 10.403, $\Delta \varepsilon$ and ε, as previously defined, are the corresponding quantities for circular dichroism and isotropic absorption. An extensive discussion of band shapes, especially as they affect electronically forbidden (weak) magnetically allowed transitions, has been given by Dekkers and Closs (43). Because emission lifetimes are frequently of the order of relaxation times for Brownian motions associated with the internal and rotational dynamics of macromolecules, orientation effects may also affect the measurements. The problem has been examined by Snir and Schellman (44), who conclude that corrections for rotatory Brownian motion are generally small. Measurements of CPL have been made by Richardson and his colleagues on camphorquinone (45) and on a variety of lanthanide complexes (46), and it appears that increasing use of this technique is likely.

10.5   SUMMARY

Of the three chiroptical phenomena that are discussed in this chapter, two are dependent on the natural chirality of the molecular systems: direct vibrational effects observable in the infrared and Raman spectra, and

PPENDIX I

ACTER TABLES AND GROUP
SFORMATION PROPERTIES

acter Tables
I. For which some or all transitions are naturally optically active.
II. For which optical activity can be induced by perturbations which split degeneracies.
III. For which no natural optical activity is possible.

sformation Properties of the Group Representations for Optical Activity

ACTER TABLES

or which some or all transitions are naturally optically active.

C_1	E
A	1

C_2	E	C_2		
A	1	1	z	R_z
B	1	-1	x,y	R_x,R_y

C_3	E	C_3	C_3^2		
A	1	1	1	z	R_z
E	$\begin{cases}1 \\ 1\end{cases}$	$\begin{matrix}\varepsilon \\ \varepsilon^*\end{matrix}$	$\begin{matrix}\varepsilon^* \\ \varepsilon\end{matrix}$	(x,y)	(R_x,R_y)

$$\varepsilon = \exp(2\pi i/3)$$

luminescent effects observable as fluorescence-detected circular dichroism or as circularly polarized luminescence. CPL is a useful structural and spectroscopic tool because of its relationship to the excited state. While not explicit in the discussion in Section 10.4, it should be clear from earlier considerations that, as in isotropic absorption and CD, it is the symmetry elements common to both states that control the electric and magnetic dipole allowedness of the transitions. In this respect, ordinary CD and CPL do not actually differ when the structure of the two states are different. However, the intensities and band shapes are also controlled by the vibrational manifolds, in the case of CD, by the excited state manifold and in the case of CPL, by the ground state manifold. Thus the emphasis on the excited state structural difference in the discussion of this phenomenon is not intended to signify that the shape of the CPL spectrum is determined by the excited state, or that the shape of the ordinary CD spectrum is determined by the ground state.

The subjects of MCD and MORD deserve more detailed discussion than is given here. We have emphasized the significance of the magnetic field perturbation in eliciting the optical activity associated with degenerate systems.

With this discussion, except for the appendices, we close this examination of chiroptical effects. The preface already disclaims our intent and ability to cover every aspect of this phenomenon, but we would add a note on two specific and important topics that have not been covered. One is the subject of optical activity induced in achiral compounds by chiral solvents (47), and the other, for which an excellent introductory summary exists, is the subject of chirality functions developed by Ernst Ruch and his associates (48). Our reason for not including them is that these chiral phenomena are not as intimately related to the spectroscopic states which is the origin of the emphasis we have given to the discussions of the other chiroptical phenomena. Measurements of optical rotatory power were made and provided useful information long before the concept of spectroscopic states was postulated and still longer before it was reasonably well understood. It seems clear, however, that our modern understanding of the phenomena, and therefore our ability to correlate observations and to predict new ones, is very dependent on this concept.

REFERENCES AND NOTES

1. A. Kastler, *C. R. Acad. Sci. (Paris)*, **191**, 565 (1930).
2. C. V. Raman, *Indian J. Phys.*, **2**, 1 (1928); C. V. Raman and K. S. Krishnan, *Indian J. Phys.*, **2**, 339 (1928); C. V. Raman, *Phys. Rev.*, **33**, 871 (1929).
3. C. W. Deutsche and A. Moscowitz, *J. Chem. Phys.*, **49**, 3257 (1963); *ibid.*, **53**, 2630 (1970).

4. E. B. Wilson, J. C. Decius, and P. Cross, *Molecular Vibrations*, McGraw-Hill Book Co., New York, 1955.

5. J. A. Schellman, *J. Chem. Phys.*, **58**, 2882 (1973).

6. S. Lifson and A. Warshel, *J. Chem. Phys.*, **49**, 5116 (1968); A. Warshel, M. Levitt, and S. Lifson, *J. Mol. Spectros.*, **33**, 84 (1970); A. Warshel and S. Lifson, *J. Chem. Phys.*, **53**, 582 (1970).

7. I. Chabay, E. C. Hsu, and G. Holzwarth, *Chem. Phys. Lett.*, **15**, 211 (1972). G. Holzwarth, E. C. Hsu, H. S. Mosher, T. R. Faulkner, and A. Moscowitz, *J. Am. Chem. Soc.*, **96**, 251 (1974).

8. J. H. Schactschneider and R. G. Snyder, *Spectros. Acta*, **19**, 117 (1963); R. G. Snyder and J. H. Schactschneider, *J. Mol. Spectros.*, **30**, 290 (1965).

9. A. B. Dempster and G. Zerbi, *J. Mol. Spectros.*, **39**, 1 (1971).

10. L. D. Barron and A. D. Buckingham, *Annu. Rev. Phys. Chem.*, **26**, 381–396 (1975); L. D. Barron in *Advances in Infrared and Raman Spectroscopy*, Vol. 4, R. J. H. Clark and R. E. Hester, eds. Heyden, London (1978).

11. L. D. Barron and A. D. Buckingham, *Mol. Phys.*, **20**, 1111 (1971).

12. L. D. Barron, M. P. Bogard, and A. D. Buckingham, *J. Am. Chem. Soc.*, **95**, 603 (1973); L. D. Barron, M. P. Bogard, and A. D. Buckingham, *Nat.* **241**, 113 (1973); L. D. Baron and A. D. Buckingham, *J. Chem. Soc. Chem. Commun.* 152 (1973); *ibid.*, 1028 (1974).

13. G. Placzek, "The Rayleigh and Raman Scattering," in *Handbuch der Radiologie*, E. Marx (Ed.), Akademische Verlag., Germany, 1934.

14. H. Boucher, T. R. Brocki, M. Moskovitz, and B. Bosnich, *J. Am. Chem. Soc.*, **99**, 6870 (1977).

15. M. Faraday, *Philos. Mag.*, **28**, 294 (1846); *ibid.*, **29**, 163 (1846); M. Faraday, *Philos. Trans.*, **136**, 1 (1846).

16. A. Cotton, *C. R.*, **195**, 561 (1932); M. Scherer, *C. R.*, **195**, 950 (1932); A Cotton and M. Scherer, *C. R.*, **195**, 1342 (1932).

17. J. R. Partington, *An Advanced Treatise on Physical Chemistry*, Vol. 4, Longmans-Green & Co., London, 1954.

18. W. H. Eberhardt and H. Renner, *J. Mol. Spectros.*, **6**, 483 (1961) and references therein. See also T. Carroll, *Phys. Rev.*, **52**, 822 (1937), for a review of earlier work, and B. Briat, in *Fundamental Aspects and Recent Developments in Optical Rotatory Disperson and Circular Dichroism*, F. Ciardelli and P. Salvadori (Eds.), Hayden & Son, Ltd., London, 1973, for an interesting exposition and discussion of work to 1971.

19. H. A. Kramers, *Proc. Acad. Sci. (Amsterdam)*, **33**, 959 (1930).

20. R. Serber, *Phys. Rev.*, **41**, 489 (1932).

21. J. T. Hougen, *J. Chem. Phys.*, **32**, 1122 (1960).

22. H. F. Hameka, *J. Chem. Phys.*, **36**, 2540 (1962).

23. P. J. Stephens, *J. Chem. Phys.*, **52**, 3489 (1970).

24. A. D. Buckingham and P. J. Stephens, *Annu. Rev. Phys. Chem.*, **17**, 399 (1966).

25. In Stephens's paper (23), the MCD expressions are given in terms of $\Delta k_{\pm}$ corresponding to the definition of the absorption coefficient: $I = I_0 \exp(-2\omega kz/c)$ where the symbols I, I_0, ω, z, and c have the definitions generally used in this monograph. To correspond with a more common usage and the one used here, the MCD is given in terms of $\Delta \varepsilon_H$ corresponding to the definition of the absorption coefficient as $I = I_0 \exp(-\varepsilon Cz)$ where the C is the concentration in moles per liter and the other symbols are as elsewhere defined.

26. D. J. Caldwell and H. Eyring (*The Theory of Op[York, 1971) have pointed out that the electroma; that the electron interacting with the fields of th three dimensions—only in a plane. This is the eq no requirement for molecular chirality in MCD.

27. This statement is true of the static magnetic field or MORD spectra are taken. The use of a modu make it possible to separate the MCD from the CI

28. R. D. Linder, E. Bunnenberg, L. Seamans, and A (1974).

29. Computed by P. J. Stephens, W. Suetaka, and P (1966), from the MORD spectrum of V. E. Shasho

30. M. Malley, G. Feher, and D. Mauzerall, *J. Mol. S₁*

31. J. S. Rosenfield, A. Moscowitz, and R. E. Linder, Rosenfield, *J. Chem. Phys.*, **66**, 921 (1977); J. S. I (1976).

32. L. Seamans, A. Moscowitz, R. E. Linder, G. Bar *Chem. Phys.*, **13**, 135 (1976); R. E. Linder, G. B Seamans, and A. Moscowitz, *Chem. Phys. Lett.*, **38**, S. Dixon, G. Barth, E. Bunnenberg, C. Djerassi, L. *Chem. Soc.*, **99**, 727 (1977).

33. J. Michl and Josef Michl, *Tetrahedron*, **30**, 4215 (Vogel, *J. Am. Chem. Soc.*, **98**, 3935 (1976), and refer

34. P. J. Stephens, in *Electronic States of Inorganic* Publishing Co., Dordrecht, Holland, 1975.

35. D. W. Miles, R. K. Robins, and H. Eyring, *Proc.* (1967).

36. W. Voelter, R. Records, E. Bunnenberg, and C. Dj (1968).

37. M. F. Maestre, D. M. Gray, and R. B. Cook, *Biopoly*

38. I. Tinoco, Jr., and D. H. Turner, *J. Am. Chem. Soc.*,

39. I. Tinoco, Jr., B. Ehrenberg, and I. Z. Steinberg, *J. C*

40. Apparently the first observation of CPL was made sodium uranyl acetate: B. N. Samojlov, *J. Exp. Theor*

41. C. A. Emeis and L. J. Oosterhoff, *Chem. Phys. Lett.*, 1

42. J. P. Riehl and F. S. Richardson, *J. Chem. Phys.*, **65**,

43. H. P. J. M. Dekkers and L. E. Closs, *J. Am. Chem. So*

44. J. Snir and J. A. Schellman, *J. Phys. Chem.*, **78**, 387 (1

45. C. K. Luk and F. S. Richardson, *J. Am. Chem. Soc.*, **9**

46. C. K. Luk and F. S. Richardson, *J. Am. Chem. Soc.*, **9**

47. D. P. Craig, E. A. Power, and T. Thirunamachandran,

48. E. Ruch, *Acc. Chem. Res.*, **5**, 49 (1972); see also E. Ruch *Acta (Berlin)*, **19**, 225 (1970).

C_4	E	C_4	C_4^3	C_2		
A	1	1	1	1	z	R_z
B	1	-1	-1	1		
E	$\left\{\begin{matrix}1\\1\end{matrix}\right.$	$\begin{matrix}i\\-i\end{matrix}$	$\begin{matrix}-i\\i\end{matrix}$	$\left.\begin{matrix}-1\\-1\end{matrix}\right\}$	(x,y)	(R_x,R_y)

C_5	E	C_5	C_5^2	C_5^3	C_5^4		
A	1	1	1	1	1	z	R_z
E_1	$\left\{\begin{matrix}1\\1\end{matrix}\right.$	$\begin{matrix}\varepsilon\\\varepsilon^*\end{matrix}$	$\begin{matrix}\varepsilon^2\\\varepsilon^{2*}\end{matrix}$	$\begin{matrix}\varepsilon^{2*}\\\varepsilon^2\end{matrix}$	$\left.\begin{matrix}\varepsilon^*\\\varepsilon\end{matrix}\right\}$	(x,y)	(R_x,R_y)
E_2	$\left\{\begin{matrix}1\\1\end{matrix}\right.$	$\begin{matrix}\varepsilon^2\\\varepsilon^{2*}\end{matrix}$	$\begin{matrix}\varepsilon^*\\\varepsilon\end{matrix}$	$\begin{matrix}\varepsilon\\\varepsilon^*\end{matrix}$	$\left.\begin{matrix}\varepsilon^{2*}\\\varepsilon^2\end{matrix}\right\}$		

$$\varepsilon=\exp(2\pi i/5)$$

C_6	E	C_6	C_6^5	C_3	C_3^2	C_2		
A	1	1	1	1	1	1	z	R_z
B	1	-1	-1	1	1	-1		
E_1	$\left\{\begin{matrix}1\\1\end{matrix}\right.$	$\begin{matrix}\varepsilon\\\varepsilon^*\end{matrix}$	$\begin{matrix}\varepsilon^*\\\varepsilon\end{matrix}$	$\begin{matrix}-\varepsilon^*\\-\varepsilon\end{matrix}$	$\begin{matrix}-\varepsilon\\-\varepsilon^*\end{matrix}$	$\left.\begin{matrix}-1\\-1\end{matrix}\right\}$	(x,y)	(R_x,R_y)
E_2	$\left\{\begin{matrix}1\\1\end{matrix}\right.$	$\begin{matrix}-\varepsilon\\-\varepsilon^*\end{matrix}$	$\begin{matrix}-\varepsilon^*\\-\varepsilon\end{matrix}$	$\begin{matrix}-\varepsilon^*\\-\varepsilon\end{matrix}$	$\begin{matrix}-\varepsilon\\-\varepsilon^*\end{matrix}$	$\left.\begin{matrix}1\\1\end{matrix}\right\}$		

$$\varepsilon=\exp(2\pi i/6)$$

D_2	E	$C_2'(x)$	$C_2''(y)$	$C_2'''(z)$		
A	1	1	1	1		
B_1	1	1	-1	-1	x	R_x
B_2	1	-1	1	-1	y	R_y
B_3	1	-1	-1	1	z	R_z

D_3	E	$2C_3$	$3C_2$		
A_1	1	1	1		
A_2	1	1	-1	z	R_z
E	2	-1	0	(x,y)	(R_x,R_y)

D_4	E	$2C_4$	C_2	$2C_2'$	$2C_2''$		
A_1	1	1	1	1	1		
A_2	1	1	1	-1	-1	z	R_z
B_1	1	-1	1	1	-1		
B_2	1	-1	1	-1	1		
E	2	0	-2	0	0	(x,y)	(R_x,R_y)

D_5	E	$2C_5$	$2C_5^2$	$5C_2$		
A_1	1	1	1	1		
A_2	1	1	1	-1	z	R_z
E_1	2	$2\cos 72°$	$2\cos 144°$	0	(x,y)	(R_x,R_y)
E_2	2	$2\cos 144°$	$2\cos 72°$	0		

D_6	E	$2C_6$	$2C_3$	C_2	$3C_2'$	$3C_2''$		
A_1	1	1	1	1	1	1		
A_2	1	1	1	1	-1	-1	z	R_z
B_1	1	-1	1	-1	1	-1		
B_2	1	-1	1	-1	-1	1		
E_1	2	1	-1	-2	0	0	(x,y)	(R_x,R_y)
E_2	2	-1	-1	2	0	0		

Class II. For which optical activity can be induced by perturbations that split degeneracies

C_{3v}	E	$2C_3$	$3\sigma_v$		
A_1	1	1	1	z	
A_2	1	1	-1		R_z
E	2	-1	0	(x,y)	(R_x,R_y)

C_{4v}	E	$2C_4$	C_2	$2\sigma_v$	$2\sigma_d$		
A_1	1	1	1	1	1	z	
A_2	1	1	1	-1	-1		R_z
B_1	1	-1	1	1	-1		
B_2	1	-1	1	-1	1		
E	2	0	-2	0	0	(x,y)	(R_x,R_y)

C_{5v}	E	$2C_5$	$2C_5^2$	$5\sigma_v$		
A_1	1	1	1	1	z	
A_2	1	1	1	-1		R_z
E_1	2	$2\cos 72°$	$2\cos 144°$	0	(x,y)	(R_x,R_y)
E_2	2	$2\cos 144°$	$2\cos 72°$	0		

C_{6v}	E	$2C_6$	$2C_3$	C_2	$3\sigma_v$	$3\sigma_d$		
A_1	1	1	1	1	1	1	z	
A_2	1	1	1	1	-1	-1		R_z
B_1	1	-1	1	-1	1	-1		
B	1	-1	1	-1	-1	1		
E_1	2	1	-1	-2	0	0	(x,y)	(R_x,R_y)
E_2	2	-1	-1	2	0	0		

D_{2d}	E	$2S_4$	C_2	$2C_2'$	$2\sigma_d$		
A_1	1	1	1	1	1		
A_2	1	1	1	-1	-1		R_z
B_1	1	-1	1	1	-1		
B_2	1	-1	1	-1	1	z	
E	2	0	-2	0	0	(x,y)	(R_x,R_y)

S_4	E	S_4	S_4^3	C_2		
A	1	1	1	1		R_z
B	1	-1	-1	1	z	
E	$\begin{cases}1 \\ 1\end{cases}$	$\begin{matrix}i \\ -i\end{matrix}$	$\begin{matrix}-i \\ i\end{matrix}$	$\begin{matrix}-1 \\ -1\end{matrix}$	(x,y)	(R_x,R_y)

C_5	E	S_8	C_4	S_8^3	C_2	S_8^5	C_4^3	S_8^7		
A	1	1	1	1	1	1	1	1		R_z
B	1	-1	1	-1	1	-1	1	-1	z	
E_1	$\begin{cases}1 \\ 1\end{cases}$	$\begin{matrix}\varepsilon \\ \varepsilon^*\end{matrix}$	$\begin{matrix}i \\ -i\end{matrix}$	$\begin{matrix}-\varepsilon^* \\ -\varepsilon\end{matrix}$	$\begin{matrix}-1 \\ -1\end{matrix}$	$\begin{matrix}-\varepsilon \\ -\varepsilon^*\end{matrix}$	$\begin{matrix}-i \\ i\end{matrix}$	$\begin{matrix}\varepsilon^* \\ \varepsilon\end{matrix}$	(x,y)	(R_x,R_y)
E_2	$\begin{cases}1 \\ 1\end{cases}$	$\begin{matrix}i \\ -i\end{matrix}$	$\begin{matrix}-1 \\ -1\end{matrix}$	$\begin{matrix}-i \\ i\end{matrix}$	$\begin{matrix}1 \\ 1\end{matrix}$	$\begin{matrix}i \\ -i\end{matrix}$	$\begin{matrix}-1 \\ -1\end{matrix}$	$\begin{matrix}-i \\ i\end{matrix}$		
E_3	$\begin{cases}1 \\ 1\end{cases}$	$\begin{matrix}-\varepsilon^* \\ -\varepsilon\end{matrix}$	$\begin{matrix}-i \\ i\end{matrix}$	$\begin{matrix}\varepsilon \\ \varepsilon^*\end{matrix}$	$\begin{matrix}-1 \\ -1\end{matrix}$	$\begin{matrix}\varepsilon^* \\ \varepsilon\end{matrix}$	$\begin{matrix}i \\ -i\end{matrix}$	$\begin{matrix}-\varepsilon \\ -\varepsilon^*\end{matrix}$		

T	E	$4C_3$	$4C_3^2$	$3C_2$		
A	1	1	1	1		
E	$\begin{cases}1 \\ 1\end{cases}$	$\begin{matrix}\varepsilon \\ \varepsilon^*\end{matrix}$	$\begin{matrix}\varepsilon^* \\ \varepsilon\end{matrix}$	$\begin{matrix}1 \\ 1\end{matrix}$		
T	3	0	0	-1	(x,y,z)	(R_x,R_y,R_z)

$$\varepsilon = \exp(2\pi i/3)$$

O	E	$8C_3$	$3C_2$	$6C_4$	$6C_2'$		
A_1	1	1	1	1	1		
A_2	1	1	1	-1	-1		
E	2	-1	2	0	0		
T_1	3	0	-1	1	-1	(x,y,z)	(R_x,R_y,R_z)
T_2	3	0	-1	-1	1		

$C_\infty V$	E	$2C_\infty^\phi$	$\cdots$	$\infty\sigma_v$		
$A_1\equiv\Sigma^+$	1	1	$\cdots$	1	z	
$A_2\equiv\Sigma^-$	1	1	$\cdots$	-1		R_z
$E_1\equiv\Pi$	2	$2\cos\phi$	$\cdots$	0	(x,y)	(R_x,R_y)
$E_2\equiv\Delta$	2	$2\cos2\phi$	$\cdots$	0		
$E_3\equiv\Phi$	2	$2\cos3\phi$	$\cdots$	0		
$\cdots$	$\cdots$	$\cdots$	$\cdots$	$\cdots$		

Class III. For which no natural optical activity is possible.

C_s	E	σ_h		
A'	1	1	x,y	R_z
A''	1	-1	z	R_x,R_y

C_i	E	i	
A_g	1	1	R
A_u	1	-1	T

C_{2v}	E	C_2	σ_v'	σ_v''		
A_1	1	1	1	1	z	
A_2	1	1	-1	-1		R_z
B_1	1	-1	1	-1	x	R_y
B_2	1	-1	-1	1	y	R_x

C_{2h}	E	C_2	i	σ_h		
A_g	1	1	1	1		R_z
B_g	1	-1	1	-1		R_x,R_y
A_u	1	1	-1	-1	z	
B_u	1	-1	-1	1	x,y	

C_{3h}	E	C_3	C_3^2	σ_h	S_3	S_3^5	
A'	1	1	1	1	1	1	R_z
A''	1	1	1	-1	-1	-1	z
E'	$\begin{cases} 1 \\ 1 \end{cases}$	$\begin{matrix} \varepsilon \\ \varepsilon^* \end{matrix}$	$\begin{matrix} \varepsilon^* \\ \varepsilon \end{matrix}$	$\begin{matrix} 1 \\ 1 \end{matrix}$	$\begin{matrix} \varepsilon \\ \varepsilon^* \end{matrix}$	$\left.\begin{matrix} \varepsilon^* \\ \varepsilon \end{matrix}\right\}$	(x,y)
E''	$\begin{cases} 1 \\ 1 \end{cases}$	$\begin{matrix} \varepsilon \\ \varepsilon^* \end{matrix}$	$\begin{matrix} \varepsilon^* \\ \varepsilon \end{matrix}$	$\begin{matrix} -1 \\ -1 \end{matrix}$	$\begin{matrix} -\varepsilon \\ -\varepsilon^* \end{matrix}$	$\left.\begin{matrix} -\varepsilon^* \\ -\varepsilon \end{matrix}\right\}$	(R_x,R_y)

$$\varepsilon = \exp(2\pi i/3)$$

C_{4h}	E	C_4	C_4^3	C_2	i	S_4^3	S_4	σ_h	
A_g	1	1	1	1	1	1	1	1	R_z
B_g	1	-1	-1	1	1	-1	-1	1	
A_u	1	1	1	1	-1	-1	-1	-1	z
B_u	1	-1	-1	1	-1	1	1	-1	
E_g	$\begin{cases} 1 \\ 1 \end{cases}$	$\begin{matrix} i \\ -i \end{matrix}$	$\begin{matrix} -i \\ i \end{matrix}$	$\begin{matrix} -1 \\ -1 \end{matrix}$	$\begin{matrix} 1 \\ 1 \end{matrix}$	$\begin{matrix} i \\ -i \end{matrix}$	$\begin{matrix} -i \\ i \end{matrix}$	$\left.\begin{matrix} -1 \\ -1 \end{matrix}\right\}$	(R_x,R_y)
E_u	$\begin{cases} 1 \\ 1 \end{cases}$	$\begin{matrix} i \\ -i \end{matrix}$	$\begin{matrix} -i \\ i \end{matrix}$	$\begin{matrix} -1 \\ -1 \end{matrix}$	$\begin{matrix} -1 \\ -1 \end{matrix}$	$\begin{matrix} -i \\ i \end{matrix}$	$\begin{matrix} i \\ -i \end{matrix}$	$\left.\begin{matrix} 1 \\ 1 \end{matrix}\right\}$	(x,y)

C_{5h}	E	C_5	C_5^2	C_5^3	C_5^4	σ_h	S_5	S_5^7	S_5^8	S_5^9	
A'	1	1	1	1	1	1	1	1	1	1	R_z
E_1'	$\begin{cases} 1 \\ 1 \end{cases}$	$\begin{matrix} \varepsilon \\ \varepsilon^* \end{matrix}$	$\begin{matrix} \varepsilon^2 \\ \varepsilon^{2*} \end{matrix}$	$\begin{matrix} \varepsilon^{2*} \\ \varepsilon^2 \end{matrix}$	$\begin{matrix} \varepsilon^* \\ \varepsilon \end{matrix}$	$\begin{matrix} 1 \\ 1 \end{matrix}$	$\begin{matrix} \varepsilon \\ \varepsilon^* \end{matrix}$	$\begin{matrix} \varepsilon^2 \\ \varepsilon^{2*} \end{matrix}$	$\begin{matrix} \varepsilon^{2*} \\ \varepsilon^2 \end{matrix}$	$\left.\begin{matrix} \varepsilon^* \\ \varepsilon \end{matrix}\right\}$	(x,y)
E_2'	$\begin{cases} 1 \\ 1 \end{cases}$	$\begin{matrix} \varepsilon^2 \\ \varepsilon^{2*} \end{matrix}$	$\begin{matrix} \varepsilon^* \\ \varepsilon \end{matrix}$	$\begin{matrix} \varepsilon \\ \varepsilon^* \end{matrix}$	$\begin{matrix} \varepsilon^{2*} \\ \varepsilon^2 \end{matrix}$	$\begin{matrix} 1 \\ 1 \end{matrix}$	$\begin{matrix} \varepsilon^2 \\ \varepsilon^{2*} \end{matrix}$	$\begin{matrix} \varepsilon^* \\ \varepsilon \end{matrix}$	$\begin{matrix} \varepsilon \\ \varepsilon^* \end{matrix}$	$\left.\begin{matrix} \varepsilon^{2*} \\ \varepsilon^2 \end{matrix}\right\}$	
A''	1	1	1	1	1	-1	-1	-1	-1	-1	z
E_1''	$\begin{cases} 1 \\ 1 \end{cases}$	$\begin{matrix} \varepsilon \\ \varepsilon^* \end{matrix}$	$\begin{matrix} \varepsilon^2 \\ \varepsilon^{2*} \end{matrix}$	$\begin{matrix} \varepsilon^{2*} \\ \varepsilon^2 \end{matrix}$	$\begin{matrix} \varepsilon^* \\ \varepsilon \end{matrix}$	$\begin{matrix} -1 \\ -1 \end{matrix}$	$\begin{matrix} -\varepsilon \\ -\varepsilon^* \end{matrix}$	$\begin{matrix} -\varepsilon^2 \\ -\varepsilon^{2*} \end{matrix}$	$\begin{matrix} -\varepsilon^{2*} \\ -\varepsilon^2 \end{matrix}$	$\left.\begin{matrix} -\varepsilon^* \\ -\varepsilon \end{matrix}\right\}$	(R_x,R_y)
E_2''	$\begin{cases} 1 \\ 1 \end{cases}$	$\begin{matrix} \varepsilon^2 \\ \varepsilon^{2*} \end{matrix}$	$\begin{matrix} \varepsilon^* \\ \varepsilon \end{matrix}$	$\begin{matrix} \varepsilon \\ \varepsilon^* \end{matrix}$	$\begin{matrix} \varepsilon^{2*} \\ \varepsilon^2 \end{matrix}$	$\begin{matrix} -1 \\ -1 \end{matrix}$	$\begin{matrix} -\varepsilon^2 \\ -\varepsilon^{2*} \end{matrix}$	$\begin{matrix} -\varepsilon^* \\ -\varepsilon \end{matrix}$	$\begin{matrix} -\varepsilon \\ -\varepsilon^* \end{matrix}$	$\left.\begin{matrix} -\varepsilon^{2*} \\ -\varepsilon^2 \end{matrix}\right\}$	

$$\varepsilon = \exp(2\pi i/5)$$

C_{6h}	E	C_6	C_6^5	C_3	C_3^2	C_2	i	S_6^5	S_6	S_3^2	S_3	σ_h	
A_g	1	1	1	1	1	1	1	1	1	1	1	1	R_z
B_g	1	-1	-1	1	1	-1	1	-1	-1	1	1	-1	
A_u	1	1	1	1	1	1	-1	-1	-1	-1	-1	-1	z
B_u	1	-1	-1	1	1	-1	-1	1	1	-1	-1	1	
E_{1g}	1	ε	ε^*	$-\varepsilon^*$	$-\varepsilon$	-1	1	ε	ε^*	$-\varepsilon^*$	$-\varepsilon$	-1	(R_x, R_y)
	1	ε^*	ε	$-\varepsilon$	$-\varepsilon^*$	-1	1	ε^*	ε	$-\varepsilon$	$-\varepsilon^*$	-1	
E_{2g}	1	$-\varepsilon$	$-\varepsilon^*$	$-\varepsilon^*$	$-\varepsilon$	1	1	$-\varepsilon$	$-\varepsilon^*$	$-\varepsilon^*$	$-\varepsilon$	1	
	1	$-\varepsilon^*$	$-\varepsilon$	$-\varepsilon$	$-\varepsilon^*$	1	1	$-\varepsilon^*$	$-\varepsilon$	$-\varepsilon$	$-\varepsilon^*$	1	
E_{1u}	1	ε	ε^*	$-\varepsilon^*$	$-\varepsilon$	-1	-1	$-\varepsilon$	$-\varepsilon^*$	ε^*	ε	1	(x, y)
	1	ε^*	ε	$-\varepsilon$	$-\varepsilon^*$	-1	-1	$-\varepsilon^*$	$-\varepsilon$	ε	ε^*	1	
E_{2u}	1	$-\varepsilon$	$-\varepsilon^*$	$-\varepsilon^*$	$-\varepsilon$	1	-1	ε	ε^*	ε	ε	-1	
	1	$-\varepsilon^*$	$-\varepsilon$	$-\varepsilon$	$-\varepsilon^*$	1	-1	ε^*	ε	ε	ε^*	-1	

$$\varepsilon = \exp(2\pi i/6)$$

D_{2h}	E	C_2'	C_2''	C_2'''	i	σ_v^{yz}	σ_v^{zx}	σ_h	
A_g	1	1	1	1	1	1	1	1	
B_{1g}	1	1	-1	-1	1	1	-1	-1	R_x
B_{2g}	1	-1	1	-1	1	-1	1	-1	R_y
B_{3g}	1	-1	-1	1	1	-1	-1	1	R_z
A_u	1	1	1	1	-1	-1	-1	-1	
B_{1u}	1	1	-1	-1	-1	-1	1	1	x
B_{2u}	1	-1	1	-1	-1	1	-1	1	y
B_{3u}	1	-1	-1	1	-1	1	1	-1	z

D_{3h}	E	$2C_3$	$3C_2$	σ_h	$2S_3$	$3\sigma_v$		
A_1'	1	1	1	1	1	1		
A_2'	1	1	-1	1	1	-1	R_z	
A_1''	1	1	1	-1	-1	-1		
A_2''	1	1	-1	-1	-1	1	z	
E'	2	-1	0	2	-1	0	(x, y)	
E''	2	-1	0	-2	1	0		(R_x, R_y)

D_{4h}	E	$2C_4$	C_2	$2C_2'$	$2C_2''$	i	$2S_4$	σ_h	$2\sigma_v$	$2\sigma_d$	
A_{1g}	1	1	1	1	1	1	1	1	1	1	
A_{2g}	1	1	1	-1	-1	1	1	1	-1	-1	R_z
B_{1g}	1	-1	1	-1	1	1	-1	1	-1	1	
B_{2g}	1	-1	1	1	-1	1	-1	1	1	-1	
A_{1u}	1	1	1	1	1	-1	-1	-1	-1	-1	
A_{2u}	1	1	1	-1	-1	-1	-1	-1	1	1	z
B_{1u}	1	-1	1	-1	1	-1	1	-1	1	-1	
B_{2u}	1	-1	1	1	-1	-1	1	-1	-1	1	
E_{1g}	2	0	-2	0	0	2	0	-2	0	0	(R_x, R_y)
E_{2u}	2	0	-2	0	0	-2	0	2	0	0	(x,y)

D_{5h}	E	$2C_5$	$2C_5^2$	$5C_2$	σ_h	$2S_5$	$2S_5^3$	$5\sigma_v$	
A_1'	1	1	1	1	1	1	1	1	
A_2'	1	1	1	-1	1	1	1	-1	R_z
E_1'	2	$2\cos 72°$	$2\cos 144°$	0	2	$2\cos 72°$	$2\cos 144°$	0	(y,z)
E_2'	2	$2\cos 144°$	$2\cos 72°$	0	2	$2\cos 144°$	$2\cos 72°$	0	
A_1''	1	1	1	1	-1	-1	-1	-1	
A_2''	1	1	1	-1	-1	-1	-1	1	x
E_1''	2	$2\cos 72°$	$2\cos 144°$	0	-2	$-2\cos 72°$	$-2\cos 144°$	0	(R_x, R_y)
E_2''	2	$2\cos 144°$	$2\cos 72°$	0	-2	$-2\cos 144°$	$-2\cos 72°$	0	

D_{6h}	E	$2C_6$	$2C_3$	C_2	$3C_2'$	$3C_2''$	i	$2S_3$	$2S_6$	σ_h	$3\sigma_d$	$3\sigma_v$	
A_{1g}	1	1	1	1	1	1	1	1	1	1	1	1	
A_{2g}	1	1	1	1	-1	-1	1	1	1	1	-1	-1	R_z
B_{1g}	1	-1	1	-1	1	-1	1	-1	1	-1	1	-1	
B_{2g}	1	-1	1	-1	-1	1	1	-1	1	-1	-1	1	
A_{1u}	1	1	1	1	1	1	-1	-1	-1	-1	-1	-1	
A_{2u}	1	1	1	1	-1	-1	-1	-1	-1	-1	1	1	z
B_{1u}	1	-1	1	-1	1	-1	-1	1	-1	1	-1	1	
B_{2u}	1	-1	1	-1	-1	1	-1	1	-1	1	1	-1	
E_{1g}	2	1	-1	-2	0	0	2	1	-1	-2	0	0	(R_x, R_y)
E_{2g}	2	-1	-1	2	0	0	2	-1	-1	2	0	0	
E_{1u}	2	1	-1	-2	0	0	-2	-1	1	2	0	0	(x,y)
E_{2u}	2	-1	-1	2	0	0	-2	1	1	-2	0	0	

D_{3d}	E	$2C_3$	$3C_2$	i	$2S_6$	$3\sigma_d$		
A_{1g}	1	1	1	1	1	1		
A_{2g}	1	1	-1	1	1	-1		R_z
A_{1u}	1	1	1	-1	-1	-1		
A_{2u}	1	1	-1	-1	-1	1	z	
E_g	2	-1	0	2	-1	0		(R_x,R_y)
E_u	2	-1	0	-2	1	0	(x,y)	

D_{4d}	E	$2S_8$	$2C_4$	$2S_8^3$	C_2	$4C_2'$	$4\sigma_d$		
A_1	1	1	1	1	1	1	1		
A_2	1	1	1	1	1	-1	-1		R_z
B_1	1	-1	1	-1	1	1	-1		
B_2	1	-1	1	-1	1	-1	1	z	
E_1	2	$\sqrt{2}$	0	$-\sqrt{2}$	-2	0	0	(x,y)	
E_2	2	0	-2	0	2	0	0		
E_3	2	$-\sqrt{2}$	0	$\sqrt{2}$	-2	0	0		(R_x,R_y)

D_{5d}	E	$2C_5$	$2C_5^2$	$5C_2$	i	$2S_{10}^3$	$2S_5$	$5\sigma_d$		
A_{1g}	1	1	1	1	1	1	1	1		
A_{2g}	1	1	1	-1	1	1	1	-1		R_z
B_{1g}	2	$2\cos 72°$	$2\cos 144°$	0	2	$2\cos 72°$	$2\cos 144°$	0		(R_x,R_y)
B_{2g}	2	$2\cos 144°$	$2\cos 72°$	0	2	$2\cos 144°$	$2\cos 72°$	0		
A_{1u}	1	1	1	1	-1	-1	-1	-1		
A_{2u}	1	1	1	-1	-1	-1	-1	1	z	
B_{1u}	2	$2\cos 72°$	$2\cos 144°$	0	-2	$-2\cos 72°$	$-2\cos 144°$	0	(x,y)	
B_{2u}	2	$2\cos 144°$	$2\cos 72°$	0	-2	$-2\cos 144°$	$-2\cos 72°$	0		

D_{6d}	E	$2S_{12}$	$2C_6$	$2S_4$	$2C_3$	$2S_{12}^5$	C_2	$6C_2'$	$6\sigma_d$		
A_1	1	1	1	1	1	1	1	1	1		
A_2	1	1	1	1	1	1	1	-1	-1		R_z
B_1	1	-1	1	-1	1	-1	1	1	-1		
B_2	1	-1	1	-1	1	-1	1	-1	1	z	
E_1	2	$\sqrt{3}$	1	0	-1	$-\sqrt{3}$	-2	0	0	(x,y)	
E_2	2	1	-1	-2	-1	1	2	0	0		
E_3	2	0	-2	0	2	0	-2	0	0		
E_4	2	-1	-1	2	-1	-1	2	0	0		
E_5	2	$-\sqrt{3}$	1	0	-1	$\sqrt{3}$	-2	0	0		(R_x,R_y)

S_6	E	C_3	C_3^2	i	S_6^5	S_0	
A_g	1	1	1	1	1	1	R_z
E_g	$\begin{cases} 1 \\ 1 \end{cases}$	$\begin{matrix} \varepsilon \\ \varepsilon^* \end{matrix}$	$\begin{matrix} \varepsilon^* \\ \varepsilon \end{matrix}$	$\begin{matrix} 1 \\ 1 \end{matrix}$	$\begin{matrix} \varepsilon \\ \varepsilon^* \end{matrix}$	$\begin{matrix} \varepsilon^* \\ \varepsilon \end{matrix}$	(R_x, R_y)
A_u	1	1	1	-1	-1	-1	z
E_u	$\begin{cases} 1 \\ 1 \end{cases}$	$\begin{matrix} \varepsilon \\ \varepsilon^* \end{matrix}$	$\begin{matrix} \varepsilon^* \\ \varepsilon \end{matrix}$	$\begin{matrix} -1 \\ -1 \end{matrix}$	$\begin{matrix} -\varepsilon \\ -\varepsilon^* \end{matrix}$	$\begin{matrix} -\varepsilon^* \\ -\varepsilon \end{matrix}$	(x, y)

T_d	E	$8C_3$	$3C_2$	$6S_4$	$6\sigma_d$	
A_1	1	1	1	1	1	
A_2	1	1	1	-1	-1	
E	2	-1	2	0	0	
T_1	3	0	-1	1	-1	(R_x, R_y, R_z)
T_2	3	0	-1	-1	1	(x, y, z)

O_h	E	$8C_3$	$6C_2'$	$6C_4$	$3C_2$	i	$8S_6$	$6\sigma_d$	$6S_4$	$3\sigma_h$	
A_{1g}	1	1	1	1	1	1	1	1	1	1	
A_{2g}	1	1	-1	-1	1	1	1	-1	-1	1	
A_{1u}	1	1	1	1	1	-1	-1	-1	-1	-1	
A_{2u}	1	1	-1	-1	1	-1	-1	1	1	-1	
E_g	2	-1	0	0	2	2	-1	0	0	2	
E_u	2	-1	0	0	2	-2	1	0	0	-2	
T_{1g}	3	0	-1	1	-1	3	0	-1	1	-1	(R_x, R_y, R_z)
T_{2g}	3	0	1	-1	-1	3	0	1	-1	-1	
T_{1u}	3	0	-1	1	-1	-3	0	1	-1	1	(x, y, z)
T_{2u}	3	0	1	-1	-1	-3	0	-1	1	1	

B. TRANSFORMATION PROPERTIES OF THE GROUP REPRESENTATIONS FOR OPTICAL ACTIVITY

The transformation properties of the representation of the rotational strength vector product $\mu \cdot \mathbf{m}$, may be summarized from the character tables in a way that permits a rapid determination of which point groups have representations that allow chiral electromagnetic transitions. The summary, in the form of the table below, gives the polarizations of the vector products. Optically active transitions in these groups are those that have repeating letters, xx, yy, zz. Transitions that may be optically active if degeneracies are lifted are those with the same repeating indices in two or more columns, and those with nonrepeating indices that appear in two columns with opposite signs. Optical activity is forbidden for all other transitions and for those groups for which there are no designated transformations (Class III). The table is adapted from Table II of reference 153, Chapter 7.

Table of Transformation Properties of the Representations of $\mu \cdot m$.

Class	Symmetry Groups	Polarizations						
		xx	yy	zz	xy	yx	xz	yz
I	C_1	xx	yy	zz	xy	yx	xz	yz
	C_2	xx	yy	zz	xy	yx	xz	yz
	$C_n (n>2)$	xx	xx	zz	xy	$-xy$		
	D_2	xx	yy	zz				
	$D_n (n>2)$	xx	xx	zz				
II	$C_{nv}(n>2<\infty)$				xy	$-xy$		
	S_4	xx	$-xx$		xy	xy		
	D_{2d}	xx	$-xx$					
	T, O	xx	xx	xx				
III	C_{1h}						xz	yz
	C_{2v}				xy	yx		
	$C_{nh}, D_{nh}(n>1)$	. . . . None						
	$S_2, S_6, D_{nd}(n>2), T_h, T_d, O_h$	. . . . . . None						

GLOSSARY

With very few exceptions, terms that may be expected to be unfamiliar or ambiguous are defined as they appear in the text. This short glossary is designed to amplify a few of these definitions and to provide a definition where none has been given.

Antipode: A synonym for enantiomer.

Basis Set: In approximate molecular orbital (MO) energy calculations, the assignment of atomic orbitals (AO) to a molecular orbital is somewhat arbitrary. Thus an MO can be constructed from the simplest obvious electronic orbital; for example, the MO for a π orbital between carbon atoms from $2p$ orbitals on the carbons or from the $1s$ orbitals of adjacent bonded hydrogens and/or the $3d$ unfilled orbitals of the two carbons, and so on. Each of these combinations constitutes a starting or basis set in a quantum calculation. Furthermore, each of these AO's may be a Slater-type orbital which is intended as a close approximation to the isolated unperturbed orbital, or, for computational ease, each of the AO's may be further approximated by Gaussian functions (Gaussian-type orbitals known as GTO's). In this case, the basis set is said to be comprised of N GTO's, where N is an arbitrary number. Presumably the larger the number (the bigger the basis set), the closer the approximation is to the truth.

Birefringence: *Linear*—the difference between the indices of refraction for plane polarized light incident along different directions (usually along principal axes) in an anisotropic crystal; *circular*—the difference between the indices of refraction for circular polarized light of opposite handedness of an optically active substance. In this case the substance need not be anisotropic, a solution for example, but if it is, then the circular birefringence will be different in different directions just as in the linear birefringence.

Chiral and Achiral: As used in this book, the words refer to a property associated with the interaction of matter with circularly polarized radiation that gives rise to the phenomenon of optical activity. Chiral molecules or chromophores cannot exhibit optical activity with un-polarized light. Achiral molecules do not exhibit optical activity with unpolarized or with circularly polarized light.

Chromophore: See Chapter 7.

Configuration: *Structural*—the absolute stereostructure of the atoms of a molecule; *quantum mechanical*—the specification of the occupied atomic orbitals in a given electronic state of a molecule or chromophore.

Conformation: One of a set of specific arrangements of bond angles and bond lengths that interchange with others of the set without breaking bonds (to distinguish from configurations). The difference in energy between conformations and the height of the energy barrier between them determines, respectively, the equilibrium distribution of conformers at a given temperature and the rate at which conformers interchange.

Consignate and Dissignate: Optical rotation or circular dichroism of the same sign as predicted from other observations or from calculations, or of opposite sign, respectively.

Dichroism: Linear and circular—the substitution of the words "extinction coefficients" for the words "refractive indices" in the foregoing definitions of the birefringence give the definitions for linear and circular dichroism, respectively.

Direct Product: A matrix in which the *mn*th element is the product of the corresponding elements of the *n*th row and *m*th column of two matrices. It has a special utility in group theory because the direct product of the matrices of group representations of a given symmetry yields directly the overall symmetry of, for example, the transition moment.

Dispersion: The wavelength dependence of ....

Dissymmetry: See Section 1.1.

Eigenfunction: Referring to wavefunctions—from the German *eigen* for characteristic or exact; arises from the fact that for a *given energy* $\mathcal{E}_i$, there is only one characteristic or exact function, ψ, which satisfies the differential equation, $\mathcal{H}\psi = \mathcal{E}\psi$ (the Schrödinger equation). The Hamiltonian function $\mathcal{H}$ is itself a function of the molecular parameters that describe the sum of the potential and kinetic energy of the atom, molecules, or groups of atomic particles whose distribution in space is to be described by ψ. If the function $\mathcal{H}$ is a correct description of the way in which the energy varies with the molecular parameters, then $\mathcal{E}_i$ and ψ_i are not only corresponding and exact, but are also the correct energy and eigenfunction. That is, ψ_i is a correct quantum mechanical description of the distribution in space (and/or time) of the particle or molecule having an energy $\mathcal{E}_i$.

Ellipticity: See the discussion of Eq. 2.212 in text and the definitions in Appendix III.

Enantiomer: A stereochemical configuration of bonded atoms for which an exact mirror image may exist. See also the definitions of "antipode" and "mirror image" and Section 1.2.

Eipmer: One of several possible stereochemical configurations (stereoisomers) of a group of atoms with more than one chiral center. If each of two epimers is the mirror image of the other, then they are also enantiomers.

Field (Electromagnetic): See the discussion in Section 2.0.

Handedness: The juxtaposition of at least four different points in space. For scientific purposes, right- and left-handed are defined by relating the position of the points to the position of the observer. See the discussion in Sections 1.2 and 2.2.

Isotropic and Anisotropic: For optical phenomena, isotropic is defined as a property of matter that is independent of its orientation relative to the direction of an incident plane polarized electromagnetic wave. Thus solutions, gases, and crystals of cubic symmetry are isotropic if optically inactive; the solution and gas, even if optically active. Correspondingly, the response of optically anisotropic matter is generally dependent on its orientation relative to the direction of an incident plane wave, although there can be unique directions for which certain classes of optically anisotropic matter are isotropic. As an example of the latter, the rotation of the plane of polarization about the direction of propagation has no observable effect if the direction of propagation is precisely parallel to the unique axis of a uniaxial crystal. This is true for both optically active and inactive crystals.

Mirror Image: The projection of an object through a plane viewed from the position of the object. It is therefore of opposite handedness to the object itself viewed from the position of the plane.

Normal Coordinate: See Section 3.6.

Operator: A designation for a mathematical procedure for transforming one function or quantity into another. Thus $\div$ is the operator for the procedure of dividing one function by another as in $x^2 \div x = x$. Symmetry operators are a special class designating the procedures for carrying one or more space-fixed points into an equivalent point or set of points. Thus $C_2(x,y,z)=(x,-y,-z)$ if the C_2 operation designates the procedure of rotating the y and z axes by 180° about the x axis (a "twofold" rotation).

Polarized Light: Linear polarization defines a plane electromagnetic wave and circular polarization defines an electromagnetic wave traced out by a vector circling about its axis of propagation. See Figs. 2.00 and 2.04 and Sections 2.0 and 3.4.

Pseudo-scalar: Used to distinguish a property that, although it has a magnitude and sign, unlike a vector, does not specify direction.

Racemization: See Section 1.1.

Representation: The word is used here in its group theoretical connotation. See Section 5.3 where it is introduced and Section 5.5 where it is explained and defined.

Selection Rules: One of the important consequences of quantum mechanics is that electromagnetic (radiative) transitions are permitted only between discrete states. The quantum mechanical and group theoretical symmetry rules which govern the allowedness of such transitions are called selection rules. An abbreviated shorthand for determining these rules is discussed in Chapter 6 and appears in the form of the character tables of Appendix I for a variety of structures having elements of group symmetry.

Stereoisomer: One of a set of two or more molecular configurations of a given assembly of atoms (i.e., of a particular molecule) with a chiral configuration different from other members of the set. While different stereoisomers may (rarely) have the same optical activity at some particular wavelength, the optical activity must be different at some wavelengths.

Transition Moment: Quantum mechanically, the magnitude of the electric or magnetic dipole (or other multipole) moment associated with a radiative transition between states. The electric and magnetic moments, respectively, are: $\int \psi_a e \mathbf{r} \psi_b \, d\tau$ and $\int \psi_a [(e/2mc)\mathbf{r} \times \mathbf{P}] \psi_b \, d\tau$. See Section 2.4 for a physical interpretation originating with Heisenberg.

Vector Potential: A property of an electromagnetic wave which specifies both the phase relationship of the magnetic field, and the directional property of the electric field of the wave. That is, the vector potential is in phase with $\mathbf{H}$ (its amplitude has precisely the same time or space dependence) and in the direction of $\mathbf{E}$. It is therefore out-of-phase with $\mathbf{E}$ since $\mathbf{E}$ is out-of-phase with $\mathbf{H}$. See the discussion of the vector potential in Section 3.4.

Vibronic: Strictly, a vibronic transition or band is a symmetry or spin *forbidden* transition which has been made allowed by vibrational coupling with an allowed electronic transition of the same molecule or chromophore. An *allowed* electronic transition may have vibrational components and vibronic components as well, the latter arising from the combination of symmetry allowed and forbidden vibrations which become allowed in the electronic symmetry of the transition. The allowed vibrational components of an allowed electronic band are not vibronic! Thus the mere existence of vibrational lines in an electronic (visible or ultraviolet band) does not justify calling them vibronic. The rules for vibronic allowedness in optical activity have a similar theoretical origin as the Herzberg-Teller rules for vibronic allowedness in achiral absorption or emission spectra. (See Section 3.6.) Vibronic bands derive their intensity by borrowing it from allowed electronic transitions, which as a consequence, are less intense since the total oscillator strength for all the electronic transitions of a molecule cannot exceed the total number of electrons. A similar rule holds for rotational strengths except that because of the laws of conservation of angular momentum, the total rotational strength must sum to zero.

OPTICAL QUANTITIES AND UNITS

The recommendations of the subcommission on the Chiroptical Techniques of the International Union of Pure and Applied Chemistry (IUPAC) are not yet available at the time of this writing. When these are finally adopted, any previous discussion of units and definitions will be superfluous. Fot this reason, this appendix is provided only as an aid to the reader with respect to this monograph and for the purpose of gauging different systems used in the literature. As a typical example of some of the difficulties which may be encountered, the symbol v almost always represents frequency, but it may be in units of cm^{-1} (i.e., cycles/cm) as it is in the paper by McLachlen and Ball [*Mol. Phys.*, **8**, 581 (1964)] or in units of sec^{-1} (i.e., cycles/sec) as it is in the paper by Tinoco [*Adv. Chem. Phys.*, **4**, 113 (1962)]. On the other hand, the symbol ω is also used for frequency, usually in units of sec^{-1} as it is in the McLachlen and Ball paper cited above. Similar problems occur with other symbols, but there is some consensus and we have tried to utilize the more commonly accepted symbols in this monograph. There is also the problem of the use of SI versus cgs units. In general throughout this monograph, without prejudice, the latter were used because they appear most frequently in the literature of optical activity. For nonchiral optical phenomena, the IUPAC has adopted a system of units and definitions which appear as the Manual of Symbols and Terminology for Physico-Chemical Quantities and Units [*Pure Appl. Chem.*, **21**, (1970)].

There are a number of different measures of the strengths of ordinary isotropic transitions and of chiral transitions. These may be placed in two categories; observational, which are extensive properties depending on the quantity of material in the absorbing path (either in terms of mass or molar concentrations and of pathlength), and intensive properties, which are intrinsic to the substance at hand and measure the cross section presented to the incident radiation per molecule, that is, the probability that the radiation of frequency, v, will produce the appropriate optical transition. In the first category are such parameters as transmission usually designated as T or I/I_0, the absorbance A, or optical density, O.D., rotation δ, and the ellipticity, θ. Extensive quantities may be transformed to quantities having the characteristics of intensive properties by specifying them in terms of unit quantities such as pathlength (usually cm or dm) or unit concentrations (per mole). A few such properties are the extinction coefficient ε, molar rotation $[\phi]$, molar ellipticity $[\theta]$, the circular dichroism $(\varepsilon_L - \varepsilon_R)$, and the amplitude A, of the Cotton effect in an ORD curve. The relationship between these quantities is as follows.

A. EXTENSIVE QUANTITIES

ACHIRAL ABSORPTION

At a fixed wavelength achiral absorption is

$$I = I_0 e^{-\alpha z} \tag{III.1}$$

or

$$I = I_0 10^{-\varepsilon C z} \tag{III.2}$$

or

$$I = I_0 e^{-4\pi k z/\lambda} \tag{III.3}$$

where, $z =$ pathlength (cm)
   $e =$ natural number 2.718
   $I =$ transmitted intensity in arbitrary units
   $I_0 =$ incident intensity in the same arbitrary units as I
   $C =$ molar concentration (moles/liter)
   $\lambda =$ wavelength of the radiation (cm)
   $\varepsilon =$ decadic molar extinction coefficient in liter/mole·cm
   $\alpha =$ exponential absorption coefficient (cm^{-1})
   $k =$ exponential absorption index (dimensionless)
(Note: Since I and I_0 must be in the same units, the factors of Eqs. III.1 to III.3 must be dimensionless.)

From Eqs. III.2 and III.3 for the same pathlength z,

$$10^{-\varepsilon C z} = e^{-4\pi k z/\lambda} \tag{III.4}$$

The natural logarithm of both sides yields

$$- \varepsilon Cz \ln 10 = \frac{-4\pi kz \ln e}{\lambda} \qquad \text{(III.5)}$$

where, since the $\ln 10 = 2.303$ and $\ln e = 1$,

$$k = \frac{2.303\lambda C\varepsilon}{4\pi} \qquad \text{(III.6)}$$

which is the equivalent of the text Eq. 2.214. Modern spectroscopic instruments calculate automatically the transmittance, $T = I/I_0$ and/or absorbance, O.D. $= \log I_0/I = \varepsilon Cz$ and polarimetric instruments calculate the ellipticity, $\theta = 33\Delta$O.D., or the rotation, $\Phi = 1800\delta/\pi$.

CHIRAL ABSORPTION (circular dichroism at a fixed wavelength)

From Eq. III.6, if the extinction coefficients are different for left- and right-circularly polarized light,

$$k_l - k_r = \frac{2.303\lambda C}{4\pi}(\varepsilon_l - \varepsilon_r) \qquad \text{(III.7)}$$

$\varepsilon_l - \varepsilon_r = \Delta\varepsilon$ is the *circular dichroism* in the same units as ε (liters/mole·cm). The measured quantity, because of the design of the measuring instruments, is more usually the ellipticity (Eq. 2.213)

$$\psi = \frac{\pi z}{\lambda}(k_l - k_r) \qquad \text{(III.8)}$$

From Eqs. III.7 and III.8,

$$\psi = \frac{2.303\,Cz}{4}\Delta\varepsilon \qquad \text{(III.9)}$$

The units of ψ are radians. In the generally preferred units of degrees, the ellipticity, θ, is obtained by multiplying by the number of degrees per radian:

$$\theta = \psi\frac{360}{2\pi} \qquad \text{(III.10)}$$

From which, by substituting Eq. III.9 for ψ,

$$\theta = 32.90\,Cz\,\Delta\varepsilon \approx 33\,\Delta\,\text{O.D.} \qquad \text{(III.11)}$$

The molar ellipticity, defined in units of degrees-cm^2/decimole, is

$$[\theta] = \frac{M.W.}{C'z'} \cdot 10^3 \theta \tag{III.12}$$

where M.W. = grams/mole, C' = grams/cm^3, and z' = 10 cm. Substituting appropriately for C', z' and for θ from Eq. III.11, the relation between the molar ellipticity and the circular dichroism in these units becomes:

$$[\theta] = 3300 \Delta\varepsilon \tag{III.13}$$

CHIRAL ROTATION (optical activity at a fixed wavelength)

From the text Eq. 2.208, the rotation is proportional to the refractive index difference for left- and right-circularly polarized light. For the rotation in units of radians,

$$\delta = \frac{\pi z}{\lambda}(\eta_l - \eta_r) \tag{III.14}$$

or in the more usual units of degrees per decimeter

$$\Phi = \frac{1800}{\pi} \cdot \delta = \frac{1800}{\lambda}(\eta_l - \eta_r) \tag{III.15}$$

To obtain the molar rotation in units of degrees-liters per centimeter-mole, the observed rotation is multiplied by $100/Cz$,

$$[\Phi] = \frac{100\Phi}{Cz} \tag{III.16}$$

A frequently used measure of the rotation, especially for analytical purposes is the specific rotation $[\alpha]$ in units of degrees-cm^3 per centimeter-mole

$$[\alpha] = \frac{10^4\Phi}{Cz \, M.W.} \tag{III.17}$$

or more commonly in units of degrees-cm^3 per decimeter-gram

$$[\alpha] = \frac{\Phi/10}{C'z} \tag{III.18}$$

where C' is concentration in grams per cm^3.

CHIRAL AMPLITUDE, A

This somewhat arbitrary quantity is especially useful in analytical determinations involving the optical rotation associated with an isolated Cotton effect. It is 0.001 of the absolute magnitude of the observed difference between the molecular rotation at the peaks of the positive and negative lobes (the "peak" and "trough") of the optical rotatory dispersion curve associated with the Cotton effect.

B. INTENSIVE QUANTITIES

Finally, there are the intrinsically intensive properties that characterize the optical transitions independently of the quantity of material involved: the electric and magnetic dipole transition moments, μ_{ab} and $\mathbf{m}_{ab}$; the dipole (electric) strength, D_{ab}; the rotational strength, R_{ab}; and the dimensionless quantity, the oscillator strength, f_{ab}. These quantities as they are used to describe radiative transitions may be obtained from experimental measurement of the extensive properties or a priori from quantum mechanical calculations. When obtained experimentally they refer to the total or integrated strength over the frequency range encompassed by the radiative transitions that, for an electronic transition, covers the entire range of associated vibrational and rotational transitions obtained from quantum calculations. The partition of the values into transitions between these substates is usually ignored, but so is the effect of environmental perturbations; the theoretical calculations almost always are for systems of independent molecules (gas state).

Of the fundamental intensive quantities, the oscillator strength and rotational strength are the most generally used measures of the ability of a molecule to absorb or rotate light. These quantities are related to the Einstein coefficients for absorption, and the Rosenfeld rotational parameter g, through equations given in the text. Here we present their relationships to experimentally observable quantities and also to the dipole strength, the latter of which is another way of specifying the cross-section for achiral absorption. (The subscripts ab have been dropped for convenience in the discussion that follows.)

THE ACHIRAL OSCILLATOR STRENGTH

The one-electron oscillator strength is a dimensionless quantity measuring the probability that a transition between two states will take place. It may have values from zero to one and could exceed one only if a fully-allowed transition were to become more intense by borrowing intensity from still another transition or, more than one electron is involved in the radiative transition. As an experimental fact, the oscillator strength of very intense transitions rarely exceeds values of a few tenths. The oscillator strength is related to the total absorption for a given transition by:

$$f = \frac{2303\nu mc^2}{\pi N_0 e^2} \int \frac{\varepsilon(\nu)\, d\nu}{\nu} \qquad \text{(III.19)}$$

where, m = electronic mass = 9.109×10^{-28} g
c = velocity of light = 2.998×10^{10} cm/sec
N_0 = Avogadro's number = 6.023×10^{23}
e = electronic charge = 4.803×10^{-10} esu
$\varepsilon(\nu)$ = molar decadic extinction coefficient as a function of frequency
ν = frequency in cm^{-1}

The value of the factor in front of the integral is therefore 4.34×10^{-9}.

The relation between the oscillator strength and the fundamental molecular property, the transition dipole moment for a given radiative transition is

$$f = \frac{8\pi^2 mc\nu}{3he^2} \mu^2 \qquad \text{(III.20)}$$

where h = Planck's constant = 6.625×10^{-27} erg-sec.

Combining Eqs. III.19 and III.20, we can obtain the value of the electric-dipole transition moment in terms of the oscillator strength and the energy (frequency) of the corresponding absorption band:

$$\mu = \pm \left[\frac{3he^2}{8\pi^2 mc} \cdot \frac{f}{\nu} \right]^{1/2} \qquad \text{(III.21)}$$

(Note: The sign of μ is indeterminate from an ordinary absorption measurement, since it is proportional to the square root of the oscillator strength. The dipole strength D is just the square of μ.)

Evaluating the constants in Eq. III.22 and dropping the sign indication for convenience:

$$\mu = 1.46 \times 10^{-15} \sqrt{\frac{f}{\nu}} \quad \text{cgs units} \qquad \text{(III.22)}$$

The Debye unit is defined as 10^{-18} cgs units. (Note that it has also been variously defined as the product of the electronic charge and the radius of the first Bohr orbit of hydrogen (0.529×10^{-8} cm) that would result in a value of 2.54×10^{-18} cgs units, and as the product of the electronic charge and a distance of one Å that would result in a value of 4.803×10^{-18} cgs units, but the current consensus appears to be as defined here.) With this value:

$$\mu = 1.46 \times 10^{3} \sqrt{\frac{f}{\nu}} \text{ debye} \qquad (III.23)$$

THE ROTATIONAL (ROTATORY) STRENGTH

Unlike the oscillator strength, the rotational strength is not dimensionless, but is nevertheless an intensive property of the transitions of optically active molecules, independent therefore of the quantity. Like the oscillator strength, it may be obtained by theoretical calculation (in this case, of the $\text{Im}\,\mu \cdot \mathbf{m}$ or the proportionate quantity, the rotatory parameter, β). Alternately it may be obtained from experimental values of the circular dichroism, from measurements of the optical activity by an approximation using Eq. III.26 below or more exactly by the use of the Kronig-Kramers transform [A. Moskowitz, *Tetrahedron*, **13**, 48 (1961); E. Emeis, L. J. Oosterhoff and G. DeVries, *Proc. Roy. Soc. (London)*, **A297**, 54 (1967).] In cgs units the rotational strength is related to the area under a circular dichroism band by:

$$R = \frac{3 \times 2303\,hc}{32\pi^{3}N_{0}} \int \frac{\Delta\varepsilon(\nu)}{\nu}\,d\nu \qquad (III.24)$$

The value of the factor in front of the integral is 22.9×10^{-49}. By approximating the area under a CD band by a Gaussian function, Eq. III.25 may be converted to

$$R \approx 0.408 \times 10^{-38} \frac{\Delta(\nu)}{\nu_{max}} \Delta\varepsilon(\nu_{max}) \qquad (III.25)$$

where $\Delta(\nu)$ is one-half the bandwidth at $\Delta\varepsilon = 0.368\Delta\varepsilon(\nu_{max})$ and ν_{max} is the frequency at the peak of the CD band.

The Rosenfeld equation relates the rotational strength to the fundamental molecular properties, the electric and magnetic transition moments

$$R = -\mu \cdot \mathbf{m}' \qquad (III.26)$$

where $\mathbf{m}'$ is the real part of $\mathbf{m}$. The magnetic transition moments $\mathbf{m}'$ is given by the classical equation (text Eq. 3.404)

$$\mathbf{m} = \frac{e}{2mc}(\mathbf{r} \times \mathbf{P}) \qquad (III.27)$$

The unit of magnetic moment utilized in rotational strength calculations is obtained by letting $\mathbf{m}$ be the real part of the moment associated with an electron rotating in a Bohr orbit at the velocity of light for which $\mathbf{r} \times \mathbf{P} = \pm h/2\pi$ so that

$$\mathbf{m}' = \pm \frac{eh}{4\pi mc} = 0.9273 \times 10^{-20} \text{ cgs units} \qquad (III.28)$$

This is defined as the Bohr magneton and is the magnetic moment associated with a fully-allowed magnetic dipole transition in which there is an increase or decrease of one quantum of angular momentum ($\pm h/2\pi$).

The magnitude of the rotational strength (in cgs units) associated with a fully-allowed electric and magnetic dipole transition is therefore (ignoring signs):

$$R = \left(1.46 \times 10^{-15}\sqrt{\frac{f}{\nu}}\right)(0.9273 \times 10^{-20}) \qquad (III.29)$$

$$R = 1.35 \times 10^{-35}\sqrt{\frac{f}{\nu}} \qquad (III.30)$$

where, since the transition is fully-allowed ($f = 1$);

$$R = 1.35 \times 10^{-35}\nu^{-1/2} \qquad (III.31)$$

THEORETICAL MAGNITUDES

Quantum calculations of R for a dissymmetric chromophore proceed from the calculation of transition moments. If the moments are specified in the dipole velocity form (see Sections 4.6 and 7.31) and computed in atomic units, the R is obtained in cgs units from the text Eq. 4.407 with the factor $[(he^2/2mc) = 6.41 \times 10^{-37})]$:

$$R = 6.41 \times 10^{-37}\frac{\boldsymbol{\mu} \cdot \mathbf{m}'}{\Delta \mathcal{E}} \qquad (III.32)$$

where $\Delta \mathcal{E}$ is the energy of the transition in units of electron volts ($1 \text{ eV} = 8.066 \times 10^3 \text{ cm}^{-1}$).

EXPERIMENTAL MAGNITUDES

Consider the magnitude of R for a fully allowed electric and magnetic transition that results in a circular dichroism band at 20,000 cm^{-1} (500 nm)

$$R = \frac{1.35 \times 10^{-35}}{1.414 \times 10^2} = 0.955 \times 10^{-37} \text{ cgs}$$

Since 1 Debye Bohr magneton (DBM) is $10^{-18} \times 0.9273 \times 10^{-20} = 0.9273 \times 10^{-38}$ cgs, the rotational strength associated with this transition is 10.3 DBM. For a one-electron transition at 20,000 cm^{-1} involving a one quantum change in angular momentum, this is the maximum possible value of the rotational strength. However, since the magnetic and electric moments are not necessarily parallel (recall that $R = -\boldsymbol{\mu} \cdot \mathbf{m} = -\mu m \cos \theta$) and since even fully-allowed transitions rarely develop moments of this magnitude, observed rotational strengths are usually much smaller except for transitions where more than one quantum change in angular momentum may be involved; for example, in optically active bands of transition metal ions. The normal range of oscillator strengths for allowed transitions is $f = 0.05$ to 0.5, and for vibronically activated forbidden transitions from 0.0001 to 0.05; rotational strengths of 0.001 to 5.0 DBM are therefore much more usual.

For convenience, still another unit of rotational strength has been introduced, the "reduced rotational strength" $[R]$. The relation between the units are: $R(\text{cgs}) = 0.9273 \times 10^{-38} \, R(\text{DBM}) = 10^{-40} \, [R]$.

ACKNOWLEDGMENTS

Permission from the following publishers and authors to reproduce in whole or in part copyrighted figures and text material is gratefully acknowledged:

Academic Press, Division of Harcourt, Brace, Jovanovich; Addison-Wesley/Benjamin Publishing Company; American Chemical Society; American Institute of Physics; Heyden & Son Ltd.; John Wiley & Sons, Inc.; Journal of Chemical Physics; Journal of Molecular Biology; Pergamon Press and the McGraw Hill Book Company and the Societè de Chimie-Physique, France. In each case, reference is made to the source. Authors who also consented to reproduction of the same material, where required, are W. Curtis Johnson, Melvin B. Robin, John A. Schellman, Otto Schnepp, Philip J. Stephens, and Ignacio Tinoco, Jr. Their enthusiastic acquiescence is appreciated.

Permission to reproduce the following noncopyrighted material is also gratefully acknowledged:

Figure 8.07, granted by Dr. Carol Cech; Figure 8.04, granted by Professor F. Ciardelli; Figure 3.04, granted by Dr. William A. Eaton, and the photograph in Figure 2.05 kindly provided by Dr. Ulrich Weiss.

INDEX